RADIATION HYDRODYNAMICS

Radiation Hydrodynamics studies the dynamics of matter interacting with radiation, when the radiation is strong enough to have a profound effect on the matter. It applies to normal stars, to exploding stars or stars with violent winds, to active galaxies, and on Earth wherever matter is very hot. This broad and up-to-date treatment provides an accessible introduction to the theory and the large-scale simulation methods currently used in radiation hydrodynamics. Chapters cover all the central topics, including: a review of the fundamentals of gas dynamics; methods for computational fluid dynamics; the theory of radiative transfer and of the dynamical coupling of matter and radiation; and quantum mechanics of matter–radiation interaction. Also covered are: the details of spectral line formation out of thermodynamic equilibrium; the theory of refraction and transfer of polarized light and current computational methods for radiation transport; and a description of some notable applications of the theory in astrophysics and laboratory plasmas. This is a valuable text for research scientists and graduate students in physics and astrophysics.

JOHN CASTOR received a Ph.D. in astronomy from the California Institute of Technology in 1967 for work involving radiation hydrodynamics in pulsating stars. He then joined the University of Colorado and the Joint Institute for Laboratory Astrophysics, where he worked on the foundations of radiation hydrodynamics and spectral line formation in high-velocity flows. This led to a very successful theory of radiatively-driven stellar winds. Since 1981 he has been a physicist at the Lawrence Livermore National Laboratory in California, working on problems in high-energy-density physics, including the interpretation of experiments with high-power lasers.

RADIATION HYDRODYNAMICS

JOHN I. CASTOR
Lawrence Livermore National Laboratory

CAMBRIDGE
UNIVERSITY PRESS

CAMBRIDGE UNIVERSITY PRESS
Cambridge, New York, Melbourne, Madrid, Cape Town, Singapore, São Paulo

Cambridge University Press
The Edinburgh Building, Cambridge CB2 2RU, UK

Published in the United States of America by Cambridge University Press, New York

www.cambridge.org
Information on this title: www.cambridge.org/9780521833097

First published 2004
This digitally printed first paperback version 2006

A catalogue record for this publication is available from the British Library

Library of Congress Cataloguing in Publication data

Castor, John I., 1943–
Radiation hydrodynamics/John I. Castor.
p. cm.
Includes bibliographical references and index.
ISBN 0-521-83309-4 – ISBN 0-521-54062-3 (pb.)
1. Radiative transfer. 2. Hydrodynamics. 3. Astrophysics. I. Title.

QC175.23.R3C37 2004
530.13′8–dc22 2004043559

ISBN-13 978-0-521-83309-7 hardback
ISBN-10 0-521-83309-4 hardback

ISBN-13 978-0-521-54062-9 paperback
ISBN-10 0-521-54062-3 paperback

Contents

Figures

Preface

Much of the material in this book originated with lectures given for the Summer School on Radiative Transfer and Radiation Hydrodynamics at the Institute of Theoretical Astrophysics of the University of Oslo during June 1–11, 1999. Those lectures focused on the specifics of the dynamic coupling of radiation and matter, and on the detailed processes of the interaction. The other lecturers were Rob Rutten of the University of Utrecht, Phil Judge of the High Altitude Observatory in Boulder, Colorado, and Mats Carlsson of the University of Oslo, the organizer of the Summer School and the Director of the Institute. Their lectures treated the introduction to radiative transfer, atomic processes and spectral line diagnostics with special reference to the sun, and numerical methods in radiative transfer and radiation hydrodynamics. For that reason the original content of these lectures was light in those areas, especially in numerical methods. In putting the lectures into the present form some effort was invested to expand the coverage of the neglected topics.

The background for the theory of radiation hydrodynamics as presented here came from work at JILA, the Joint Institute for Laboratory Astrophysics of the University of Colorado and the National Bureau of Standards, as they were called then, in the late 1960s and 1970s. It originated with the need to treat radiation–matter coupling correctly in stellar pulsations and other areas of astrophysical fluid dynamics. The theory of radiatively-driven stellar winds developed out of that work. At this same time the Boulder School of radiative transfer was flourishing through the efforts of L. Auer, D. Hummer, J. Jefferies, D. Mihalas, R. Thomas, and many others. Some of the knowledge absorbed from these people, and their colleagues E. Avrett, W. Kalkofen, and G. Rybicki at the Harvard-Smithsonian Center for Astrophysics, made it into this book. The years in JILA also provided an education in theoretical atomic physics, and the mentors in this area included D. G. Hummer again, and R. Garstang, D. Norcross, J. Cooper, M. Seaton (when he visited), and H. Nussbaumer. The leader of stellar pulsation theory was J. Cox, and his influence was felt in several areas, as was that of his good friend A. Cox of Los Alamos National Laboratory. R. McCray provided an outstanding example of insightful analytic theory complementing computational astrophysics.

Later on the opportunity to work directly with A. Cox provided additional exposure to the methods current at that time in numerical hydrodynamics and radiation hydrodynamics. Since the beginning of the 1980s the Lawrence Livermore National Laboratory has provided many lessons in numerical techniques for radiation hydrodynamics, and A. Winslow, E. Garelis, and P. Crowley were notable teachers. Their friend G. Pomraning, when he was visiting, was a fount of knowledge as well. G. Zimmerman, then and now, is the master of this subject. R. Klein has been a valued colleague for many years, beginning with stellar winds in Boulder and continuing today in Livermore, where he is a recognized authority in precision hydrodynamic techniques. Two of today's leading experts in S_N methods for radiation transport, M. Zika and P. Nowak, have helped with ideas in that part of the book.

Every work on radiation hydrodynamics since the 1980s stands in the shadow of the monumental *Foundations of Radiation Hydrodynamics* by D. Mihalas and B. Mihalas. The present author owes a great debt of gratitude to D. Mihalas for his teaching, support, and encouragement over the years. The help and encouragement of D. Hummer, L. Auer, and R. Klein are also greatly appreciated. The colleagues D. Abbott, D. Band, D. Friend, G. Olson, S. Owocki, and D. Van Blerkom may find their fingerprints here too.

Finally, special gratitude is extended to Professor Mats Carlsson of the Institute of Theoretical Astrophysics in Oslo for the opportunity to present the original lectures.

1

Introduction

1.1 Philosophy

The discipline of radiation hydrodynamics is the branch of hydrodynamics in which the moving fluid absorbs and emits electromagnetic radiation, and in so doing modifies its dynamical behavior. That is, the net gain or loss of energy by parcels of the fluid material through absorption or emission of radiation is sufficient to change the pressure of the material, and therefore change its motion; alternatively, the net momentum exchange between radiation and matter may alter the motion of the matter directly. Ignoring the radiation contributions to energy and momentum will give a wrong prediction of the hydrodynamic motion when the correct description is radiation hydrodynamics.

Of course, there are circumstances when a large quantity of radiation is present, yet can be ignored without causing the model to be in error. This happens when radiation from an exterior source streams through the problem, but the latter is so transparent that the energy and momentum coupling is negligible. Everything we say about radiation hydrodynamics applies equally well to neutrinos and photons (apart from the Einstein relations, specific to bosons), but in almost every area of astrophysics neutrino hydrodynamics is ignored, simply because the systems are exceedingly transparent to neutrinos, even though the energy flux in neutrinos may be substantial.

Another place where we can do "radiation hydrodynamics" without using any sophisticated theory is deep within stars or other bodies, where the material is so opaque to the radiation that the mean free path of photons is entirely negligible compared with the size of the system, the distance over which any fluid quantity varies, and so on. In this case we can suppose that the radiation is in equilibrium with the matter locally, and its energy, pressure, and momentum can be lumped in with those of the rest of the fluid. That is, it is no more necessary to distinguish photons from atoms, nuclei, and electrons than it is to distinguish hydrogen atoms from helium atoms, for instance. They are all just components of a mixed fluid in this case.

So why do we have a special subject called "radiation hydrodynamics", when photons are just one of the many kinds of particles that comprise our fluid? The reason is that photons couple rather weakly to the atoms, ions, and electrons, much more weakly than those particles couple with each other. Nor is the matter–radiation coupling negligible in many problems, since the star or nebula may be millions of mean free paths in extent. Radiation hydrodynamics exists as a discipline to treat those problems for which the energy and momentum coupling terms between matter and radiation are important, and for which, since the photon mean free path is neither extremely large nor extremely small compared with the size of the system, the radiation field is not very easy to calculate.

In the theoretical development of this subject, many of the relations are presented in a form that is described as approximate, and perhaps accurate only to order of v/c. This makes the discussion cumbersome. Why are we required to do this? It is because we are using Newtonian mechanics to treat our fluid, yet its photon component is intrinsically relativistic; the particles travel at the speed of light. There is a perfectly consistent relativistic kinetic theory, and a corresponding relativistic theory of fluid mechanics, which is perfectly suited to describing the photon gas. But it is cumbersome to use this for the fluid in general, and we prefer to avoid it for cases in which the flow velocity satisfies $v \ll c$. The price we pay is to spend extra effort making sure that the source-sink terms relating to our relativistic gas component are included in the equations of motion in a form that preserves overall conservation of energy and momentum, something that would be automatic if the relativistic equations were used throughout.

Some general references on the subject of radiation hydrodynamics are these:

- The most comprehensive general reference is *Foundations of Radiation Hydrodynamics*, by Mihalas and Mihalas (1984). This provides all the needed background in statistical physics, hydrodynamics, and radiative transfer, as well as a thorough discussion of the nonrelativistic and relativistic formulations of radiation hydrodynamics, mostly in one space dimension. The applications include several of the more important topics as of 1984.
- The book *The Equations of Radiation Hydrodynamics* by Pomraning (1973) reflects the viewpoint of the neutron transport community, in that the $O(v/c)$ effects are discussed largely in the fixed frame, and the distinction between fixed-frame diffusion and comoving-frame diffusion is not made sufficiently clear. Considerable space is devoted to Compton scattering with frequency redistribution, a small effect in astrophysics (except for hard x-rays) although the analogous problem in neutron transport is important.
- The NATO workshop *Astrophysical Radiation Hydrodynamics*, edited by Winkler and Norman (1982), contains useful articles on the fundamental theory by Mihalas,

on advanced computational methods in hydrodynamics by Norman and Winkler and by Woodward, on particle methods by Eastwood and on finite-element methods by Griffiths.

- *Physics of Shock Waves and High-Temperature Hydrodynamic Phenomena* by Zel'dovich and Raizer (1967) is one of the most valuable references for radiation hydrodynamics in general. The treatment of radiation here is simplified, but the insightful analysis of complicated shock phenomena is outstanding.
- *Stellar Atmosphere Modeling*, edited by Hubeny, Mihalas, and Werner (2003), is the most recent (in 2003) conference devoted to the advanced methods in numerical radiation transport for astrophysics.
- *A Guide to the Literature on Quantitative Spectroscopy in Astrophysics* by Mihalas (2003) is a comprehensive bibliography on astrophysical radiative transfer and related topics, and includes a 30-page historical review of the field.

1.2 Outline

The succeeding chapters of the book start with an introduction to gas dynamics that covers the essential elements: Euler's equations, the Lagrangian equations, and the especially important topic for numerical calculations, arbitrary Lagrangian Eulerian (ALE). Then comes viscosity and the Navier–Stokes equation, Bernoulli's equation, and some of the important topics like sound waves, shocks, and the self-similar Taylor–Sedov blast wave. The following chapter is a fairly up-to-date survey of methods for numerical hydrodynamics, which is divided into Lagrangian, Eulerian, and ALE, with special notice for methods such as Godunov's and the weighted-essentially-nonoscillatory (WENO) methods.

Next we go on to three chapters on radiation and radiative transfer at increasing levels of complexity. The first chapter gives the basic definitions of things like the intensity, angle and frequency moments, the idea of diffusion, and the initial, more naive view of how radiation affects the state of the gas. The second radiation chapter deals with the simple methods for steady-state radiative transfer, and introduces some key ideas like Milne's equations and the Eddington–Barbier relation. The third radiation chapter in this group introduces the special-relativistic picture of radiation transport in all its complexity, and then attempts to wash away the dross and leave a simple enough picture of the dynamics of radiation and matter when the matter does not move *too* fast ($v \ll c$) that it can be incorporated into practical calculations.

Once we are fully apprised of the true way in which matter is coupled to radiation we return to some analytical examples of gas dynamics coupled now with radiation, including the modifications to wave motion due to energy and momentum coupling.

The following chapter deals with the atomic details of the processes that actually couple radiation and matter; these are the source-sink terms that appear in the conservation equations for matter and radiation separately. We can talk *about* the quantum mechanics that enters the calculations of these interactions, but the real work is done in computer codes such as OPAL that supply radiative process data to hydrodynamic simulations.

Next is a chapter on the most detailed methods for calculating the transport of spectral line radiation through a gas, when the radiative processes are dominant over the effects of electron–atom collisions to such a degree that the detailed quantum state of the matter is altered by this radiation which is not representative of the local thermal equilibrium distribution. This is the topic that is called non-LTE, where LTE stands for local thermodynamic equilibrium. The non-LTE condition is prevalent in many astrophysical and laboratory plasmas. The chapter describes the general features of non-LTE calculations, and then examines some interesting details of non-LTE line transport associated with frequency redistribution or the lack of it.

The next chapter concerns a subject often overlooked in treatments of radiative transfer, namely the effects of refraction and polarization on the radiation. Polarization is an important aspect of the measurements of radiation from the sun, and it may have increasing importance in observations of other bodies and in laboratory experiments as the quality of the data improves.

The next-to-last chapter is a survey of numerical methods of various kinds that are applied to radiation, either by itself or coupled with hydrodynamics. In a survey like this the names of the methods, some pointers into the literature, and summaries of what they are about is the limit of what can be presented in the space available, but it hoped that the reader will be enticed into reading more deeply.

The final chapter gives a short selection of radiation-dominated examples that either are important in themselves or bring out some interesting aspects of the theory.

2

A quick review of gas dynamics

Before beginning a discussion of the special effects brought about by the energy and momentum of radiation, we review ideal gas dynamics as it exists without these sources. We will define the variables we use and present the main equations that will be modified later.

A quite good introduction to fluid mechanics is the volume *Fluid Mechanics* in the *Course of Theoretical Physics* by Landau and Lifshitz (1959). This does not spend much time on the microscopic picture of fluids, but is very strong on the physical applications. The approach is entirely analytic. Mihalas and Mihalas (1984) describe kinetic theory in some detail, and the basis of viscosity, in addition to some of the basic results of gas dynamics. The chapters on viscous effects and relativistic flows are valuable.

2.1 Ideal fluid description: ρ, p, $\mathbf{u}$, and e

A fluid is, as the name suggests, free to flow, which distinguishes it from an elastic solid. The solid can deform, but as it does stresses are produced that depend on the displacements. In a fluid the stress is primarily (i.e., apart from a small correction due to viscosity) an isotropic pressure, and this depends on the local temperature and density of the matter, and is independent of how far a parcel may have moved from its starting point. So density is the parameter that expresses how the kinematics will change the state of the matter. The density evolves as the fluid moves, and the volume occupied by a parcel of material changes. The motion is described by the fluid velocity, $\mathbf{u}$. This is a vector which is equal to the mass-weighted average of the individual velocities of all the particles contained in a microscopic cell surrounding the point in question. A coarse-graining assumption is applied in the fluid picture. The assumption is made that a cell size can be chosen that is infinitesimal compared with the scale of the fluid region, yet so large that the statistical uncertainty of averages over the atoms in the cell is negligible. Furthermore, we assume that, if the cell is carried along with the fluid velocity, it has just the same atoms in it at the end of some time of interest as it has at the start. In other words, even though the individual atoms are flying around with velocities that may be large compared with the mean – the fluid velocity – the amount of net progress an atom makes is negligible compared with the cell size. Clearly, this requires that the mean free path between atom collisions is negligible compared with the cell size. Both because of the atom statistics and because the mean free path varies inversely with the density, the fluid approximation is valid at high densities and breaks down at low density. It fails, for example, in high-vacuum laboratory experiments and in the outer parts of the solar wind.

The ideal fluid we have been describing is characterized by its mass density ρ, its pressure p, which gives the momentum flux of the particles across any infinitesimal element of area, and by the internal energy per unit mass, e. This includes the kinetic energies of all the particles, but, since kinetic energy may be stored temporarily in atomic excitation, the excitation energy is included as well, to make a conserved quantity. Density, pressure, and internal energy are three of the thermodynamic functions. For a system with a fixed mass such as the cell we are considering, only two of these are independent, for example density and internal energy. The pressure can be calculated from the other two using the known equation of state of the material, and all the other thermodynamic functions such as temperature, entropy, Helmholtz free energy, etc., can be as well.

2.2 Euler's equations

The mass of all the particles in a cell is thus one conserved quantity. A second conserved quantity is the total momentum of the particles in the cell. This is conserved in the sense that its rate of change equals the net gain in momentum due to particles entering or leaving across the cell boundary, plus any body force that may exist such as gravity. The first of these terms is the surface integral of the pressure over the boundary. The third conservation law is for the internal energy in the cell. This is expressed by the first law of thermodynamics, which says that the change in the internal energy is the negative of the pressure times the rate of volume increase of the cell, plus the rate of any addition of heat to the cell by external agents. By applying a little calculus to the conservation laws, they can be expressed as differential equations, Euler's equations for an ideal fluid.

The mathematical form of the conservation laws differs depending on whether the time derivative is taken at a fixed point in inertial space (Eulerian picture) or following a given parcel of fluid (Lagrangian picture). The manipulation is somewhat easier with the Eulerian form, but then the expression of a conservation law must include terms for the flux of the conserved quantity across a fixed cell boundary. For conservation of mass, conservation says that the rate of change of the mass in a space-fixed volume, V, is just the negative of the integral of the mass flux, $\rho\mathbf{u}$, over the surface S of V, i.e.,

$$\frac{d}{dt}\int_V \rho dV = -\int_S \rho\mathbf{u}\cdot d\mathbf{A}. \tag{2.1}$$

By making use of the divergence theorem, and requiring the relation to be true for every cell volume, we find the first Euler equation

$$\frac{\partial\rho}{\partial t} + \nabla\cdot(\rho\mathbf{u}) = 0. \tag{2.2}$$

The momentum conservation law follows in a similar way. Two additions are needed, however. The surface integral contains not only the flux of momentum, $-(\rho\mathbf{u})\mathbf{u}\cdot d\mathbf{A}$, being carried across the boundary, but also the momentum flux associated with pressure, $-p d\mathbf{A}$. Also, if there is a body force, it gives a momentum source rate F per unit volume. The result is the second Euler equation

$$\frac{\partial\rho\mathbf{u}}{\partial t} + \nabla\cdot(\rho\mathbf{u}\mathbf{u}) + \nabla p = F. \tag{2.3}$$

The divergence in the second term on the left-hand side needs a comment. This is the divergence of a tensor and the result is a vector. The tensor has two indices, one from each factor $\mathbf{u}$, and the divergence is taken by assigning first 1 to the second index, and forming the ordinary divergence of $\rho\mathbf{u}u_1$, then repeating for index 2 and index 3, thus obtaining the three components of the result.

The derivation of Euler's energy conservation law from the first law of thermodynamics is more roundabout. We have to begin with the notion of the time derivative taken following the motion of the fluid, the Lagrangian time derivative D/Dt, which is given by $Df/Dt = \partial f/\partial t + \mathbf{u}\cdot\nabla f$ for any function f. Secondly, we recognize that the volume occupied by a unit mass of material is $1/\rho$. The time rate of change of the volume of this fixed parcel of mass is $D(1/\rho)/Dt$. According to the first law of thermodynamics, the Lagrangian rate of change of the specific internal energy, plus the rate per unit mass at which the pressure is doing work, equals the rate per unit mass at which heat is being deposited from external sources, q. Thus the Lagrangian internal energy equation is

$$\frac{De}{Dt} + p\frac{D(1/\rho)}{Dt} = q \tag{2.4}$$

or

$$\frac{\partial e}{\partial t} + \mathbf{u}\cdot\nabla e + p\frac{\partial 1/\rho}{\partial t} + p\mathbf{u}\cdot\nabla(1/\rho) = q. \tag{2.5}$$

In order to make this look more like an Eulerian conservation law we make use of this handy identity for any function f:

$$\rho \frac{Df}{Dt} = \frac{\partial \rho f}{\partial t} + \nabla \cdot (\rho \mathbf{u} f), \tag{2.6}$$

which is easily proved by expanding the derivatives of the products and using the Eulerian mass conservation equation, (2.2). When this identity is applied to (2.4) multiplied by ρ, this form of the internal energy conservation law results:

$$\frac{\partial \rho e}{\partial t} + \nabla \cdot (\rho \mathbf{u} e) + p \nabla \cdot \mathbf{u} = \rho q. \tag{2.7}$$

This is still not an expression of "energy conservation" because of the pressure work term, $p\nabla \cdot \mathbf{u}$. In order to eliminate that, we first derive a mechanical energy conservation law by forming the dot product of the velocity with the momentum conservation equation, (2.3), and making use of mass conservation again. The result is

$$\frac{\partial}{\partial t}\left(\frac{1}{2}\rho u^2\right) + \nabla \cdot \left(\frac{1}{2}\rho \mathbf{u} u^2\right) + \mathbf{u} \cdot \nabla p = \mathbf{u} \cdot F. \tag{2.8}$$

The sum of the internal energy equation (2.7) and the mechanical energy equation (2.8) is the total energy equation

$$\frac{\partial}{\partial t}\left(\rho e + \frac{1}{2}\rho u^2\right) + \nabla \cdot \left(\rho \mathbf{u} e + \frac{1}{2}\rho \mathbf{u} u^2 + p\mathbf{u}\right) = \rho q + \mathbf{u} \cdot F. \tag{2.9}$$

This is the form usually considered as the third Euler equation. The conserved quantity is the sum of the internal and the kinetic energies, and its flux has a part from advection with the fluid and another part due to the pressure. The effect of this pressure term is the same as replacing the internal energy flux by the flux of enthalpy, $h = e + p/\rho$.

The way that the three Euler equations are used to solve a hydrodynamic initial value problem is conceptually something like the following. The current values of the three conserved quantities, ρ, $\rho\mathbf{u}$, and $\rho e + 1/2\rho u^2$ are solved for ρ, $\mathbf{u}$, and e. Given ρ and e, the equation of state supplies the pressure p. These are all the variables needed to determine the fluxes in the three conservation laws and the source terms, if any. From these values the three time derivatives are calculated, and the conserved quantities can be advanced to the next time step. Refined calculations may differ from this scheme in detail, but the concept is the same.

2.3 Lagrangian equations

If we imagine that at an initial time we draw a spatial mesh on the material that makes up our problem, and that subsequently this mesh is dragged along with the material as it moves, then we have the Lagrangian picture. The appropriate time derivative, at a fixed point in the Lagrangian mesh, is D/Dt, which we used above. When spatial derivatives are required in the equations they are with respect to ordinary fixed coordinates, and they have to be calculated first with respect to the dragged-along Lagrangian mesh, then transformed using the chain rule to the fixed-space variables. When the flow is highly rotational, not to mention turbulent, the Lagrangian mesh is increasingly distorted as time goes by, and eventually this transformation introduces so much error that the Lagrangian mesh can no longer be used. Thus in rotational problems (in the sense $\nabla \times \mathbf{u} \neq 0$) Lagrangian coordinates are limited-life components. In some cases, for example spherical symmetry, this problem never arises, and the Lagrangian mesh can be used indefinitely.

The main equations of the Lagrangian method have been mentioned already. Mass conservation deserves special mention. Because the mesh follows the material, conservation of mass is guaranteed. What is needed, however, is the formula to compute the density from the current mapping from Lagrangian space to fixed space. If we call our Lagrangian coordinates ξ, η, ζ, and the Eulerian coordinates x,y, and z, then the true volume of a cell corresponding to a Lagrangian volume $d\xi d\eta d\zeta$ is $dV = d\xi d\eta d\zeta\, \partial(x, y, z)/\partial(\xi, \eta, \zeta)$. The last factor is the Jacobian of the mapping. Since the mass of the cell is constant, the density varies exactly as the reciprocal of the Jacobian. It is not hard to show that the logarithmic time derivative of the Jacobian is exactly $\mathbf{\nabla} \cdot \mathbf{u}$, as expected from $D\rho/Dt + \rho\mathbf{\nabla} \cdot \mathbf{u} = 0$, another form of (2.2). The Lagrangian momentum equation is found by applying the identity (2.6) in reverse to the Eulerian momentum equation (2.3) to give

$$\frac{D\mathbf{u}}{Dt} + \frac{1}{\rho}\nabla p = \frac{F}{\rho}. \tag{2.10}$$

In practical calculations the total force is found by combining the specified volume force F and the pressure gradient term, of which the latter is obtained using the chain rule for differentiation as mentioned above. Then the total force, converted to an acceleration, is used to update the velocity of the Lagrangian mesh. The velocity is used to move the mesh to its position for the next time step. After the mesh is moved, the Jacobian of the mapping can be recalculated, and with it the densities of all the Lagrangian cells.

The energy update is done with the Lagrangian internal energy equation (2.4), which we have discussed already. The pdV term can be evaluated since the change in density is now known, and the result is a new internal energy for every cell.

This is enough information to enable the new pressures to be evaluated, and the calculation can proceed to the next time step.

2.4 Moving mesh – ALE

There are complementary advantages of the Eulerian and Lagrangian pictures. The Eulerian picture has the advantage of a regular mesh – often Cartesian – which makes it possible to construct relatively sophisticated finite-difference or finite-element numerical representations of the Euler equations. The truncation terms can be bounded, and adjustments to the time step or to the mesh can be made if the estimated error is too great. A disadvantage is that the advection terms in the equations (terms like $\mathbf{u} \cdot \nabla\rho$), absent in the Lagrangian picture, are hard to represent accurately, and Eulerian methods in the past often produced unacceptable smearing of contact discontinuities, interfaces carried along with the fluid that separate two different materials, or regions of different entropy and temperature or transverse component of velocity. In the Lagrangian picture contact discontinuities are not a problem, since the mesh follows the matter. The serious problem is that the mesh becomes progressively distorted as the calculation proceeds, and the accuracy of differencing in the distorted mesh is of a lower order. Eventually the distortion results in zones being turned inside out or otherwise grossly disturbed, and the calculation simply stops. At this point, a new mesh has to be created by hand, and all the fluid variables must be interpolated from the old, distorted mesh to the new one. This interpolation is by no means very accurate, not least since the operation is not applied until the mesh distortion is already severe. The accuracy in following contact discontinuities normally possessed by the Lagrangian method may all be lost in the interpolation process.

Modern techniques have improved both Lagrangian and Eulerian methods. Advances in numerical algorithms have produced treatments of the advection problem that allow sharp definition of contact discontinuities even as they propagate across many cells of the Eulerian mesh, with the result that advection is no longer the Achilles heel of Eulerian methods. Sophisticated rezoning techniques that are applied automatically within the code enable Lagrangian codes to keep running long after they would formerly have crashed. A general formulation that amounts to applying a rezoning operation every time step is called arbitrary Lagrangian Eulerian, or ALE. The idea is that the computational mesh moves with respect to fixed space with a velocity $\mathbf{v}_g$, a function of space and time that is whatever the hydrodynamics code, through an adaptive procedure, decides to make it. If $\mathbf{v}_g$ were zero, the method would be Eulerian; if $\mathbf{v}_g$ were the fluid velocity $\mathbf{u}$, the method would be Lagrangian. Since $\mathbf{v}_g$ is arbitrary, we have the name of the method. The ALE equations are found by replacing the Eulerian time derivative $\partial/\partial t$ with

$(\partial/\partial t)_g - \mathbf{v}_g \cdot \nabla$, where $(\partial/\partial t)_g$ is the time derivative at a fixed point in the mesh. The result is a set of Eulerian-like equations including extra terms involving $\mathbf{v}_g$:

$$\left(\frac{\partial \rho}{\partial t}\right)_g + \nabla \cdot [\rho(\mathbf{u} - \mathbf{v}_g)] = -\rho \nabla \cdot \mathbf{v}_g, \tag{2.11}$$

$$\left(\frac{\partial \rho \mathbf{u}}{\partial t}\right)_g + \nabla \cdot [\rho(\mathbf{u} - \mathbf{v}_g)\mathbf{u}] + \nabla p = F - \rho \mathbf{u} \nabla \cdot \mathbf{v}_g, \tag{2.12}$$

$$\left(\frac{\partial}{\partial t}\right)_g \left(\rho e + \frac{1}{2}\rho u^2\right) + \nabla \cdot \left[\rho(\mathbf{u} - \mathbf{v}_g)\left(e + \frac{1}{2}u^2\right) + p\mathbf{u}\right]$$
$$= \rho q + \mathbf{u} \cdot F - \rho \left(\frac{1}{2}u^2 + e\right) \nabla \cdot \mathbf{v}_g. \tag{2.13}$$

We see that the advection fluxes, instead of being proportional to the mass flux $\rho \mathbf{u}$, as in the Eulerian method, are proportional to the flux $\rho(\mathbf{u} - \mathbf{v}_g)$ that crosses the moving zone boundary. The residual terms in $\nabla \cdot \mathbf{v}_g$ are due to the changing volumes of the ALE zones.

We can make a useful transformation of these equations by introducing the definition $J = \partial(x, y, z)/\partial(\xi, \eta, \zeta)$ for the Jacobian of the mapping from ALE coordinates to Cartesian coordinates. The volume of an ALE zone varies with time in proportion to the local value of J. The product ρJ is proportional to the zone mass. The identity relating J and $\mathbf{v}_g$ is

$$\frac{1}{J}\left(\frac{\partial J}{\partial t}\right)_g = \nabla \cdot \mathbf{v}_g. \tag{2.14}$$

By introducing J as a factor in the time derivatives in (2.11)–(2.13) the $\nabla \cdot \mathbf{v}_g$ terms are absorbed, which gives the ALE equations:

$$\frac{1}{J}\left(\frac{\partial \rho J}{\partial t}\right)_g + \nabla \cdot [\rho(\mathbf{u} - \mathbf{v}_g)] = 0, \tag{2.15}$$

$$\frac{1}{J}\left(\frac{\partial \rho J \mathbf{u}}{\partial t}\right)_g + \nabla \cdot [\rho(\mathbf{u} - \mathbf{v}_g)\mathbf{u}] + \nabla p = F, \tag{2.16}$$

$$\frac{1}{J}\left(\frac{\partial}{\partial t}\right)_g \left[\rho J\left(e + \frac{1}{2}u^2\right)\right] + \nabla \cdot \left[\rho(\mathbf{u} - \mathbf{v}_g)\left(e + \frac{1}{2}u^2\right) + p\mathbf{u}\right] = \rho q + \mathbf{u} \cdot F. \tag{2.17}$$

There is a great deal of freedom in choosing $\mathbf{v}_g$, and ALE codes include sophisticated routines for making an optimum choice in which the goals of following the material and keeping the mesh regular and orthogonal, if possible, are balanced. The algorithms for doing this vary with the ALE code.

The numerical methods used to solve this set of equations are much like those for the Eulerian method, which are too varied to discuss here. A strategy often used for both Eulerian and ALE codes is to break the time step into two parts: a first part in which the advection fluxes are dropped, i.e., a pure Lagrangian step, followed by a remap process that accounts for the amount of advection during the time step.

2.5 Transport terms: viscosity and heat conduction

So far we have considered ideal fluids for which we can completely ignore the possibility that individual atoms may migrate from a fixed position with respect to the mean fluid. In reality, of course, the atoms travel a nonzero distance between collisions, and so the atoms that populate any small parcel of fluid had their last collisions some finite distance away, in various random directions. This means that the local velocity distribution function may not be exactly isotropic, but will include fluctuations related to the gradients of temperature and fluid velocity. For example, if the fluid velocity component u_y is increasing in the x direction, an atom that arrives in the parcel traveling generally in the $+x$ direction will have a u_y value that is biased negatively compared with the average value in the parcel. The amount of bias is of the order of the mean free path times the x-gradient of u_y. Alternatively, if the temperature is increasing in the x direction, that atom that arrives in the parcel traveling generally toward $+x$ will have a small negative bias in its kinetic energy. When these biases are evaluated using an accurate kinetic theory model, what result are corrections (viscous stress) to the ideal pressure and a nonzero heat flux.

The viscous stress and the conductive heat flux are expressed in terms of coefficients that are either empirical constants or derived from kinetic theory. For terrestrial gases and liquids the values are tabulated in handbooks; for plasmas and for conditions very different from those realizable on earth, theoretical values must be used. Viscosity results in the replacement of the isotropic pressure with a stress tensor:

$$p\delta_{ij} \to P_{ij} = p\delta_{ij} - \sigma_{ij}, \tag{2.18}$$

in component notation (i and j are free indices running from 1 to 3), and δ_{ij} is the Kronecker delta, 1 if $i = j$ and 0 otherwise. The viscous stress σ_{ij} is determined by the rate-of-strain tensor according to

$$\sigma_{ij} = \mu \left(u_{i,j} + u_{j,i} - \frac{2}{3} u_{k,k}\delta_{ij} \right) + \zeta u_{k,k}\delta_{ij}. \tag{2.19}$$

(In tensor component equations like this a subscript following a comma indicates differentiation by the coordinate with that index, and a repeated index, such as k in this case, is to be summed over.) The coefficient μ is the (ordinary) coefficient of viscosity, and ζ is the coefficient of bulk viscosity. The inclusion of the ζ term often causes a lengthy discussion, since there are some good reasons for thinking it should be zero, and the kinetic theory models generally give a zero value. Experimental measurements of the bulk viscosity are elusive, because its effect vanishes in an incompressible flow where $u_{k,k} = \nabla \cdot \mathbf{u} = 0$, and compressible flows are generally high-velocity flows, for which the Reynolds number is large and viscous effects are therefore unimportant. The order of magnitude of μ is ρ times the mean thermal speed of an atom times the atomic mean free path. Because the mean free path is proportional to the reciprocal of the atomic cross section times the number density of atoms, the density factors cancel out in the coefficient of viscosity, which should therefore be nearly independent of density, though varying somewhat with temperature.

Including the viscous stress in Euler's momentum conservation equation leads to the first form of the Navier–Stokes equation:

$$\frac{\partial \rho \mathbf{u}}{\partial t} + \nabla \cdot (\rho \mathbf{u}\mathbf{u}) + \nabla p = F + \nabla \cdot \{\mu[\nabla \mathbf{u} + (\nabla \mathbf{u})^{\mathrm{T}}]\} + \nabla \left[\left(\zeta - \frac{2}{3}\mu \right) \nabla \cdot \mathbf{u} \right]. \tag{2.20}$$

The notation for the tensor components that is used here is that if i is the row index and j is the column index, so that the divergence operation involves summing on i, then the components of $\nabla \mathbf{u}$ are $u_{i,j}$ and those of its transpose, $\nabla \mathbf{u}^{\mathrm{T}}$, are $u_{j,i}$.

As mentioned above, the viscosity coefficients are not too strongly dependent on the state of the material. If this dependence is neglected, then the coefficients can be taken outside the spatial derivatives to give the second form of the Navier–Stokes equation:

$$\frac{\partial \rho \mathbf{u}}{\partial t} + \nabla \cdot (\rho \mathbf{u}\mathbf{u}) + \nabla p = F + \mu \nabla^2 \mathbf{u} + \left(\zeta + \frac{1}{3}\mu \right) \nabla \nabla \cdot \mathbf{u}, \tag{2.21}$$

and since we are frequently interested in viscous effects when the flow is incompressible ($\nabla \cdot \mathbf{u} = 0$), we have the third form:

$$\frac{\partial \rho \mathbf{u}}{\partial t} + \nabla \cdot (\rho \mathbf{u}\mathbf{u}) + \nabla p = F + \mu \nabla^2 \mathbf{u}. \tag{2.22}$$

There are no viscous effects for mass conservation, since the fluid velocity is defined as the mean mass flux divided by the density. The viscous term in the energy equation will be considered after heat conduction is discussed. Thermal conduction is very simple; Fourier's law is

$$F_{\text{heat}} = -K\nabla T. \tag{2.23}$$

The thermal conductivity K is tabulated in handbooks for terrestrial materials, and estimated theoretically using methods of kinetic theory or condensed matter theory for others. The magnitude of K is approximately the mass density times the specific heat at constant volume times the atomic mean free path. So for gases, the thermal conductivity and viscosity are closely related. Anticipating some of the later discussion, we note that the ratio of the diffusivity of momentum, μ/ρ, to the diffusivity of heat, $K/(\rho C_P)$, defines the Prandtl number, and that for a hard-sphere monatomic gas it has the value 2/3. The effect of heat conduction on the energy equation is to add a volume energy source $-\nabla \cdot F_{\text{heat}}$ to the right-hand side. The effect of viscosity on the energy equation comes because the rate of doing work by the viscous stress subtracts from the internal energy. The term $p\nabla \cdot \mathbf{u}$ in (2.7) is modified by subtracting $\Phi = \sigma_{ij}u_{i,j}$, where a sum over i and j is implied. This quantity Φ is called the dissipation function, and can be written as

$$\Phi = \frac{1}{2}\mu(u_{i,j} + u_{j,i})(u_{i,j} + u_{j,i}) + \left(\zeta - \frac{2}{3}\mu\right)(\nabla \cdot \mathbf{u})^2. \tag{2.24}$$

The dissipation function is never negative. The μ part is nonnegative, and vanishes if and only if the strain rate is isotropic, i.e., uniform dilation. The ζ part is also nonnegative (for $\zeta > 0$) and vanishes if and only if the bulk expansion $\nabla \cdot \mathbf{u}$ vanishes. The internal energy equation (2.7) is therefore modified to

$$\frac{\partial \rho e}{\partial t} + \nabla \cdot (\rho \mathbf{u} e) + p\nabla \cdot \mathbf{u} = \rho q + \Phi + \nabla \cdot (K\nabla T). \tag{2.25}$$

The left-hand side can also be written $\rho[De/Dt + pD(1/\rho)/Dt]$, which is equal to $\rho Ds/Dt$, where s is the specific entropy. From this we see that the viscous dissipation function Φ always acts to increase the entropy. With the inclusion of both viscosity and heat conduction, the total energy conservation equation becomes

$$\begin{aligned} \frac{\partial}{\partial t}\left(\rho e + \frac{1}{2}\rho u^2\right) &+ \nabla \cdot \left\{\rho \mathbf{u} e + \frac{1}{2}\rho \mathbf{u} u^2 + p\mathbf{u}\right. \\ &\left. -\mu\left[\mathbf{u} \cdot \nabla \mathbf{u} + \nabla\left(\frac{1}{2}u^2\right)\right] - \left(\zeta - \frac{2}{3}\mu\right)(\nabla \cdot \mathbf{u})\mathbf{u} - K\nabla T\right\} \\ &= \rho q + \mathbf{u} \cdot F, \end{aligned} \tag{2.26}$$

where now q refers to heat deposition other than from thermal conduction, and F continues to be the externally applied force per unit volume.

2.6 Bernoulli's equation and applications

A couple of simple manipulations of Euler's equations give results that are very useful for applications. The first is Bernoulli's equation. We begin by observing this identity for the velocity gradient term in the acceleration equation:

$$\mathbf{u} \cdot \nabla \mathbf{u} = \nabla \frac{1}{2} u^2 - \mathbf{u} \times \nabla \times \mathbf{u}. \tag{2.27}$$

Furthermore, suppose that the external force F is a body force derived from a potential, V, viz., $F = -\rho \nabla V$. Then the acceleration equation becomes

$$\frac{\partial \mathbf{u}}{\partial t} + \nabla \left(\frac{1}{2} u^2 \right) - \mathbf{u} \times \nabla \times \mathbf{u} + \frac{1}{\rho} \nabla p + \nabla V = 0. \tag{2.28}$$

In the case of steady flow, when all the variables are time-independent, we can drop the $\partial \mathbf{u}/\partial t$ term, then form the dot product of this equation with $\mathbf{u}$, and obtain

$$\mathbf{u} \cdot \nabla \left(\frac{1}{2} u^2 + \int \frac{dp}{\rho} + V \right) = 0, \tag{2.29}$$

which says that the quantity in parentheses has a constant value on any given streamline. This is the weak form of Bernoulli's law. It is useful if there is a functional relation between p and ρ (barotropic law) such as is provided in adiabatic flow or isothermal flow. In the adiabatic case the integral $\int dp/\rho$ becomes the enthalpy and in the isothermal case it becomes the Gibbs free energy, for example.

A stronger form of Bernoulli's law results for irrotational flows, i.e., flows for which the vorticity, $\nabla \times \mathbf{u}$, vanishes. As we know from electrostatics, when $\nabla \times \mathbf{u}$ vanishes globally $\mathbf{u}$ can be derived from a potential, $\mathbf{u} = \nabla \phi$. Making this replacement and dropping the vorticity term in (2.28) leads to

$$\nabla \left(\frac{\partial \phi}{\partial t} + \frac{1}{2} u^2 + \int \frac{dp}{\rho} + V \right) = 0, \tag{2.30}$$

which says that the quantity in parentheses is constant in all space, though it may vary with time. If the flow is both steady and irrotational, the $\partial \phi/\partial t$ term may be dropped, which then says that the same Bernoulli constant as in the weak form is in fact uniform over all space, not just on a streamline.

Potential flow is the name for flows that are both irrotational and incompressible. By virtue of the first assumption we can write $\mathbf{u} = \nabla \phi$; by virtue of the second we can require $\nabla \cdot \mathbf{u} = 0$, so the velocity potential ϕ must satisfy Laplace's equation $\nabla^2 \phi = 0$. The problem is virtually solved at that point, since we can use the methods of potential theory, boundary-value problems and the like, to find the solution for ϕ. The pressure need not even be considered in this procedure, depending on the boundary conditions. As a last step, Bernoulli's equation in its strong form is used to find the pressure. Incompressible flow is always barotropic since $\int dp/\rho = p/\rho$ when ρ is uniform and constant.

As mentioned above, the vorticity is defined by $\boldsymbol{\omega} = \nabla \times \mathbf{u}$. A helpful equation for understanding vorticity is found by taking the curl of the Eulerian acceleration equation (2.28):

$$\frac{\partial \boldsymbol{\omega}}{\partial t} - \nabla \times (\mathbf{u} \times \boldsymbol{\omega}) - \frac{\boldsymbol{\nabla}\rho \times \boldsymbol{\nabla}p}{\rho^2} = \nabla \times \frac{F}{\rho}. \tag{2.31}$$

This is an interesting equation, since it says that vorticity cannot be produced in a barotropic flow that has an external body force derived from a potential, if any. This is because the barotropic relation $p = f(\rho)$ implies that $\boldsymbol{\nabla}p$ is parallel to $\boldsymbol{\nabla}\rho$, and therefore the cross product term, called the baroclinic term, vanishes. The integral Γ of the vorticity over a surface S bounded by a curve C is called the circulation around C, since by Stokes's theorem,

$$\Gamma = \int_S \boldsymbol{\omega} \cdot d\mathbf{A} = \oint_C \mathbf{u} \cdot d\mathbf{l}. \tag{2.32}$$

It is a calculus exercise to show that the equation

$$\frac{\partial \boldsymbol{\omega}}{\partial t} = \nabla \times (\mathbf{u} \times \boldsymbol{\omega}) \tag{2.33}$$

obtained by dropping the baroclinic term and the external force term implies that Γ is constant in time for any curve C that is carried along by the fluid. In this sense, vorticity is a conserved quantity in the absence of the generation terms.

Vorticity is also the basis of numerical methods for incompressible hydrodynamics. Using the analogy $\mathbf{u}$ is to $\boldsymbol{\omega}$ as magnetic induction $\mathbf{B}$ is to current density $\mathbf{J}$, appropriate for the incompressible case, $\mathbf{u}$ can be reconstituted from $\boldsymbol{\omega}$. Thus we can regard $\boldsymbol{\omega}$ as the basic unknown and derive $\mathbf{u}$ from it, then use the vorticity equation to update the vorticity, i.e., move the vortices. Vortex dynamics is the name for the method that uses this approach with a finite set of line vortices. This approach is outstandingly successful in modeling Jupiter's Great Red Spot, to pick one example. (See Marcus (1993).)

Even in a flow that has quite a high Reynolds number, and for which, therefore, the viscous effects should be quite small, there may be viscous boundary layers, since the flow solution in the interior of the problem, which is essentially inviscid, may not obey the true boundary conditions. An example is flow over a surface, where the flow velocity must be zero at the surface, but is some finite value a moderate distance away. In this case a rapid transition occurs in a layer next to the surface to join the interior solution to the required boundary condition. The thickness of the transition layer is proportional to the small viscosity, since only by having a large gradient can the small viscosity term have a macroscopic effect.

These boundary layers can act as producers of vorticity, which then is transported into the bulk downstream flow.

Deposition of heat is another mechanism by which vorticity can be generated. The addition of heat can easily have a spatial dependence that results in the pressure no longer being strictly a function of the density. Once this occurs there is baroclinic vorticity production. As we shall see, the same is true of shock waves. A shock wave that is not uniform in the transverse direction results in an entropy increase that has a transverse gradient, and therefore a nonzero baroclinic term.

As an example of the use of the method of potential flow and the strong form of Bernoulli's law, we consider the case of two superposed incompressible fluids, possibly of different densities, which may also be in relative horizontal motion, and acted upon by vertical gravity. When the fluid interface is perturbed slightly, surface waves, or possibly instabilities may be produced. The waves are deep-water waves, and the instabilities are Rayleigh–Taylor or Kelvin–Helmholtz, depending on the setup.

The unperturbed state is of density ρ_1 for $z > 0$ and density ρ_2 for $z < 0$, and a velocity in the x direction of U_1 for $z > 0$ and U_2 for $z < 0$. The perturbed surface is displaced by a small amount z_s in the z direction, which we take to be $\zeta \exp(i\mathbf{k}_\mathrm{h} \cdot \mathbf{r} - i\omega t)$. The vector $\mathbf{k}_\mathrm{h}$ has only x and y components, and we are looking for the dispersion relation that gives ω in terms of $\mathbf{k}_\mathrm{h}$. A real ω means surface waves; a complex ω with a positive imaginary part means an unstable interface.

In the unperturbed state, the pressure is uniform across the plane $z = 0$, and has the same value on either side of the interface. We will take this as the zero of pressure. When the surface is perturbed, a flow exists in both $z > 0$ and $z < 0$, but we expect the flow in each region to be irrotational. If U_1 and U_2 are different, there is a vortex sheet at $z = 0$ in the unperturbed state, and we will see that even when U_1 and U_2 are equal, vorticity may develop on the perturbed interface. So the solution we are looking for is a potential flow in each region, for which we then require matching on the interface. The quantities that have to match at the interface are the normal component of velocity and the pressure.

We let ϕ_1 and ϕ_2 be the velocity potentials in the respective regions. Each of these satisfies Laplace's equation. Since we have chosen the interface perturbation to have the horizontal variation $\exp(i\mathbf{k}_\mathrm{h} \cdot \mathbf{r})$, the appropriate harmonic functions are $\exp(\pm k_h z + i\mathbf{k}_\mathrm{h} \cdot \mathbf{r})$, where k_h is the vector magnitude of $\mathbf{k}_\mathrm{h}$. Since the fluctuations should vanish far from the interface, we choose the $-$ sign for $z > 0$ and the $+$ sign for $z < 0$. To these are added the potentials corresponding to the unperturbed uniform flows, $U_1 x$ or $U_2 x$. Thus we put

$$\phi = \begin{cases} U_1 x + \psi_1 \exp(-k_h z + i\mathbf{k}_\mathrm{h} \cdot \mathbf{r} - i\omega t) = \phi_1, & z > 0, \\ U_2 x + \psi_2 \exp(k_h z + i\mathbf{k}_\mathrm{h} \cdot \mathbf{r} - i\omega t) = \phi_2, & z < 0. \end{cases} \tag{2.34}$$

The normal velocity matching proceeds as follows. The direction of the surface normal is the vector

$$-ik_x\zeta\exp(i\mathbf{k}_{\mathrm{h}}\cdot\mathbf{r}-i\omega t)\mathbf{e}_x-ik_y\zeta\exp(i\mathbf{k}_{\mathrm{h}}\cdot\mathbf{r}-i\omega t)\mathbf{e}_y+\mathbf{e}_z. \tag{2.35}$$

The vectors $\mathbf{e}_x$, $\mathbf{e}_y$ and $\mathbf{e}_z$ are unit vectors in the three coordinate directions. The component along the normal of the velocity of the interface is just $\partial z_s/\partial t$ to first order, or

$$-i\omega\zeta\exp(i\mathbf{k}_{\mathrm{h}}\cdot\mathbf{r}-i\omega t). \tag{2.36}$$

The fluid velocity in region 1 at the interface is

$$(U_1+ik_x\psi_1\exp(i\mathbf{k}_{\mathrm{h}}\cdot\mathbf{r}-i\omega t))\mathbf{e}_x+ik_y\psi_1\exp(i\mathbf{k}_{\mathrm{h}}\cdot\mathbf{r}-i\omega t)\mathbf{e}_y$$
$$-k_h\psi_1\exp(i\mathbf{k}_{\mathrm{h}}\cdot\mathbf{r}-i\omega t)\mathbf{e}_z. \tag{2.37}$$

Taking the component of this along the normal vector gives a part from U_1 due to the tilt of the surface in addition to the vertical component of the fluid velocity, i.e.,

$$-ik_x\zeta\exp(i\mathbf{k}_{\mathrm{h}}\cdot\mathbf{r}-i\omega t)U_1-k_h\psi_1\exp(i\mathbf{k}_{\mathrm{h}}\cdot\mathbf{r}-i\omega t). \tag{2.38}$$

Matching this to the velocity of the interface gives

$$-i\omega\zeta=-k_h\psi_1-ik_xU_1\zeta. \tag{2.39}$$

Repeating the argument for region 2 gives

$$-i\omega\zeta=k_h\psi_2-ik_xU_2\zeta. \tag{2.40}$$

Next we have to evaluate the pressure approaching the interface from each of the two sides, and match them. In each region the pressure follows from the strong form of Bernoulli's law, which gives

$$p=-\rho\left(gz+\frac{\partial\phi}{\partial t}+\frac{1}{2}u^2-\frac{1}{2}U_\infty^2\right) \tag{2.41}$$

in each region separately. Here U_∞ is U_1 or U_2, as appropriate. This additive constant has been chosen to ensure that p matches the unperturbed value $-\rho gz$ far from the interface. We put in (2.34) for ϕ and $\mathbf{u}=\nabla\phi$, and expand, dropping the second order terms. Then setting $z=0$ we get the upper-side and lower-side interface pressures:

$$p(0+)=-\rho_1(g\zeta-i\omega\psi_1+U_1ik_x\psi_1)\exp(i\mathbf{k}_{\mathrm{h}}\cdot\mathbf{r}-i\omega t), \tag{2.42}$$
$$p(0-)=-\rho_2(g\zeta-i\omega\psi_2+U_2ik_x\psi_2)\exp(i\mathbf{k}_{\mathrm{h}}\cdot\mathbf{r}-i\omega t). \tag{2.43}$$

We next equate these two expressions to each other, and substitute for ψ_1 and ψ_2 from (2.39) and (2.40). Canceling the factor ζ and the exponential then gives the dispersion relation

$$(\rho_2 - \rho_1)g - \rho_1 \frac{(\omega - k_x U_1)^2}{k_h} - \rho_2 \frac{(\omega - k_x U_2)^2}{k_h} = 0. \tag{2.44}$$

When this is arranged as a quadratic equation in ω, it becomes

$$(\rho_1 + \rho_2)\omega^2 - 2k_x(\rho_1 U_1 + \rho_2 U_2)\omega + k_x^2(\rho_1 U_1^2 + \rho_2 U_2^2) - (\rho_2 - \rho_1)k_h g = 0. \tag{2.45}$$

Solving it, we get

$$\omega = k_x \frac{\rho_1 U_1 + \rho_2 U_2}{\rho_1 + \rho_2} \pm \sqrt{\frac{\rho_2 - \rho_1}{\rho_1 + \rho_2} k_h g - k_x^2 \frac{\rho_1 \rho_2 (U_1 - U_2)^2}{(\rho_1 + \rho_2)^2}}. \tag{2.46}$$

This is the main result. We can directly apply it to three different problems. The first is deep-water waves, for which ρ_1 is negligible (since region 1 is air) and U_1 and U_2 are both zero. The dispersion relation gives real frequencies $\omega = \pm\sqrt{k_h g}$. These are dispersive waves, and the phase velocity $V_p = \omega/k_h = \sqrt{g/k_h}$ increases with the wavelength. The group velocity $d\omega/dk_h = 1/2\sqrt{g/k_h}$ is half the phase velocity. So for a wave packet of water waves, the wave crests appear at the rear of the packet, ride up over the top, and disappear at the front.

The second case is Rayleigh–Taylor instability, for which $\rho_1 > \rho_2$ and $U_1 = U_2 = 0$. The values of ω are imaginary, and in particular one root is $\omega = i\sqrt{(\rho_1 - \rho_2)k_h g/(\rho_1 + \rho_2)}$. This root leads to exponential growth in time with a growth rate given by the square root factor. We see that the rate depends on the Atwood number $\alpha = (\rho_1 - \rho_2)/(\rho_1 + \rho_2)$, which is a positive number less than 1. The growth rate depends only on the vector magnitude of k_h, so any planform z_h that satisfies $(\nabla_h^2 + k_h^2)z_h = 0$, whether rolls, checkerboard or Bessel function, will lead to the same growth rate.

The third case is the Kelvin–Helmholtz instability, with $\rho_1 = \rho_2$ and $U_1 \neq U_2$. The roots are imaginary, and the growth rate is $(1/2)k_x|\Delta U|$. The analysis here, it must be remembered, has neglected viscosity and surface tension, among other things. These dissipative effects will stabilize the instabilities if the wavenumbers are high, for this increases their relative importance.

A final remark concerns the vorticity in the Rayleigh–Taylor problem. All the vorticity lies in the interface itself; it is a vortex sheet. But at the initial time the amount of vorticity is negligible. As the instability grows, however, a shear develops across the interface since ψ_1 and ψ_2 have opposite signs, as we see from (2.39) and (2.40). So we have zones of positive and negative vorticity in the interface that exponentiate in magnitude with time, along with the size of the interface perturbation. The total vorticity may remain nearly zero, but the positive and negative accumulations both increase. Eventually this shear helps amplify the instability, as in the Kelvin–Helmholtz case. This example helps illustrate the point that vorticity can be created in a problem through the action of discontinuities or the boundaries.

2.7 Sound waves

We will illustrate how to derive linear waves from the Euler equations. As more physical processes are added to the equations, the character of the waves changes, and perhaps new wave modes appear; exploring these is a way of gaining insight into the consequences of those new processes. A systematic presentation of the interaction of waves with radiation is contained in Chapter 8 of Mihalas and Mihalas (1984).

For the simplest possible waves, suppose we have a base state consisting of a uniform medium at rest, and we suppose that this state is perturbed by small amounts in density, velocity, and pressure, and suppose further that a barotropic relation $p \sim \rho^\gamma$ describes the variations. Let ρ' and p' be the fluctuations in density and pressure, and $\mathbf{u}$ itself is the fluctuation in velocity since the base state is at rest. We linearize the continuity and momentum equations by Taylor-expanding around the base state, and discard all terms of second or higher order in the fluctuations. The flow variables for the base state will keep their usual names after linearizing, since there is no possibility of confusing them with the fluctuations. The continuity equation becomes

$$\frac{\partial \rho'}{\partial t} + \rho \nabla \cdot \mathbf{u} = 0, \tag{2.47}$$

and the momentum equation becomes

$$\frac{\partial \mathbf{u}}{\partial t} + \frac{\gamma p}{\rho^2} \nabla \rho' = 0. \tag{2.48}$$

Now we seek sinusoidal wave solutions that are proportional to the complex exponential factor $\exp(i\mathbf{k} \cdot \mathbf{r} - i\omega t)$. That is, we can replace the time derivative operator by $-i\omega$ and the gradient operator by $i\mathbf{k}$. After doing this, our two equations turn into this 4×4 linear system for ρ' and the three components of $\mathbf{u}$:

$$\begin{pmatrix} -\omega & \rho \mathbf{k}^{\mathrm{T}} \\ (\gamma p/\rho^2)\mathbf{k} & -\omega \mathsf{I} \end{pmatrix} \begin{pmatrix} \rho' \\ \mathbf{u} \end{pmatrix} = 0, \tag{2.49}$$

where $\mathbf{k}$ stands for the wave vector as a column vector, $\mathbf{k}^{\mathrm{T}}$ is its transpose, a row vector, and I is the 3×3 identity matrix. In order for a wave mode to exist, this homogeneous linear system must have a nonzero solution, which means that its determinant, the dispersion function $D(\mathbf{k}, \omega)$, must vanish. The determinant is evaluated by adding $\rho k_x/\omega$ times the second row plus $\rho k_y/\omega$ times the third row plus $\rho k_z/\omega$ times the fourth row to the first row, which eliminates all the elements of $\rho \mathbf{k}^T$ in that row. Then expanding by the elements in the first row, as modified, gives

$$D(\mathbf{k}, \omega) = \left(-\omega + \frac{\gamma p}{\rho}\frac{\mathbf{k}\cdot\mathbf{k}}{\omega}\right)(-\omega^3) = \omega^2\left(\omega^2 - \frac{\gamma p}{\rho}k^2\right). \tag{2.50}$$

We write scalar k for the magnitude of the vector $\mathbf{k}$. We use this dispersion relation in the following way. Let $\mathbf{k}$ be any real vector. Then solve $D(\mathbf{k}, \omega) = 0$ for the roots, ω. Since D is fourth order in ω, there are four roots, corresponding to the four flow variables ρ, u_x, u_y, and u_z. Each root gives a wave mode. However, the pair $\omega = \pm\sqrt{\gamma p/\rho}k$ represent the two directions of propagation for this $\mathbf{k}$ and are thus considered together as one mode. The wave speed is the barotropic sound speed $c = \sqrt{\gamma p/\rho}$. The degenerate $\omega = 0$ modes exist because the two possible transverse polarizations do not produce any density fluctuation and therefore no restoring force. We can find the mode shape, i.e., the proportionality of ρ', u_x, u_y, and u_z for that mode, by substituting values of ω and $\mathbf{k}$ that obey the dispersion relation into (2.49) and solving for the amplitudes. For the nondegenerate modes in this case we get

$$\mathbf{u} = c^2\frac{\rho'}{\rho}\frac{\mathbf{k}}{\omega} = c\frac{\rho'}{\rho}\frac{\mathbf{k}}{k}, \tag{2.51}$$

so the mode is longitudinal and the velocity fluctuation is the sound speed times the fractional density fluctuation.

For a given mode, the phase velocity vector is in the direction of $\mathbf{k}$ with the magnitude of ω/k. For the case of the positive root, this is c. The group velocity is the gradient of ω with respect to $\mathbf{k}$, where ω is considered as a function of $\mathbf{k}$ and is evaluated on a consistent branch of the multivalued function. For the positive ω branch, this also is a vector of magnitude c in the direction of $\mathbf{k}$, the same as the phase velocity. They agree because these sound waves are not dispersive.

For a second example we will include the energy equation instead of assuming the barotropic relation, and suppose that there is a Newton's cooling type of coupling to an external heat bath, perhaps by means of radiation. (See Mihalas and Mihalas (1984), Section 101.) This means that the heat deposition term is proportional to the negative of the temperature fluctuation. We will use the ideal gas relation $e = p/[(\gamma - 1)\rho]$, and take q to be given by

$$q = \frac{e}{\tau}\left(\frac{\rho'}{\rho} - \frac{p'}{p}\right) = \frac{1}{(\gamma - 1)\tau}\left(\frac{p\rho'}{\rho^2} - \frac{p'}{\rho}\right). \tag{2.52}$$

The quantity τ, with dimension time, is the time constant for the temperature to relax to that of the heat bath.

The linearization of the internal energy equation (2.7) multiplied by $\gamma - 1$ gives

$$-i\omega\left(\frac{p'}{\rho} - \frac{p}{\rho^2}\rho'\right) + i\omega\frac{\rho'}{\rho^2}p(\gamma - 1) = (\gamma - 1)q = \frac{1}{\tau}\left(\frac{p\rho'}{\rho^2} - \frac{p'}{\rho}\right). \quad (2.53)$$

We can add this as a fifth equation to the previous set of four, and treat p' and ρ' as independent variables. Expanding the 5×5 determinant leads to a dispersion function that is fifth order in ω and second order in $\mathbf{k}$. A somewhat shorter route to the dispersion relation is to solve the linearized energy equation for the ratio p'/ρ', namely

$$\frac{p'}{\rho'} = \frac{p}{\rho}\frac{\gamma\omega\tau + i}{\omega\tau + i}, \quad (2.54)$$

in other words, exactly as if γ were replaced by the rational function

$$\frac{\gamma\omega\tau + i}{\omega\tau + i}. \quad (2.55)$$

The physical significance of this function is that the effective γ tends to 1 when the wave period is long compared with the cooling time constant, since then the oscillations are nearly isothermal, and it tends to the usual value when the period is much less than the cooling time constant, since there is too little time for any cooling to occur, and the oscillations are nearly adiabatic. When the effective γ is substituted into the dispersion relation for the acoustic modes we find, after clearing the linear function of ω in the denominator,

$$(\omega\tau + i)\omega^2 - \frac{c^2}{\gamma}(\gamma\omega\tau + i)k^2 = 0. \quad (2.56)$$

This cubic equation for ω has two roots that belong to the acoustic mode, and a single root that represents the cooling mode. We make this into a real, nondimensional equation by introducing a variable $\xi = i\omega\tau$. The equation for ξ is then

$$\xi^3 - \xi^2 + (kc\tau)^2\xi - \frac{(kc\tau)^2}{\gamma} = 0. \quad (2.57)$$

The exact roots of this cubic are messy (illustrated in Figure 2.1), but the roots are easily discussed in the limits of large and small $kc\tau$. When $kc\tau$ is large, which is the high-frequency limit, then one approximate root will be found by balancing the third and fourth terms in the polynomial, and the other two come from approximately balancing the first and third terms. The first gives $\xi \approx 1/\gamma$, or $\omega \approx -i/(\gamma\tau)$. The other two are $\xi \approx \pm ikc\tau$, or $\omega \approx \pm kc$. So this is an adiabatic sound wave, and a cooling mode in which the specific heat at constant pressure,

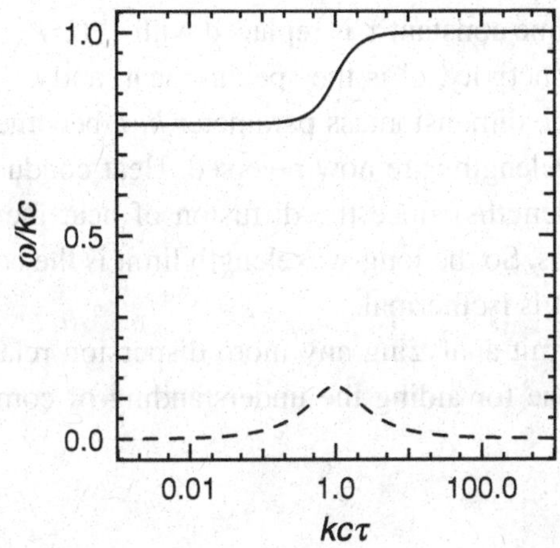

Fig. 2.1 Real and imaginary frequencies for sound waves with Newton's cooling. Ordinate, ω scaled by kc; abscissa, cooling time in units of $1/kc$. Solid line, real part; dashed line, imaginary part.

not constant volume, gives the decay time. This mode occurs at constant pressure since it is slow compared with the sound propagation time.

In the limit of small $kc\tau$ one approximate root comes from balancing the first and second terms, and the other two come from balancing the second and fourth. The first is thus $\xi \approx 1$, or $\omega \approx -i/\tau$, and the other two are $\xi \approx \pm ikc\tau/\sqrt{\gamma}$, or $\omega \approx \pm kc/\sqrt{\gamma}$. So in this low-frequency limit, the sound wave is isothermal, and the cooling mode uses the specific heat at constant volume, since cooling occurs too quickly for any motion to affect it.

The cooling term causes the sound wave to be damped; the roots for ω have a small negative imaginary part. To see this, we apply one step of Newton–Raphson iteration to account for the terms in the cubic that were neglected in the previous paragraph. For the $kc\tau$ large case, we get

$$\omega \approx kc - i\frac{\gamma - 1}{\gamma\tau}, \tag{2.58}$$

while for the small $kc\tau$ case the result is

$$\omega \approx \frac{kc}{\sqrt{\gamma}} - i\frac{\gamma - 1}{2\gamma}(kc)^2\tau. \tag{2.59}$$

Interestingly, the damping is small in both the case of no cooling and the case of strong cooling. Maximum damping occurs when the cooling time and the wave period are comparable.

All this analysis can be used virtually intact if we replace Newton's cooling, due perhaps to coupling with a radiation bath, with thermal conductivity. The only difference is that the time constant τ is replaced with $1/(k^2\kappa) = \rho C/(k^2 K)$. Here K is the thermal conductivity, C is the specific heat, and κ is the corresponding thermal diffusivity. The dimensionless parameter $kc\tau$ becomes $c/(k\kappa)$. The roles of long and short wavelengths are now reversed. Heat conduction has the largest effect for short wavelengths, since the diffusion of heat increases faster with k than the frequency does. So the long-wavelength limit is the adiabatic one, and the short-wavelength limit is isothermal.

Space does not permit analyzing any more dispersion relations, but this technique is a powerful one for aiding the understanding of complicated hyperbolic systems.

2.8 Characteristics and simple waves

The method of characteristics for solving hyperbolic partial differential equations (PDEs) is primarily helpful for analytic studies of the properties of the systems, and not so much for practical computations. (The exception is Godunov's method – see below.) It is also a 1-D method. Characteristics in two and three dimensions are of a different nature than those in one dimension, and will not be discussed. Characteristics are very well explained by Courant and Friedrichs (1948), as are deflagration and detonation waves, which are analogous to ionization fronts, to be discussed in the last chapter of the present work.

A 1-D hyperbolic system of partial differential equations is represented as follows:

$$\frac{\partial U}{\partial t} + \frac{\partial F(U)}{\partial x} = G, \tag{2.60}$$

where U is a vector of unknowns, and F is a vector whose elements are functions of the elements of U. The definition of a characteristic is a curve in space-time along which a total differential equation involving elements of U is obeyed.

Here is how we find the characteristics of a system like (2.60). We expand the derivative of F using the chain rule:

$$\frac{\partial U}{\partial t} + A\frac{\partial U}{\partial x} = G, \tag{2.61}$$

where now A is the square Jacobian matrix $\partial F/\partial U$. Suppose we can find a left eigenvector $\mathbf{m}^{\mathrm{T}}$ of A, a row vector, so that

$$\mathbf{m}^{\mathrm{T}} A = c\mathbf{m}^{\mathrm{T}}, \tag{2.62}$$

where c is a scalar that has dimensions of velocity. Multiplying (2.61) from the left by $\mathbf{m}^{\mathrm{T}}$ gives

$$\mathbf{m}^{\mathrm{T}}\left(\frac{\partial U}{\partial t}+c\frac{\partial U}{\partial x}\right)=\mathbf{m}^{\mathrm{T}}G. \tag{2.63}$$

This is our characteristic equation. It says that along the curve $dx/dt = c$ in space-time, the total differentials of the components of U obey

$$\mathbf{m}^{\mathrm{T}}dU=\mathbf{m}^{\mathrm{T}}G\,dt. \tag{2.64}$$

Each different eigenvector of A gives a different characteristic. If A has a full set of eigenvectors, as many as the components of U, then that is the number of characteristics. If A is defective and has fewer independent eigenvectors then the system is in fact not hyperbolic. The fan of characteristics going back into the past from a point (x, t) in space time covers all the points that directly influence the flow at (x, t). Adding on all the fans from those points as well fills out a cone-like region, called the domain of dependence of the point (x, t). What happened in the past at a point outside that cone cannot possibly affect conditions at (x, t).

If the system of PDEs is not in conservation-law form, we can still find the characteristics. Suppose the system is

$$M\frac{\partial U}{\partial t}+N\frac{\partial U}{\partial x}=G. \tag{2.65}$$

Then we look for generalized eigenvectors $\mathbf{m}^{\mathrm{T}}$ that obey

$$\mathbf{m}^{\mathrm{T}}N=c\mathbf{m}^{\mathrm{T}}M. \tag{2.66}$$

Given such an eigenvector, we again left-multiply the system by $\mathbf{m}^{\mathrm{T}}$ and obtain in this case

$$\mathbf{m}^{\mathrm{T}}M\left(\frac{\partial U}{\partial t}+c\frac{\partial U}{\partial x}\right)=\mathbf{m}^{\mathrm{T}}G, \tag{2.67}$$

or

$$\mathbf{m}^{\mathrm{T}}M\,dU=\mathbf{m}^{\mathrm{T}}G\,dt, \tag{2.68}$$

which, like (2.64), is a total differential equation along $dx/dt = c$.

Let's take the particular case of the Eulerian equations not in conservation-law form, viz.,

$$\frac{\partial \rho}{\partial t} + u\frac{\partial \rho}{\partial x} + \rho\frac{\partial u}{\partial x} = 0, \tag{2.69}$$

$$\frac{\partial u}{\partial t} + u\frac{\partial u}{\partial x} + \frac{1}{\rho}\frac{\partial p}{\partial x} = 0, \tag{2.70}$$

$$\frac{\partial s}{\partial t} + u\frac{\partial s}{\partial x} = 0. \tag{2.71}$$

The momentum and energy sources have been omitted for simplicity. The energy equation has been replaced by the entropy equation. (Algebraically equivalent systems of PDEs give rise to exactly the same characteristics, so we are free to do this.) If we take ρ, u, and s as the components of U, then we still need to expand the pressure derivative to get the form (2.65). The thermodynamic relations tell us that

$$dp = c^2 d\rho + (\gamma - 1)\rho T\, ds \tag{2.72}$$

for an ideal gas, where c is the adiabatic sound speed, $\sqrt{\gamma p/\rho}$.

After substituting relation (2.72) for $\partial p/\partial x$, we can write the system in the form

$$\frac{\partial}{\partial t}\begin{pmatrix} \rho \\ u \\ s \end{pmatrix} + \begin{pmatrix} u & \rho & 0 \\ c^2/\rho & u & (\gamma - 1)T \\ 0 & 0 & u \end{pmatrix} \frac{\partial}{\partial x}\begin{pmatrix} \rho \\ u \\ s \end{pmatrix} = 0. \tag{2.73}$$

We need the eigenvalues of the matrix N in the second term. Expanding the determinant of $N - v\mathsf{I}$, where I is the 3×3 identity matrix, yields $(u - v)[(u - v)^2 - c^2]$. So the eigenvalues, which we will now call v to avoid confusion with the sound speed, are $v = u$ and $v = u \pm c$. The left eigenvector for $v = u$ is $(0, 0, 1)$, the one for $v = u + c$ is $(c^2, \rho c, (\gamma - 1)\rho T)$, and the one for $v = u - c$ is $(c^2, -\rho c, (\gamma - 1)\rho T)$. So our characteristic equations are

$$c^2 d\rho + \rho c\, du + (\gamma - 1)\rho T\, ds = 0 \qquad \text{on } C_+\text{: } \frac{dx}{dt} = u + c, \tag{2.74}$$

$$c^2 d\rho - \rho c\, du + (\gamma - 1)\rho T\, ds = 0 \qquad \text{on } C_-\text{: } \frac{dx}{dt} = u - c, \tag{2.75}$$

$$ds = 0 \qquad \text{on } C_0\text{: } \frac{dx}{dt} = u. \tag{2.76}$$

We recognize the differential of the pressure in the first two equations, so we can write the system in the simpler form

$$du + \frac{dp}{\rho c} = 0 \qquad \text{on } C_+\text{: } \frac{dx}{dt} = u + c, \tag{2.77}$$

$$du - \frac{dp}{\rho c} = 0 \qquad \text{on } C_-\text{: } \frac{dx}{dt} = u - c, \tag{2.78}$$

$$ds = 0 \qquad \text{on } C_0\text{: } \frac{dx}{dt} = u. \tag{2.79}$$

In an isentropic flow the entropy is not only constant following a parcel of fluid, which is what adiabatic means and which is expressed by the C_0 characteristic equation, but is spatially uniform as well, so the entropy is everywhere and always the same (barring shocks). If this is the case, then $dp/(\rho c)$ is a total differential of the thermodynamic function

$$\int \frac{dp}{\rho c} = \frac{2c}{\gamma - 1}, \tag{2.80}$$

where the equality is valid for gamma-law gases. For this special case the first two characteristic equations are integrable, and take the form

$$r \equiv u + \frac{2c}{\gamma - 1} = \text{constant} \qquad \text{on } C_+\text{: } \frac{dx}{dt} = u + c, \tag{2.81}$$

$$l \equiv u - \frac{2c}{\gamma - 1} = \text{constant} \qquad \text{on } C_-\text{: } \frac{dx}{dt} = u - c. \tag{2.82}$$

These two functions, r and l, are called the Riemann invariants. The royal road to solving ideal gas dynamics problems of this class is to select a point where you want to know the flow variables, trace a C_+ characteristic back to the start time and evaluate r there, and this is therefore the value at the desired point. Also trace back a C_- characteristic to the start time and get l. (This tracing back may be easier said than done in some cases!) The average of r and l is the velocity at the desired point, and their difference determines the sound speed. Through the adiabatic relations, since we know the value of the entropy, we can get density, pressure, and so on.

This prescription actually works in the case of what are called simple waves. These are isentropic flows of an ideal gas for which one of the two Riemann invariants is the same everywhere. To be specific, suppose l is the same everywhere. Then consider what happens along a particular C_+ characteristic. This characteristic will have its own value of r, which will be constant along it. So on this characteristic both l and r are constant, which means that u and c are also constant, which means that the characteristic velocity is constant too. That is, this characteristic is a straight line in space-time. With different C_+ characteristics having different values of r their slopes will differ, so they will form a fan. If the fan opens with time it is a simple rarefaction wave; if it is converging with time it is a simple compression wave. In the latter case, the characteristics eventually intersect forming a cusp, and a discontinuity, a shock, is generated at that point. This situation in which one of the Riemann invariants is constant everywhere arises when a large area has uniform density, temperature, and entropy at the initial time. The characteristics that emanate from there will carry a constant Riemann invariant when they cross into an adjacent region. The result is the theorem, "The flow adjacent to a region of constant state is a simple wave" (Courant and Friedrichs, 1948).

A simple centered rarefaction wave is an illustration of this. Suppose that a uniform block of ideal gas located at $x \leq 0$ is confined by a membrane that is removed at $t = 0$. What happens is a rarefaction wave that eats into the gas at $x < 0$ with a velocity of $-c_0$, where c_0 is the sound speed in the undisturbed gas. The gas that is disturbed moves in the $+x$ direction with a velocity that increases with x. The density drops from the undisturbed value to very small values at positive x. In this problem there is a region of constant state on the left, so the Riemann invariant r is constant everywhere, and has the value $r = 2c_0/(\gamma - 1)$. The whole region of rarefaction belongs to the fan of C_- characteristics that emanate from the point $(x, t) = (0, 0)$. This is what makes it a centered rarefaction. The slope of the C_- characteristic through the point (x, t) is the slope of the line connecting this point to $(0, 0)$, namely x/t. Thus

$$\frac{x}{t} = u - c. \tag{2.83}$$

Combining this relation with

$$r = u + \frac{2c}{\gamma - 1} = \frac{2c_0}{\gamma - 1} \tag{2.84}$$

gives

$$u = \frac{2}{\gamma + 1}\left(\frac{x}{t} + c_0\right), \tag{2.85}$$

$$c = \frac{\gamma - 1}{\gamma + 1}\left(\frac{2}{\gamma - 1}c_0 - \frac{x}{t}\right), \tag{2.86}$$

from which we can also get the density profile

$$\rho = \rho_0 \left[\frac{\gamma - 1}{\gamma + 1}\left(\frac{2}{\gamma - 1} - \frac{x}{c_0 t}\right)\right]^{2/(\gamma-1)}, \tag{2.87}$$

with a similar relation for the pressure. We see that at any given time $t > 0$ the velocity increases linearly from 0 at the head of the rarefaction at $x = -c_0 t$ to a maximum value at the tail, $x = (2/(\gamma - 1))c_0 t$, at which point the velocity is

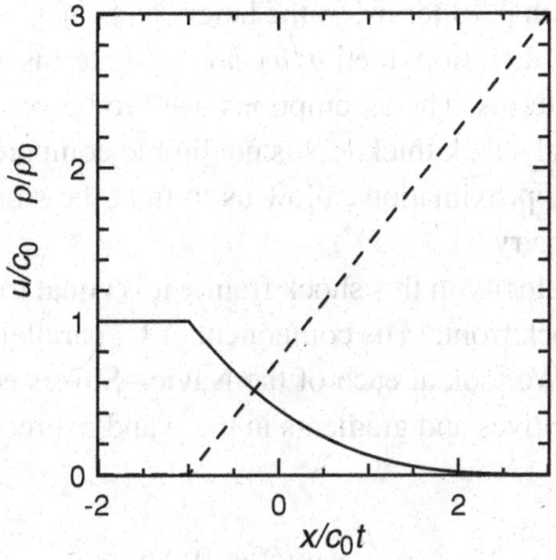

Fig. 2.2 Structure of a centered rarefaction wave. Solid line, density in units of undisturbed density; dashed line, velocity in units of undisturbed sound speed; abscissa, similarity coordinate x/c_0t.

$2c_0/(\gamma - 1)$, the same as x/t. The sound speed decreases linearly from c_0 to 0 over the same interval. The density varies as c raised to the $2/(\gamma - 1)$ power; this exponent is 3 for a monatomic gas. It means the density reaches very low values as the tail is approached. These results are shown in Figure 2.2. In a Lagrangian calculation of a free rarefaction such as this, the tail may be unphysically truncated because the remaining mass is less than the mass of a single zone. The tail velocity for the free rarefaction in a monatomic gas is $3c_0$, which is a useful rule of thumb to keep in mind.

2.9 Shock waves: Rankine–Hugoniot relations

Shocks are surfaces of discontinuity of the inviscid flow equations, or represent regions interior to the flow where the viscosity and heat conduction terms are locally important – like internal boundary layers. Courant and Friedrichs (1948) develop the theory of shock waves in considerable detail. An aerodynamic rather than physical point of view is contained in Liepmann and Roshko (1957). This has a useful analysis of the shock tube problem. The excellent standard reference on hydrodynamic waves is *Linear and Nonlinear Waves* by Whitham (1974). For the outer inviscid solution shock waves are fully described by the jumps across them. Here we discuss these jump conditions – the Rankine–Hugoniot relations. We will go back to the Navier–Stokes equations and look at these in the small, in a region that includes a patch of the shock front. In a coordinate system that is comoving with the shock for a short time, the flow is almost steady. The reason is that $\mathbf{u} \cdot \nabla$ terms are large compared with $\partial/\partial t$ terms, so the latter can be neglected. Similarly, if the shock normal is the x direction, then $\partial/\partial y$ and $\partial/\partial z$ terms are locally negligible compared with $\partial/\partial x$ terms. The assumptions need to be verified after the fact by showing that the actual shock thickness is negligible compared with length scales in the problem. The approximations allow us to treat the small region around the front using slab symmetry.

Let $\mathbf{v}$ be the flow velocity in this shock frame; it is equal to $\mathbf{u} - \mathbf{U}_\mathrm{s}$, where $\mathbf{U}_\mathrm{s}$ is the velocity of the shock front. (The component of $\mathbf{U}_\mathrm{s}$ parallel to the front will turn out to be irrelevant.) We look at each of the Navier–Stokes equations in turn, and discard the time derivatives and gradients in the y and z directions. The continuity equation gives simply

$$\frac{d}{dx}(\rho v_x) = 0. \tag{2.88}$$

The Navier–Stokes momentum equation turns into

$$\frac{d}{dx}(\rho v_x \mathbf{v}) + \frac{dp}{dx}\mathbf{e}_x - \frac{d}{dx}\left[\mu\left(\frac{d\mathbf{v}}{dx} + \frac{dv_x}{dx}\mathbf{e}_x\right) + \left(\zeta - \frac{2}{3}\mu\right)\frac{dv_x}{dx}\mathbf{e}_x\right] = 0. \tag{2.89}$$

The total energy equation becomes

$$\frac{d}{dx}\left\{\rho v_x e + \frac{1}{2}\rho v_x v^2 + p v_x - \mu\left[v_x\frac{dv_x}{dx} + \frac{d}{dx}\left(\frac{1}{2}v^2\right)\right]\right.$$
$$\left. - \left(\zeta - \frac{2}{3}\mu\right)v_x\frac{dv_x}{dx} - K\frac{dT}{dx}\right\} = 0. \tag{2.90}$$

The source terms on the right-hand side of the energy and momentum equations are neglected since they are small compared with the large d/dx terms. As we see, the three conservation equations imply that three functions are constant in the flow (five, if we count the three components of momentum). In other words,

$$\rho v_x = \text{constant} = C_1, \tag{2.91}$$

$$\rho v_x^2 + p - \left(\frac{4}{3}\mu + \zeta\right)\frac{dv_x}{dx} = \text{constant} = C_2, \tag{2.92}$$

$$\rho v_x v_y - \mu\frac{dv_y}{dx} = \text{constant} = C_3, \tag{2.93}$$

$$\rho v_x v_z - \mu\frac{dv_z}{dx} = \text{constant} = C_4, \tag{2.94}$$

$$\rho v_x e + \frac{1}{2}\rho v_x v^2 + p v_x - \left(\frac{4}{3}\mu + \zeta\right)v_x\frac{dv_x}{dx}$$
$$-\mu\left(v_y\frac{dv_y}{dx} + v_z\frac{dv_z}{dx}\right) - K\frac{dT}{dx} = \text{constant} = C_5. \tag{2.95}$$

The first constant is simply the mass flux through the shock. The third and fourth equations imply that v_y and v_z are themselves constant, with the values C_3/C_1 and C_4/C_1, respectively. That enables us to discard the v_y and v_z gradient terms in the fifth equation and absorb the constant term $(1/2)\rho v_x(v_y^2 + v_z^2)$ into C_5. The modified fifth equation is

$$\rho v_x e + \frac{1}{2}\rho v_x^3 + p v_x - \left(\frac{4}{3}\mu + \zeta\right) v_x \frac{dv_x}{dx} - K\frac{dT}{dx} = \text{constant} = C_5. \quad (2.96)$$

The internal structure of shock waves is beyond the scope of these remarks, but a rough guide to the relevant length scale is provided by examining these equations. The second equation suggests that the scale in x is of order $\mu/(\rho v_x)$. Since v_x is of order the sound speed (as we will see), and μ is roughly ρ times the mean thermal speed times the mean free path, the scale in x is very roughly just the mean free path. We get the same result from the viscous term in the energy equation. If the Prandtl number is of order unity, the conclusion from the heat conduction term is the same. Now, if the shock thickness is comparable with the mean free path then *ipso facto* the fluid approximation is not accurate. The true shock structure is a problem in particle transport. In plasmas, the transport processes result in a separation of charge within the shock and a substantial electric field is created. All these complications can be significant if atomic processes such as ionization and photorecombination take place at an appreciable rate within the front. We will consider these problems no further here.

Let us then move the observation points a small distance away from the shock front on each side, so that they are far enough away from the shock for the viscosity and heat conduction terms to have become negligible, but still close compared with the main length scales of the flow. Then when we compare the three functions whose values are C_1, C_2, and C_5 between these two points, one upstream and one downstream of the shock, we get the jump conditions

$$\Delta(\rho v_x) = 0, \quad (2.97)$$

$$\Delta(p + \rho v_x^2) = 0, \quad (2.98)$$

$$\Delta\left(e + \frac{p}{\rho} + \frac{1}{2}v_x^2\right) = 0, \quad (2.99)$$

to which we can add the two we have seen already, namely $\Delta(v_y) = \Delta(v_z) = 0$. These are the Rankine–Hugoniot relations.

We might count the number of relations and compare with the number of variables. The upstream state is defined by five variables, ρ, u_x, u_y, u_z, and e. (Remember that the $\mathbf{v}$ components are $\mathbf{u} - \mathbf{U}_\mathrm{s}$.) There are five more downstream variables, and the shock velocity U_s in the normal direction is an additional variable. So the Rankine–Hugoniot relations give five conditions constraining the eleven unknowns. Here is how it works out that the flow is fully determined. The shock moves supersonically into the material in front of it, as we will see, so the state of that material is fully determined without any information about, or from behind, the shock. Thus the preshock values of ρ, u_x, u_y, u_z, and e are known. The jump conditions then give all the postshock values, if one more piece of information is provided. The shock velocity U_s itself is not known, and in fact the shock may speed up or slow down in response to what the flow is doing. The extra piece of information is the one that is carried along the forward characteristic in the postshock region; this characteristic overtakes the shock from behind since the shock moves subsonically with respect to the region behind it. Thus the preshock conditions, plus the one piece of information from the postshock flow, determine all the postshock flow variables, and also the shock velocity.

A number of important results emerge from manipulation of the jump conditions, such as the property mentioned already that the shock propagates supersonically with respect to the gas in front and subsonically with respect to the gas behind. In the presentation below we use the notation v_0, ρ_0, p_0, etc., for quantities in front of the shock, and a subscript 1 for quantities behind the shock. The specific volume, $1/\rho$, will be denoted by V. The mass conservation relation becomes

$$v_0 = C_1 V_0 \quad \text{and} \quad v_1 = C_1 V_1, \tag{2.100}$$

so

$$\frac{V_0}{V_1} = \frac{v_0}{v_1}. \tag{2.101}$$

Since

$$p_0 + C_1^2 V_0 = p_1 + C_1^2 V_1, \tag{2.102}$$

the mass flux is given by

$$C_1 = \sqrt{\frac{p_1 - p_0}{V_0 - V_1}}, \tag{2.103}$$

and therefore the pre- and post-shock flow speeds with respect to the shock are

$$v_0 = V_0 \sqrt{\frac{p_1 - p_0}{V_0 - V_1}} \tag{2.104}$$

and

$$v_1 = V_1 \sqrt{\frac{p_1 - p_0}{V_0 - V_1}}. \tag{2.105}$$

Therefore the velocity jump is

$$v_0 - v_1 = \sqrt{(p_1 - p_0)(V_0 - V_1)}, \tag{2.106}$$

and the kinetic energy jump is

$$\frac{1}{2}v_0^2 - \frac{1}{2}v_1^2 = \frac{1}{2}(p_1 - p_0)(V_0 + V_1), \tag{2.107}$$

which, because of the energy jump condition, also gives the enthalpy jump:

$$h_1 - h_0 = \frac{1}{2}(p_1 - p_0)(V_0 + V_1). \tag{2.108}$$

Subtracting $p_1 V_1 - p_0 V_0$ from both sides gives the internal energy jump:

$$e_1 - e_0 = \frac{1}{2}(p_1 + p_0)(V_0 - V_1). \tag{2.109}$$

(Notice the $+$ and $-$ switched!) The set of all thermodynamic states P_1, V_1 that satisfy either of the last two equations for a specific initial state p_0, V_0 defines the Hugoniot curve for that state.

Using the ideal gas law all the relations can be made explicit. For the ideal gas,

$$e = \frac{PV}{\gamma - 1}, \tag{2.110}$$

so the Hugoniot equation (2.109) can be solved for the pressure ratio to give

$$\frac{p_1}{p_0} = \frac{(\gamma + 1)V_0 - (\gamma - 1)V_1}{(\gamma + 1)V_1 - (\gamma - 1)V_0}. \tag{2.111}$$

This is the equation of a hyperbola in the P, V diagram. The vertical asymptote is at

$$\frac{V_0}{V_1} = \frac{\rho_1}{\rho_0} = \frac{\gamma + 1}{\gamma - 1}, \tag{2.112}$$

which therefore is the maximum possible compression in a single shock. This ratio is 4 for a monatomic gas, for which $\gamma = 5/3$. The horizontal asymptote is

$$\frac{p_1}{p_0} = -\frac{\gamma - 1}{\gamma + 1}. \tag{2.113}$$

The curve crosses the V axis at

$$\frac{V_1}{V_0} = \frac{\gamma + 1}{\gamma - 1}. \tag{2.114}$$

Of course, values of V_1 that are larger than V_0 represent expansion, which cannot occur in a shock. The Mach number equations are

$$M_0^2 \equiv \left(\frac{v_0}{c_0}\right)^2 = \frac{(\gamma - 1) + (\gamma + 1)p_1/p_0}{2\gamma} \tag{2.115}$$

and

$$M_1^2 \equiv \left(\frac{v_1}{c_1}\right)^2 = \frac{(\gamma - 1) + (\gamma + 1)p_0/p_1}{2\gamma}. \tag{2.116}$$

Since $p_1 > p_0$, $M_0 > 1$ and $M_1 < 1$, substantiating the earlier claim that preshock flow is supersonic and postshock flow is subsonic, at least for ideal gases. The relation is also true for a general equation of state, although the proof is more subtle in that case.

The entropy jump is

$$s_1 - s_0 = \frac{e_0}{T_0} \log \left[\frac{p_1}{p_0} \left(\frac{(\gamma - 1)p_1/p_0 + (\gamma + 1)}{(\gamma + 1)p_1/p_0 + (\gamma - 1)} \right)^{\gamma} \right]. \tag{2.117}$$

This is positive for all shocks. It is 0 for $p_1 = p_0$, the limit of a weak shock, and increases indefinitely as $p_1/p_0 \to \infty$. It remains very small for modest values of $p_1/p_0 - 1$, and is proportional to the cube of $p_1/p_0 - 1$ (or of $M_0 - 1$) for weak shocks.

The jump conditions for a strong shock are often used as a limiting case. Besides the density jump

$$\rho_1 = \frac{\gamma + 1}{\gamma - 1} \rho_0 \tag{2.118}$$

that we saw before, the other two relations are

$$v_1 = \frac{\gamma - 1}{\gamma + 1} v_0 \tag{2.119}$$

and

$$p_1 = \frac{2\gamma}{\gamma + 1} \frac{v_0^2}{c_0^2} p_0 = \frac{2}{\gamma + 1} \rho_0 v_0^2. \tag{2.120}$$

The internal energy behind a strong shock is

$$e_1 = \frac{p_1}{(\gamma - 1)\rho_1} = \frac{4}{(\gamma + 1)^2} \frac{1}{2} v_0^2; \tag{2.121}$$

the enthalpy is

$$h_1 = \frac{4\gamma}{(\gamma+1)^2}\frac{1}{2}v_0^2. \tag{2.122}$$

Thus a large fraction of the kinetic energy entering the shock is turned into heat, but not all of it, since there is still residual kinetic energy in the flow behind the shock. For the $\gamma = 5/3$ gas, the postshock internal energy is 9/16 of the incoming kinetic energy, and the enthalpy is 15/16 of the incoming kinetic energy.

2.10 The Taylor–Sedov blast wave

This is the classical example that demonstrates a self-similar solution of a gas dynamic problem. The discussion here follows Landau and Lifshitz (1959), but the reader should be beware of typographical errors in the first edition, mentioned below. There is a good discussion, lacking the mathematical details, in Zel'dovich and Raizer (1967). The blast wave is also discussed, without typographical errors, in Sedov (1959), the classic of similarity methods. Consider an infinite space filled with a gas at rest at a uniform density, ρ_0, with negligible pressure. At $t = 0$ a certain amount of energy E is deposited in a tiny region at the origin. The details of how the energy is distributed between thermal and kinetic, and the spatial distribution of the deposited energy within the tiny volume, are soon forgotten. One significant point is that the amount of extra mass in the tiny volume, above the amount expected from the ambient density, is not large enough to be important. What follows is an expanding spherical shock front, behind which is a radial distribution of density, pressure, and velocity. The idea of self-similarity is that the radial distributions at different times are scalable. That is, they are the same if the radial scale and the scale of the variable are stretched appropriately. The scale transformations in self-similar problems often follow from dimensional analysis, as they do in this case.

The only data for this problem are the ambient density ρ_0 and the energy, E. At a certain time t the three quantities ρ_0, E, and t can be used to construct combinations with the dimensions of radius, velocity, and pressure. These are

$$r_1 = \left(\frac{Et^2}{\rho_0}\right)^{1/5}, \tag{2.123}$$

$$u_1 = \frac{2}{\gamma+1}\xi_0\frac{dr_1}{dt} = \frac{4}{5}\frac{\xi_0}{\gamma+1}\left(\frac{E}{t^3\rho_0}\right)^{1/5}, \tag{2.124}$$

$$p_1 = \frac{2}{\gamma+1}\rho_0\left(\xi_0\frac{dr_1}{dt}\right)^2 = \frac{8}{25}\frac{\xi_0}{(\gamma+1)^2}\left(\frac{E^2\rho_0^3}{t^6}\right)^{1/5}. \tag{2.125}$$

The reason for the extra numerical factors in these expressions will become apparent shortly. The similarity assumption is that the scaled variables

$$\tilde{u} = \frac{u}{u_1}, \tag{2.126}$$

$$\tilde{p} = \frac{p}{p_1}, \tag{2.127}$$

$$\tilde{\rho} = \frac{\gamma - 1}{\gamma + 1}\frac{\rho}{\rho_0} \tag{2.128}$$

are functions only of the scaled radius

$$\xi = \frac{r}{r_1} \tag{2.129}$$

and not of radius and time separately.

These substitutions are made into the Euler equations and, if we have done our work correctly and the problem is really self-similar, the factors involving E, ρ_0, and t all cancel out, and we are left with three ordinary differential equations in ξ for the unknowns $\tilde{u}$, $\tilde{p}$, and $\tilde{\rho}$. The location of the shock in the ξ coordinate, ξ_0, is to be determined. Given a value for ξ_0, the jump conditions (for a strong shock $p_1 \gg p_0$) determine the values of $\tilde{u}$, $\tilde{p}$, and $\tilde{\rho}$ at that point, just behind the shock. The factors introduced in the definitions of u_1 and p_1 above, and in the definition of $\tilde{\rho}$, are based on the jump conditions, and lead to the values at $\xi = \xi_0$

$$\tilde{u} = \tilde{p} = \tilde{\rho} = 1. \tag{2.130}$$

Inward radial integrations give a provisional structure. But the integral of the total kinetic and internal energy within the shocked region does not necessarily equal E; the value of ξ_0 is adjusted to make it so. The details of the analytic integration of the equations are given by Landau and Lifshitz (1959). (Users of the first edition of the Landau and Lifshitz reference should be aware of typographical errors: the exponent ν_5 in [99.10] should be $2/(\gamma - 2)$, not $1/(\gamma - 2)$, and the power of ξ multiplying p' in [99.11] should be 4 not 9. Furthermore, the variable ξ for all the terms in the integral in [99.11], and the integration variable, should be ξ/ξ_0.)

The results for the $\gamma = 5/3$ case are that $\xi_0 = 1.152$, and the distributions of u, p, and ρ are as shown in Figure 2.3. Some interesting features of the solution are that the velocity is almost linear in r, which is the ballistic relation $v = r/t$. The pressure drops steeply behind the shock, but only by a factor of about 2.5, and is nearly constant over most of the sphere. The density has a very large dynamic range. The inner part of the sphere is almost totally evacuated, and the mass that formerly occupied the volume of the sphere has been packed into a thin shell behind the shock. The temperature and entropy, not shown in the figure, are much larger near the center than just behind the shock. That is because each parcel of

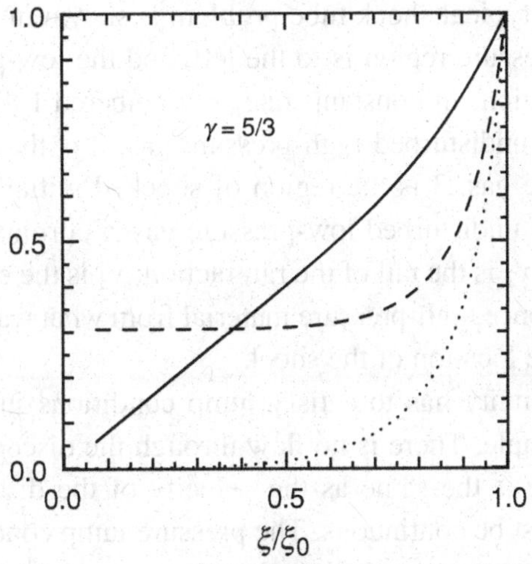

Fig. 2.3 Structure of the $\gamma = 5/3$ blast wave. The ordinates are: solid curve, scaled velocity; dashed curve, the scaled pressure; dotted curve, the scaled density, as defined in (2.126)–(2.128) in the text. The abscissa is the scaled radius ξ defined in (2.129).

material remembers how strong the shock was when it passed through the shock. (The entropy of a parcel is preserved in inviscid flow, apart from shocks.) At early time the shock is stronger and the entropy increase is correspondingly larger than at later time; this maps into an increase of entropy, and therefore of temperature, toward the center.

2.11 The shock tube

One of the best shock wave examples, often used as a test problem, is the shock tube. This is a model of a laboratory experiment in which a long gas cell initially contains a high-pressure gas on one side of a membrane and a much lower-pressure gas on the other side, possibly of a different temperature or composition. At time 0 the membrane is removed. What results is a shock wave propagating into the low-pressure gas, followed by a region of constant state made up of the initially low-pressure gas. This joins another constant-state region made up of initially high-pressure gas, but these two constant-state regions have the same pressure and velocity at this point. Next is a rarefaction progressing into the high-pressure region. Beyond the rarefaction on one side and beyond the shock on the other side are the undisturbed high- and low-pressure gases. The analysis of the shock tube demonstrates several of the methods of compressible gas dynamics.

The solution for a typical shock tube problem is shown in Figure 2.4. We suppose that the high-pressure region is to the left, and the low-pressure region is to the right. The four regions of constant state are numbered 1 through 4 going from left to right: 1 is the undisturbed high-pressure gas, 2 is the region of expanded initially high-pressure gas, 3 is the region of shocked initially low-pressure gas, and 4 is the region of undisturbed low-pressure gas. Coordinate x_1 is the head of the rarefaction wave; x_2 is the tail of the rarefaction; x_3 is the contact discontinuity separating what was once high-pressure material from what was once low-pressure material; and x_4 is the location of the shock.

A contact discontinuity has to satisfy jump conditions just as a shock does, but these are very simple. There is no flow through the discontinuity, so the fluid velocity on each side is the same as the velocity of the discontinuity; the fluid velocity therefore must be continuous. The pressure jump condition reduces in the case that there is no flow across the interface to the statement that the pressure is continuous. The energy jump condition is automatically satisfied if the first two conditions are obeyed.

For the shock tube this means that constant-state regions 2 and 3 share the same velocity and pressure. The densities and sound speeds can be, and indeed will be, different, however. The first step in the procedure for solving a shock tube is to find the pressure and velocity in regions 2 and 3. The pressure will be intermediate between the initial values p_1 and p_4, and the velocity will be some positive value. The method of solution is to use the centered rarefaction relations to express the velocity difference across the rarefaction, which will be u_2, as a function of the unknown pressure ratio p_1/p_2. Likewise the shock jump conditions are used to express the velocity jump across the shock, which will be u_3, as a function of the pressure ratio p_3/p_4. But u_2 and u_3 must be the same, and p_2 and p_3 are the same, so p_3/p_4 can be expressed as $(p_1/p_4)/(p_1/p_2)$. The ratio p_1/p_4 is one of the data for the problem, so there is a single unknown, p_1/p_2. The equation $u_2 = u_3$ becomes an algebraic equation for p_1/p_2. This has to be solved numerically. Here is the equation when we allow the two gases to have different γs:

$$\frac{2c_1}{\gamma_1 - 1}\left[1 - (p_2/p_1)^{(\gamma_1-1)/(2\gamma_1)}\right] = \frac{\sqrt{2}c_4(p_3/p_4 - 1)}{\sqrt{\gamma_4[\gamma_4 - 1 + (\gamma_4 + 1)p_3/p_4]}}, \quad (2.131)$$

where p_3/p_4 is to be replaced by $(p_1/p_4)/(p_1/p_2)$. In practice we use a root-finder method such as Newton–Raphson to find the solution. Once this equation has been solved, the left- or the right-hand side gives the value of u_2, and the adiabatic relation across the rarefaction gives ρ_2 and c_2. Other shock jump conditions give ρ_3 and the shock velocity U_s since p_3/p_4 is known.

The spatial structure of the shock tube solution is self-similar in time, since all the variables are functions of x/t, assuming that the initial membrane location was

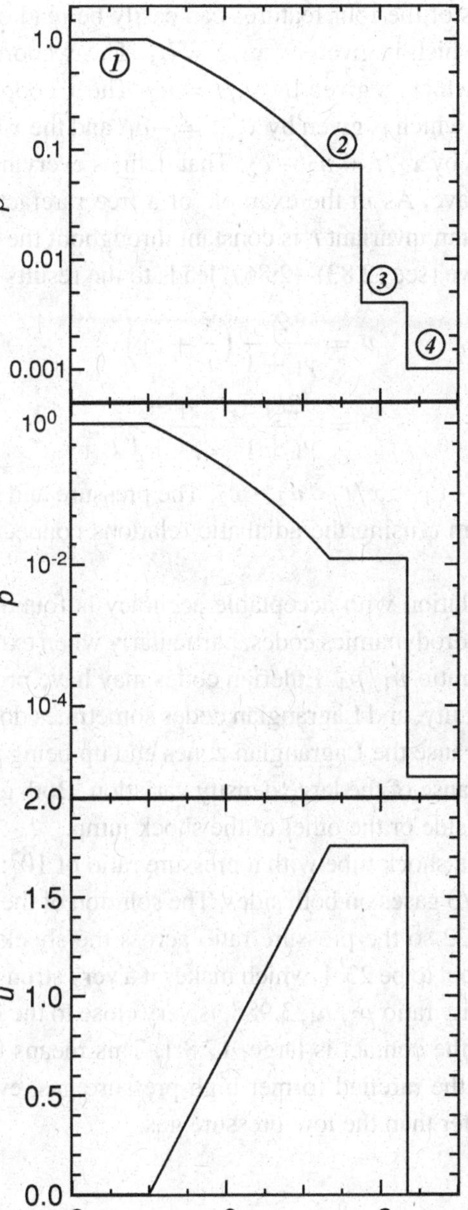

Fig. 2.4 Structure for the shock tube described in the text. Abscissa, x/t; ordinate, density (top), pressure (middle) and velocity (bottom). The four regions discussed in the text are indicated on the density curve.

$x = 0$. The locations of the four features can easily be read off. The x coordinate of the shock is x_4, which is given by $x_4/t = U_s$. The x coordinate of the contact discontinuity is x_3, which is given by $x_3/t = u_2$. The x coordinate of the head of the rarefaction is x_1, which is given by $x_1/t = -c_1$, and the x coordinate of the tail is x_2, which is given by $x_2/t = u_2 - c_2$. That defines everything except the shape of the rarefaction wave. As in the example of a free rarefaction, this is a simple wave, and the Riemann invariant r is constant throughout the wave. Following the same algebra as above (see (2.83)–(2.86)) leads to the results

$$u = \frac{2}{\gamma_1 + 1}\left(\frac{x}{t} + c_1\right), \tag{2.132}$$

$$c = \frac{2c_1}{\gamma_1 + 1} - \frac{\gamma_1 - 1}{\gamma_1 + 1}\frac{x}{t}, \tag{2.133}$$

for x/t in the range $-c_1 < x/t < u_2 - c_2$. The pressure and density distributions are then derived from c using the adiabatic relations connecting each point with region 1.

Obtaining this solution with acceptable accuracy is found to be a challenging test for numerical hydrodynamics codes, particularly when extreme values are chosen for the pressure ratio p_1/p_4. Eulerian codes may have problems smearing out the contact discontinuity, and Lagrangian codes sometimes do not do well with the rarefaction wave because the Lagrangian zones end up being poorly distributed in the x coordinate because of the large density variation. Both codes sometimes will have ringing on one side or the other of the shock jump.

Figure 2.4 shows a shock tube with a pressure ratio of 10^5:1 and a density ratio of 10^3:1 with $\gamma = 5/3$ gases on both sides. The solution of the shock tube equation gives $p_2/p_1 = 0.0122$ so the pressure ratio across the shock is 1222. The shock Mach number turns out to be 23.4, which makes it a very strong shock. As expected in this case, the density ratio ρ_3/ρ_4, 3.988, is very close to the limiting value 4. The density ratio across the contact is large, 17.8:1. This means that the shocked gas is much hotter than the rarefied former high-pressure gas, even though the latter began 100 times hotter than the low-pressure gas.

3
Numerical hydrodynamics

In Chapter 2 we have already discussed the Eulerian and Lagrangian formulations of the equations of fluid mechanics. Now we want to describe some of the numerical solution methods that have been developed to solve them, and in particular, in the more challenging 2-D and 3-D cases. The present discussion will focus on hydrodynamics without radiation, and the methods for treating coupled hydrodynamics and radiation will be mentioned later, in Chapter 11.

There are excellent references on this subject, among which are the classic book by Richtmyer and Morton (1967), the text by Bowers and Wilson (1991), the collection of papers by Norman and coworkers (Stone and Norman, 1992a,b; Stone, Mihalas and Norman, 1992), van Leer (1979), Colella and Woodward (1984), Caramana and coworkers (Caramanā and Whalen, 1998; Caramana and Shashkov, 1998; Caramana, Shashkov, and Whalen, 1998; Caramana, Burton, Shashkov, and Whalen, 1998) and Jiang and Shu (1996).

3.1 Lagrangian methods

3.1.1 Staggered-mesh hydrodynamics for 1-D slab geometry

The mother of all numerical hydrodynamics methods has the name von Neumann–Richtmyer staggered-mesh hydrodynamics. In its simplest incarnation, for 1-D slab geometry, it is described as follows. The material of the problem is divided into N zones with fixed masses, divided by $N - 1$ material interfaces. Including the outer boundaries, that makes $N + 1$ interfaces in all. The basic set of unknowns is the list of z coordinates of these interfaces: z_I, $I = 1, \ldots, N + 1$. Time is discretized as well, and these interface positions are to be found at a succession of times: $t_1, t_2, \ldots$. The position of interface I at time $t = t_n$ is denoted by z_I^n. The zone between interface I and interface $I + 1$ carries the index $I + 1/2$, and its fixed amount of mass (per unit area) is $m_{I+1/2}$. From the zone thickness and the mass comes the mass density, $\rho_{I+1/2}^n = m_{I+1/2}/(z_{I+1}^n - z_I^n)$. There are also zone-centered values of internal energy $e_{I+1/2}^n$, temperature $T_{I+1/2}^n$ and pressure $p_{I+1/2}^n$.

The other basic set of unknowns consists of the interface velocities. These carry the same spatial indexing as the interfaces themselves, but are staggered in time by half a time step relative to the positions, so that $u_I^{n+1/2}$ represents the velocity of interface I in the time interval t_n to t_{n+1}. Thus one of the time-integration equations is

$$\frac{z_I^{n+1} - z_I^n}{t_{n+1} - t_n} = u_I^{n+1/2}. \tag{3.1}$$

In order to find the evolution of u_I we need to relate it to the pressure gradient, etc. Because of the staggering in time, the acceleration that changes $u_I^{n-1/2}$ to $u_I^{n+1/2}$ is centered in time at t_n, and therefore it can be calculated from the information available at that time:

$$2\frac{u_I^{n+1/2} - u_I^{n-1/2}}{t_{n+1} - t_{n-1}} = 2\frac{p_{I-1/2}^n - p_{I+1/2}^n}{m_{I-1/2} + m_{I+1/2}}. \tag{3.2}$$

This has assumed that the velocity $u_I^{n+1/2}$ is centered in the middle of the time interval $[t_n, t_{n+1}]$, and that the baseline for the pressure gradient is from the midpoint of zone $I - 1/2$ to the midpoint of zone $I + 1/2$. This equation has a flaw that will be mentioned shortly. The pressure $p_{I+1/2}^n$ must be calculated from the equation of state using the density $\rho_{I+1/2}^n$ and either the internal energy or temperature, and therefore the energy equation will be needed.

The time-centered internal energy equation is (from (2.4))

$$\left(e_{I+1/2}^{n+1} - e_{I+1/2}^n\right) + \frac{1}{2}\left(p_{I+1/2}^n + p_{I+1/2}^{n+1}\right)\left(\frac{1}{\rho_{I+1/2}^{n+1}} - \frac{1}{\rho_{I+1/2}^n}\right) = 0. \tag{3.3}$$

This is an implicit equation for $e_{I+1/2}^{n+1}$ since $p_{I+1/2}^{n+1}$ depends on it, or, equivalently, both depend on $T_{I+1/2}^{n+1}$. The work flow is the following. The pressure data at t_n are used to update the velocities to $t_{n+1/2}$; the new velocities are then used to update the interface positions and densities to t_{n+1}. The densities being known, the internal energy equations may be solved for the $e_{I+1/2}^{n+1}$ values. With a gamma-law equation of state $p_{I+1/2}^{n+1} = (\gamma - 1)\rho_{I+1/2}^{n+1} e_{I+1/2}^{n+1}$ this can be done directly, otherwise a Newton–Raphson procedure may be needed. Once the new internal energies are known, the new pressures follow, and the time step is complete.

In the limit of infinitesimal time steps the internal energy equation above becomes the equation for an isentrope, and in that limit the entropy of zone $I + 1/2$ would never change. This causes a problem. We know that the Euler equations can produce shocks, and that the entropy must increase across a shock. Therefore the finite-difference equations as written so far will give the wrong answer for a flow containing a shock. Richtmyer and Morton (1967) illustrate the numerical results in this case: in what would be the postshock region there are huge zone-to-zone fluctuations of the flow velocity and pressure. The Navier–Stokes equations can give a smooth transition across what would be a shock for the Euler equations, and this is the key to the successful numerical method. A dissipative term, a pseudo-viscous pressure q, must be added to the momentum and energy equations. In those cases, the majority, for which the Euler equations are a good approximation, the pseudo-viscous pressure is much larger than the true viscous stress. It does not matter very much exactly what q is, provided that: (1) it is positive, (2) it is negligible away from shocks, (3) it is included in the momentum and energy equations in such a way that total energy is still conserved. It then acts, like true viscosity, to turn kinetic energy of fluid motion into heat. The von Neumann–Richtmyer prescription is

$$q^n_{I+1/2} = C_Q \rho^n_{I+1/2} \max\left(u_I^{n-1/2} - u_{I+1}^{n-1/2}, 0\right)^2. \tag{3.4}$$

The factor C_Q is a constant that may be 1 or 2 or whatever value seems to give good results. It may be noted that the velocity is half a time step behind the other factors in the expression for $q^n_{I+1/2}$. This is a necessity since the advanced velocity is not available when q must be computed. Lagging the velocity is found not to have a deleterious effect. The pseudo-viscosity spreads a shock over about $\pi\sqrt{C_Q}$ zones, so a larger C_Q means a smoother, less noisy calculation, while a smaller C_Q gives improved resolution.

When pseudo-viscosity is added the momentum and internal energy equations become

$$2\frac{u_I^{n+1/2} - u_I^{n-1/2}}{t_{n+1} - t_{n-1}} - 2\frac{p^n_{I-1/2} + q^n_{I-1/2} - p^n_{I+1/2} - q^n_{I+1/2}}{m_{I-1/2} + m_{I+1/2}} = 0, \tag{3.5}$$

$$\left(e^{n+1}_{I+1/2} - e^n_{I+1/2}\right) + \frac{1}{2}\left(p^n_{I+1/2} + q^n_{I+1/2} + p^{n+1}_{I+1/2} + q^{n+1}_{I+1/2}\right) \times \left(\frac{1}{\rho^{n+1}_{I+1/2}} - \frac{1}{\rho^n_{I+1/2}}\right) = 0. \tag{3.6}$$

It can be shown that with this finite-difference system the discrete total energy of the problem

$$\mathcal{E}_{\text{tot}} = \sum_I \left[\frac{1}{2}(m_{I-1/2} + m_{I+1/2})\frac{1}{2}(u_I)^2\right] + \sum_I m_{I+1/2} e_{I+1/2} \tag{3.7}$$

is precisely conserved in the limit of short time steps, provided the boundary conditions are suitable: rigid boundaries or zero-pressure boundaries.

When we speak of boundary conditions, we recall that they must, of course, be included. When the finite-difference equations would incorporate data that do not exist, since a boundary intervenes, either the missing data are generated from the specified boundary values, or that particular difference equation is replaced by a constraint. Thus at a rigid boundary the acceleration need not be computed since the velocity is forced to be zero. At a free boundary, where the pressure vanishes, the pressure of the phantom zone outside the boundary is obtained by extrapolating from the zone adjacent to the boundary on the inside and using the condition that the pressure at the boundary interface should vanish.

A very important limitation of all the methods of numerical hydrodynamics we will discuss is that the computed solution diverges wildly from the right answer if the time step is excessive in comparison with the zone sizes. This is the Courant–Friedrichs–Lewy condition, Courant or CFL for short (Courant, Friedrichs and Lewy, 1928). A linearized 1-D Lagrangian problem can help motivate the condition. Consider this idealized system:

$$\frac{z_I^{n+1} - z_I^n}{\Delta t} = u_I^{n+1/2}, \tag{3.8}$$

$$\frac{u_I^{n+1/2} - u_I^{n-1/2}}{\Delta t} = a^2 \left(\frac{1}{z_I^n - z_{I-1}^n} - \frac{1}{z_{I+1}^n - z_I^n} \right), \tag{3.9}$$

in which a is the effective speed of sound. Now we perturb around a state in which the material is at rest and the zone size is uniform and equals Δz. Denoting the perturbations to the interface positions by δz_I^n leads to these linearized equations:

$$\frac{\delta z_I^{n+1} - \delta z_I^n}{\Delta t} = u_I^{n+1/2}, \tag{3.10}$$

$$\frac{u_I^{n+1/2} - u_I^{n-1/2}}{\Delta t} = \frac{a^2}{(\Delta z)^2} \left(\delta z_{I-1}^n - 2\delta z_I^n + \delta z_{I+1}^n \right). \tag{3.11}$$

We now look for a solution of this homogeneous linear system in which both the velocities and the interface positions are proportional to $g^n \exp(i\theta I)$, where θ is a specified angle between $-\pi$ and π, and g is an amplification ratio to be found. When the factors are eliminated this quadratic equation is found for g:

$$g^2 - 2\left(1 - 2C^2 \sin^2 \frac{\theta}{2}\right) g + 1 = 0, \tag{3.12}$$

where C stands for the Courant ratio $a\Delta t/\Delta z$. This equation has a pair of complex conjugate roots with $|g| = 1$ provided $C \sin\theta/2 \le 1$, but it has real roots, one of which has a magnitude > 1, if $C|\sin\theta/2| > 1$. Since $|\sin\theta/2|$ can be as large as 1, in the case $\theta = \pm\pi$, there will be instability if $C > 1$, the CFL result. The most unstable mode is the one in which the fluctuations alternate in sign at consecutive zones. In practice, a smaller limit is set for C, such as 0.5. This accounts for non-linear instability effects. Richtmyer and Morton (1967) demonstrate how the upper limit for C depends on shock strength.

The solution for the amplification factor in the continuum limit $\Delta z \to 0$, $\Delta t \to 0$ can be used as a check on the accuracy of the numerical solution. The value of g^n for the analytic solution is $\exp(i\omega t) = \exp(ikat) = \exp(ika\Delta tn)$, and therefore g should be $\exp(ika\Delta t) = \exp(iC\theta)$ when we identify θ with $k\Delta z$. In other words, for the analytic solution

$$g = 1 + iC\theta - \frac{1}{2}C^2\theta^2 - \frac{1}{6}iC^3\theta^3 + O(C^4\theta^4). \qquad (3.13)$$

The expansion of the numerical amplification factor in powers of θ is

$$g = 1 + iC\theta - \frac{1}{2}C^2\theta^2 - iC\frac{3C^2+1}{24}\theta^3 + O(\theta^4). \qquad (3.14)$$

The expansions agree through terms of order θ^2, and so the numerical method is second order accurate. The third order terms differ except in the case $C = 1$. In that marginally-stable circumstance the numerical method is actually exact.

This is the essence of 1-D staggered-mesh Lagrangian hydrodynamics. The variations come in slight changes in the way the time-centered factors are constructed, and variations in the pseudo-viscosity. Sometimes a linear pseudo-viscosity is used:

$$q^n_{I+1/2} = C_Q a^n_{I+1/2}\rho^n_{I+1/2} \max\left(u_I^{n-1/2} - u_{I+1}^{n-1/2}, 0\right), \qquad (3.15)$$

in which the factor $a^n_{I+1/2}$ is of the order of the sound speed in the zone. The linear pseudo-viscosity is quite a bit more dissipative for weak shocks, and for no shock at all, than the quadratic one. This makes the calculation quite smooth, perhaps spuriously so. The two examples of pseudo-viscosity we have seen so far are nonzero for zones being compressed (as in a shock) and vanish for zones in expansion. A further step in the direction of smoothness is to let q for a zone in expansion be the negative of the q with the same velocity difference in compression. This is usually undesirable. We will see a type of q later that vanishes except in shocks of at least moderate strength; this is the so-called monotonic q.

3.1.2 Staggered-mesh hydrodynamics in two dimensions – quadrilateral zones

In 2-D Lagrangian hydrodynamics with quadrilateral zones the mesh consists of a more-or-less distorted map of a uniform Cartesian grid. This underlying logical mesh has the coordinates k and ℓ, and the mesh spacing in each coordinate is unity. The nodes of the logical mesh have integer values of k and ℓ, the mesh consists of k-lines with integer values of k and variable ℓ and ℓ-lines with the reverse. The zones have half-integer values of both k and ℓ. Each zone is supposed to have a fixed mass. The fluid velocity is centered at the nodes, and staggered in time from the other quantities. The time advancement proceeds by: moving the nodes according to their current velocities, computing the new area of all the zones and thereby getting the new densities; next applying the internal energy equation to find the new internal energy and pressure in each zone; finally using the pressures in the zones surrounding a given node to compute the acceleration of the node and obtain the time-advanced nodal velocity.

The first respect in which the 2-D case is harder than the 1-D case is computing the zone area, which is needed to obtain the density. The following is one way to obtain the result. We suppose that our problem has xy geometry, i.e., it has translational invariance in the z direction. Our coordinates are x and y, and we can speak of the position vector $\mathbf{r} = (x, y)$. The zone area is given by

$$A = \iint_Z dxdy = \iint_R \left| \frac{\partial(x, y)}{\partial(k, \ell)} \right| dkd\ell, \tag{3.16}$$

in which Z represents the physical zone, and R represents the corresponding logical zone. Here we are formally using the coordinates in logical space, the indices, as real variables, and the Jacobian of the mapping from logical space to physical space appears. Of course, we do not actually know that mapping except at the integer logical mesh nodes. For definiteness let us consider the zone bounded by the mesh lines $k = 0$, $k = 1$, $\ell = 0$ and $\ell = 1$. Within this zone let us also assume a bilinear mapping

$$\mathbf{r} = \mathbf{r}_{00}(1-k)(1-\ell) + \mathbf{r}_{10}k(1-\ell) + \mathbf{r}_{01}(1-k)\ell + \mathbf{r}_{11}k\ell, \tag{3.17}$$

where the vectors $\mathbf{r}_{00}$, $\mathbf{r}_{10}$, $\mathbf{r}_{01}$, and $\mathbf{r}_{11}$ are the four corners of the quadrilateral. The bilinear mapping is consistent with the quadrilateral shape since the $k = 0$ and $k = 1$ lines and $\ell = 0$ and $\ell = 1$ lines will be straight lines connecting the corners. The Jacobian is $J = \mathbf{r}_k \times \mathbf{r}_\ell$, where the z component of the vector cross product is needed, and the subscripts k and ℓ indicate partial differentiation by that variable. But since $\mathbf{r}_k$ is a linear function of ℓ alone, and likewise $\mathbf{r}_\ell$ is a linear function of k alone, the average of J is the cross product of the average $\mathbf{r}_k$ with the average $\mathbf{r}_\ell$. This gives

$$A = \left| \frac{1}{4}(\mathbf{r}_{10} - \mathbf{r}_{00} + \mathbf{r}_{11} - \mathbf{r}_{01}) \times (\mathbf{r}_{01} - \mathbf{r}_{00} + \mathbf{r}_{11} - \mathbf{r}_{10}) \right| \\ = \frac{1}{2} |(\mathbf{r}_{10} - \mathbf{r}_{01}) \times (\mathbf{r}_{11} - \mathbf{r}_{00})|. \tag{3.18}$$

This can be arranged in a variety of ways; the last expression is half the cross product of the two diagonals of the quadrilateral. There is a fallacy in this calculation, which pops up if the sign of the Jacobian changes within the zone. Such zones are called "bowtie" zones, since they look like two triangles that touch at a single point. The calculation above gives the *net* area of the two triangles, treating one as positive and the other as negative, not the sum of the areas. Lagrangian hydrodynamics codes generally crash when a bowtie zone occurs, which is not infrequent unless steps are taken to prevent it.

The next challenge is to calculate the accelerations of the nodes, which depend on a pressure gradient. One way to do this is the following. We focus our attention on a particular node and construct around it a "dual zone" made up of pieces of all the zones surrounding that node. In a quadrilateral mesh there are four surrounding zones in general. For example, we can construct a contour C that joins the midpoint of one zone edge with the center of the adjacent zone, then on to the midpoint of the nest edge around, then the next zone center, and so on back to the starting point. This contour encloses roughly one quarter of each of the surrounding zones. We call this enclosed region the dual zone and denote it by Z'. Its area and mass can be found in the same way as the area calculation above. Next we apply a form of the divergence theorem to this dual zone:

$$\iint_{Z'} \nabla p \, dA = \oint_C p\mathbf{n} \, ds = -\mathbf{e}_z \times \oint_C p \, d\mathbf{s}, \tag{3.19}$$

in which $\mathbf{n}$ is the unit outward normal vector to the zone at points along C, ds is the element of arc length along C, $\mathbf{e}_z$ is the unit vector in the z direction and $d\mathbf{s}$ is the vector path element for traversing C counterclockwise. If the last integral is broken into pieces belonging to each of the surrounding zones, and if we approximate p as constant within each of these zones, then the integral for each piece is just the pressure of the zone times the vector joining the midpoint of one edge to the midpoint of the next edge. So finally we get the integral of ∇p over the whole dual zone. The negative of this divided by the dual zone mass is the nodal acceleration.

For a uniform or very smooth mesh this approximation for ∇p is second order. Unfortunately, when the mesh is severely distorted the accuracy is not at all good.

Specifically, if a Taylor expansion of the pressure is made in the region surrounding a node, and this is used to evaluate the gradient using the line integral formula, then it is found that the second order terms introduce an error in the derived gradient unless the mesh is orthogonal and regular, where "regular" means that the zone shape and dimensions are slowly varying in the mesh; i.e., the changes of zone dimensions between adjacent zones are themselves second order. This is a limitation that many Lagrangian codes suffer with, although some improvements have been made (*viz*., the work of Caramana and coworkers (Caramana and Whalen, 1998; Caramana and Shashkov, 1998; Caramana, Shashkov, and Whalen, 1998; Caramana, Burton, Shashkov, and Whalen, 1998).

The final complicated ingredient is the pseudo-viscosity. There is considerable ambiguity involved in the centering of q in multidimensional problems. The difference between the vector velocities of the nodes (k, ℓ) and $(k+1, \ell)$, projected on the edge connecting those nodes, leads to a q that is naturally centered at the midpoint of that edge. This would have to be averaged with the similar quantity defined for the $\ell + 1$ side of the quadrilateral to obtain a zone-centered q that can be added to the normal pressure in the zone. Clearly, there are many alternatives to this prescription. It is also possible to use a different q in computing the x acceleration than when computing the y acceleration.

3.1.3 Unstructured meshes

The logic for calculating ∇p above is not critically dependent on having exactly four zones around each node, nor of each zone having exactly four edges. In an unstructured mesh code the mesh consists of an arbitrary assortment of polygons (in two dimensions). There may be any number of edges meeting at a node, and a zone will have as many neighbors as it has edges, whatever that may be. The area calculation is simple enough if the zone is first broken into triangles. The algorithm for finding ∇p works well once the list of zones surrounding the node is constructed.

In codes of this kind the mesh connectivity may be dynamic: as the mesh becomes distorted some of the mesh lines may be removed and/or created. Zones and nodes are not indexed by k and ℓ, which implies quadrilaterals, but form two long lists. Associated with each node is its current list of neighbors, and the list of zones that share the mesh line connecting this node with the neighbors. As connections are made and broken in the course of the calculation, these lists are updated. The indexing overhead and the cost of dealing with many-sided polygons make unstructured mesh codes more costly than quadrilateral-based codes, but there can be advantages in ameliorating mesh distortion.

3.2 Eulerian methods

While in Lagrangian hydrodynamics the coordinates of the mesh are dynamic variables, from which the density is calculated, in Eulerian hydrodynamics the mesh is fixed and the density is one of the primary variables. In each case the other variables are fluid velocity and internal energy or temperature.

The Eulerian methods fall into two groups: finite-difference methods and finite-volume methods. In the former group the flow variables are conceived as being samples at certain points in space and time, and from these sampled values the partial derivatives are computed that are required to obey the Euler equations. For methods of the finite-volume class the unknowns are understood to be average values over certain finite volumes – the zones – and these must obey the conservation laws in integral form. The difference is subtle and the final equations are quite similar for the different viewpoints.

3.2.1 Finite-difference methods; Lagrangian plus advection

This approach is described very well in Bowers and Wilson (1991). The Euler equations are first written in this way

$$\frac{\partial \rho}{\partial t} = -\nabla \cdot (\rho \mathbf{u}), \tag{3.20}$$

$$\frac{\partial \rho \mathbf{u}}{\partial t} = -\nabla p - \nabla \cdot (\rho \mathbf{u}\mathbf{u}), \tag{3.21}$$

$$\frac{\partial \rho e}{\partial t} = -p\nabla \cdot \mathbf{u} - \nabla \cdot (\rho e \mathbf{u}). \tag{3.22}$$

In the method of Bowers and Wilson a time step consists of a "Lagrangian" step followed by an advection step. The idea of operator splitting is applied here, which means that for each of these substeps just *part* of the expression for the time derivative of each variable is applied. The unknowns ρ, $\rho\mathbf{u}$ and ρe are updated using those parts of their time derivatives, and the updated values are the starting point for the next partial time step. The philosophy of operator splitting is described further in Section 11.1. For the Lagrangian step the density is left alone, the momentum density is updated using just the pressure gradient, after which a new flow velocity is calculated by division. As in staggered-mesh Lagrangian hydrodynamics, the velocity (and momentum density) components are centered in time at $t_{n+1/2}$. The new velocities are used to calculated $\nabla \cdot \mathbf{u}$ and the internal energy density is updated using just the first term on the right-hand side of (3.22). That completes the Lagrangian step.

The acceleration of the material by the pressure gradient is more easily calculated in this method than in the Lagrangian codes for two reasons. One is the regularity of the mesh, which lends itself to a very natural expression for ∇p. The other is the frequent choice to center the velocity (and momentum density) components on cell edges. The x momentum is centered at the midpoints of the edges bounding the zones in the x direction, and likewise for the y momentum. This means, for example, that $(p_{k+1/2,\ell+1/2} - p_{k-1/2,\ell+1/2})/\Delta x$ is the natural expression for the gradient to accelerate the x velocity at point $k, \ell + 1/2$.

For the advection step these relations are used:

$$\frac{d}{dt}(\rho V) = -\iint_S \rho \mathbf{u} \cdot d\mathbf{A}, \tag{3.23}$$

$$\frac{d}{dt}(\rho \mathbf{u} V') = -\iint_{S'} \rho \mathbf{u}\mathbf{u} \cdot d\mathbf{A}, \tag{3.24}$$

$$\frac{d}{dt}(\rho e V) = -\iint_S \rho \mathbf{u} e \cdot d\mathbf{A}. \tag{3.25}$$

Here V is a zonal volume and V' is a dual zonal volume, since the velocity components are staggered in space from the quantities like density and internal energy. The surfaces S and S' represent the boundaries of V and V'. The surface integrals become sums over the edges of each zone or dual zone of the advection flux through each edge. The challenge for Eulerian hydrodynamics, to be accurate and minimize numerical diffusion, is the construction of these advection fluxes. Since the flow is in one direction or the other through each edge, there will be a donor zone and a receiver zone. The advection flux is the appropriate edge-centered normal component of velocity multiplied by an appropriately averaged value of the conserved quantity: mass density, momentum density or internal energy per unit volume. Stability requires that the edge-centered conserved density be calculated from the properties of the donor zone. If this is just set equal to the average density of the donor zone, the result is called donor-cell advection. Alas, this method, while very simple, is very diffusive.

The improvement on donor-cell advection is to notice that the contents of the donor zone within a distance $\mathbf{u} \cdot \mathbf{n}\Delta t$ will be advected out of the zone during the time step, where $\mathbf{n}$ is the unit vector normal to the zone edge. If we can construct an interpolation function representing the variation of ρ or the other conserved quantities *within* the zone, then the integral of this interpolation function over the volume that will be advected gives a much better estimate of the advection flux. This concept is now used in virtually all Eulerian codes. One very common interpolation scheme is the monotonic piecewise-linear scheme introduced by van Leer (1977). This is a 1-D interpolation method. Indeed, common practice, as described by Bowers and Wilson (1991), is to operator-split the advection by direction: first advect in the x direction, then in the y direction. (Or the reverse. Or first one way then the other.) Constructing the piecewise-linear interpolation function in one coordinate, say x, for, let us say, ρ, comes down to finding the slope within each zone of ρ vs x. The goal of monotonic interpolation is to ensure that the zone-edge values of the interpolant do not fall outside the range of ρ between the given zone and its neighbor in that direction. If the given zone is in fact a local extremum, then the slope must be zero. The formula can be illustrated easily if the zoning is uniform. The slope $d\rho/dx$ in zone $k + 1/2$ multiplied by Δx will be $\Delta\rho$ given by $\Delta\rho = 0$ if zone $k + 1/2$ is an extremum, otherwise

$$\Delta\rho = S \min(2|\rho_{k+1/2} - \rho_{k-1/2}|, 2|\rho_{k+3/2} - \rho_{k+1/2}|, 0.5|\rho_{k+3/2} - \rho_{k-1/2}|), \tag{3.26}$$

where S is 1 if $\rho_{k+1/2} - \rho_{k-1/2} > 0$ and -1 if $\rho_{k+1/2} - \rho_{k-1/2} < 0$. The formula for variable zone sizes is only a little more complicated. The interpolation function for zone $k + 1/2$ is then the straight line that passes through $\rho_{k+1/2}$ at zone center with this slope. The advected flux during the time step is then the integral of this function over a distance $u_x\Delta t$ adjacent to the zone edge, as described above.

Staggered-mesh finite-difference Eulerian hydrodynamics, like Lagrangian hydrodynamics, cannot calculate shocks successfully unless pseudo-viscosity is included. The prescriptions for pseudo-viscosity are very much the same, apart from the differences owing to the centering of the velocity components on zone edges rather than at vertices. Van Leer interpolation lends itself to a very useful expression for the pseudo-viscosity, called the monotonic q. The x components of velocity, centered at $k, \ell + 1/2$, can be interpolated in the van Leer fashion, where now the zones become the dual zones that extend from the center of one regular zone to the center of the next. The piecewise interpolants in these dual zones will in general not be continuous at the dual zone edges, i.e., at the regular zone centers. The amount of the discontinuity will typically be very small in smooth regions of the flow, and become large where the velocity has a sharp gradient, i.e., near a shock. The monotonic q is proportional to the zone density times the square of this velocity jump. Of course, the van Leer interpolation is done one direction at a time, and this is an instance in which q definitely differs for the two directions. Thus the Lagrangian step momentum update must be split into an x update and a y update.

The CFL condition applies to Eulerian methods just as to Lagrangian methods, except that the expression for C is somewhat modified:

$$C \equiv \max\left(\frac{(a + |u_x|)\Delta t}{\Delta x}, \frac{(a + |u_y|)\Delta t}{\Delta y}\right) < 1. \tag{3.27}$$

In fact, a smaller upper limit to C may be set, such as 0.5, as mentioned earlier.

3.2.2 Godunov methods; PPM

The numerical methods described so far are quite elderly, dating in large part to the 1940s. Now we come to the methods of the 1970s and 1980s. Godunov's method (1959) and its descendants are among these. These are finite-volume methods. This means that we now consider all the flow variables to be zonal averages, and we find their evolution by determining the gains and losses of conserved quantities by each zone. The natural place to begin is with the Euler equations written in conservation law form, as in (2.60), which for our purpose becomes

$$\frac{\partial U}{\partial t} + \nabla \cdot \mathbf{F}(U) = 0. \tag{3.28}$$

The vector U contains the conserved densities ρ, ρu_x, ρu_y, and $\rho e + \rho u^2/2$. The components of the flux vector are $\rho\mathbf{u}$, $\rho\mathbf{u}u_x + p\mathbf{e}_x$, $\rho\mathbf{u}u_y + p\mathbf{e}_y$, and $\rho\mathbf{u}e + \rho\mathbf{u}u^2/2 + p\mathbf{u}$. In order to have the Euler equations in conservation-law form we must use the total energy equation (2.9), not the internal energy equation (2.7).

All our flow variables are zone-centered quantities now. Furthermore, all are specified at the same time points, t_n. The evolution equation is found by averaging (3.28) over a zone volume V and over a time interval $[t_n, t_{n+1}]$:

$$V\frac{\Delta\langle U\rangle}{\Delta t} = -\iint_S \langle\mathbf{F}\rangle \cdot \mathbf{n}\, dA. \tag{3.29}$$

The expression $\langle U\rangle$ represents the zone average of U at the beginning or end of the time step; these are our unknowns. The quantity $\langle\mathbf{F}\rangle \cdot \mathbf{n}$ represents the flux of a conserved quantity through the edge of a zone averaged over the time step. If we can construct accurate values of these edge fluxes from the beginning-of-step values of $\langle U\rangle$, then the advanced values follow immediately. From the components of U it is simple arithmetic to find $\mathbf{u}$ and e, and then the pressure can be calculated. The question remains: What are the fluxes?

Godunov's original method is described as follows. At the beginning of the time step we imagine that each zone is uniform, with its average fluid quantities ρ, $\mathbf{u}$, and p. At an interface between two zones there will thus be a discontinuity, as in a shock tube at the initial time. The name for the situation in which two regions of constant fluid properties meet at a discontinuity is a *Riemann problem;* a shock tube is a special case in which the velocities vanish. At the next instant this discontinuity will be resolved into a pair of shocks, a shock and a rarefaction wave, or two rarefaction waves, one traveling into each of the two zones. At the material interface, which at the beginning of the time step coincides with the zone boundary, there will be continuous values of pressure and velocity. Godunov's method is to use these conditions, derived from the solution of the Riemann problem, to evaluate the edge fluxes, then to use the conservation law to update the conserved quantities to the next time step.

The general Riemann problem can be solved graphically as follows. Let the initial states be state 0 on the left of the interface and state 1 on the right. Plot the locus of points in the (u, p) diagram that can be reached from state 0 either by a shock ($p > p_0$) or a rarefaction ($p < p_0$) using these relations, assuming gamma-law gases:

$$u - u_0 = \begin{cases} \dfrac{2c_0}{\gamma_0 - 1}\left[1 - (p/p_0)^{(\gamma_0-1)/(2\gamma_0)}\right] & p < p_0, \\ -\dfrac{\sqrt{2}c_1(p/p_0 - 1)}{\sqrt{\gamma_1[\gamma_1 - 1 + (\gamma_1 + 1)p/p_0]}} & p > p_0. \end{cases} \tag{3.30}$$

Then plot the locus of points that can be reached from state 1 using

$$u - u_1 = \begin{cases} -\dfrac{2c_1}{\gamma_1 - 1}\left[1 - (p/p_1)^{(\gamma_1-1)/(2\gamma_1)}\right] & p < p_1, \\ \dfrac{\sqrt{2}c_1(p/p_1 - 1)}{\sqrt{\gamma_1[\gamma_1 - 1 + (\gamma_1 + 1)p/p_1]}} & p > p_1. \end{cases} \tag{3.31}$$

The intersection of the loci is the solution of the Riemann problem. This is illustrated in Figure 3.1 for $u_0 = 2$, $u_1 = 1$, $p_0 = 15$, $p_1 = 2$, $c_0 = 2$, $c_1 = 1$. This case produces a shock moving right into material 1 and a rarefaction moving left into material 0, as in a normal shock tube. The figure also shows, as a dashed curve, the release path if the relation for a shock were used instead of the correct

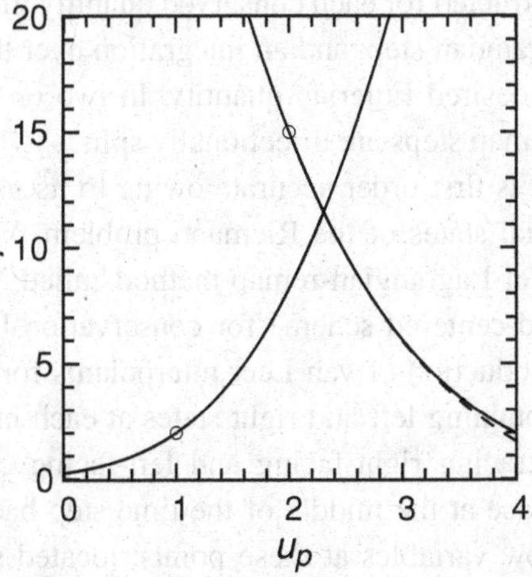

Fig. 3.1 Solid curves: states reached from left (descending curve) and right (ascending curve) initial states by a single shock or rarefaction. Circles: initial states. Dashed curve: approximation in which rarefaction is treated as a shock.

relation for a rarefaction. The locus is almost exactly the same, and this can be a useful approximation, as discussed by Colella and Glaz (1985). The solution for the Riemann problem contains a contact discontinuity, which emanates from the initial interface. The velocity of this discontinuity is u. Thus the stationary interface lies on the material 0 side of the contact if $u > 0$ and on the material 1 side of the contact if $u < 0$. This influences the correct choice of ρ at the interface, which enters the expressions for the fluxes. Efficient methods for solving the Riemann problem, including in some cases tabular equation of state data, have been discussed by van Leer (1979) and Colella and Glaz (1985).

Godunov's original method and some of its successors are actually operator-split Lagrangian-remap methods, as discussed above. The Lagrangian method still fits within the conservation-law framework, except that, in one dimension, the space coordinate is the areal mass m, the solution vector is $U = (\tau, u, E)$, with $\tau = 1/\rho$ and $E = e + u^2/2$, and the flux vector is $F = (-u, p, up)$, *viz.*,

$$\frac{\partial}{\partial t}\begin{pmatrix}\tau \\ u \\ E\end{pmatrix} + \frac{\partial}{\partial m}\begin{pmatrix}-u \\ p \\ up\end{pmatrix} = 0. \tag{3.32}$$

The reason for doing separate Lagrangian and remap steps in Godunov's method, rather than a direct Eulerian step as implied above (see also Godunov, Zabrodyn, and Prokopov (1962)), is that it is assured in the Lagrangian case that one characteristic approaches each interface from the left and the right; if the flow is supersonic this is not true for Eulerian characteristics. This is important for the higher-order Godunov methods. The remap step can make use of van Leer interpolation: the interpolant is constructed for each conserved quantity in the distorted zone that follows from the Lagrangian step, and an integration over the fixed Eulerian zone volume produces the desired Eulerian quantity. In two or three dimensions both the Lagrangian and remap steps are directionally split.

Godunov's method is first order accurate owing to its use of the zone-average quantities for the initial states of the Riemann problem. Van Leer (1979) extended it to a second order Lagrangian-remap method called MUSCL, which stands for monotonic upwind-centered scheme for conservation laws. The key element in MUSCL is the introduction of van Leer interpolants for the beginning-of-step flow variables, and obtaining left and right states at each interface for the middle of the time step by tracing right-facing and left-facing characteristics, respectively, from the interface at the middle of the time step back to the beginning of the time step. The flow variables at these points, located somewhere within the left and right zones, determine via the characteristic equations the left and right states that are used for the Riemann problem. Thus a first order accurate solution of the flow equations in characteristic form provides the data that, through the

Riemann solution, determine the fluxes that go into the conservative update for the Lagrangian step. The remap step in MUSCL is the same as just described, a van Leer interpolation on the distorted mesh followed by integration over the Eulerian zones. In two dimensions the operator-splitting follows the scheme $XYYX$, where X and Y represent Lagrangian + remap steps for the indicated directions. Doing the steps twice with the order reversed the second time, suggested by Strang (1968), eliminates the second order error associated with directional splitting. The MUSCL code is second order accurate, except near shocks, where it, and almost all others, become first order.

The next developments of Godunov-type methods were introduced by Colella and Woodward. Colella (1985) provided a second order method comparable to MUSCL, but based it on a single Eulerian step rather than Lagrangian + remap. The necessity of converting back and forth between Eulerian and Lagrangian quantities, and interpolants in mass vs space, were eliminated, and some linearizations that van Leer found necessary in the Riemann solution were also avoided. The added complication is dealing with the alternatives of two characteristics from the left, one each from left and right, and two characteristics from the right, in finding the Riemann initial states. This method is now referred to as piecewise-linear MUSCL direct Eulerian, or PLMDE, to distinguish it from the common implementation of the next method we will discuss, piecewise-parabolic MUSCL Lagrangian-remap, or PPMLR, called PPM for short.

Coming soon after Colella's second order MUSCL was the introduction of piecewise-parabolic monotonic interpolation, replacing van Leer's piecewise-linear monotonic interpolation. This makes the spatial accuracy third order, but the accuracy of the characteristic tracing and of the directional splitting in multidimensional problems reduces the overall order to second. The piecewise parabolic method (PPM) is described by Colella and Woodward (1984) and Woodward and Colella (1984). Both Lagrangian + remap and direct Eulerian versions are presented. The Lagrangian + remap version has seen considerably more use in the intervening years.

The details of PPM are rather intricate, and the reader is referred to the basic papers. The essential elements are these. Zone average data at the beginning of the time step for a specified zone, its two neighbors on the left and its two neighbors on the right, are combined to form a quadratic interpolation function in the zone that is exactly consistent with the specified zone average. The coefficients are then modified in ways described by Colella and Woodward to produce monotonicity and contact discontinuity steepening. Next the regions of influence in the left zone and the right zone for the conditions on the zone edge through the time step are found by tracing characteristics back from the end of the time step. This determines a portion of the left zone, at the beginning of the time step, that influences the left-hand side zone edge conditions, and likewise a portion of the right-hand side. The fluid variables are averaged in these regions of influence to give the left-hand side and right-hand side conditions for the Riemann problem. The space average over the region of influence gives the correct time average along the zone edge apart from terms arising from the variation of the characteristic slopes during the time step. The fluxes in the conservative Lagrangian equations (3.32) are computed from these time-average edge conditions. In most cases no additional dissipation such as pseudo-viscosity needs to be included in the method. There are exceptions, however. If a shock is nearly stationary with respect to the mesh, oscillations will develop unless the mesh is jiggled slightly, as Woodward and Colella describe. This jiggling is comparable to artificial viscosity or artificial heat conduction in its effect, but is an order of magnitude smaller than the pseudo-viscosity usually used in the staggered-mesh methods.

The test cases shown by Woodward and Colella (1984) demonstrate the improved accuracy with PPM relative to MUSCL for calculations on the same mesh. Despite the considerable cost in CPU time per zone per cycle, PPM compares well with other methods in the cost to obtain a specified accuracy. The widths of shocks computed with PPM are about one zone, not a few zones as with pseudo-viscosity methods.

3.2.3 WENO

The next class of Eulerian hydrodynamics schemes is the essentially non-oscillatory (ENO) class and its descendant weighted essentially non-oscillatory (WENO). To explain what these are it is helpful to step back and discuss the Lax–Wendroff (1960) method. It is a method for a system of conservation laws in one space dimension, such as the Euler equations, in the form of (3.28). The two-step Lax–Wendroff method is a predictor–corrector method:

$$\begin{aligned} U_{j+1/2}^{n+1/2} &= \frac{1}{2}\left(U_{j+1}^n + U_j^n\right) - \frac{\Delta t}{2\Delta x}\left(F_{j+1}^n - F_j^n\right) \\ U_j^{n+1} &= U_j^n - \frac{\Delta t}{\Delta x}\left(F_{j+1/2}^{n+1/2} - F_{j-1/2}^{n+1/2}\right). \end{aligned} \tag{3.33}$$

To conform to the notation in Richtmyer and Morton (1967) and the WENO literature (Jiang and Shu, 1996) the zone centers are given integer spatial indices and the edges have half-integer indices. We see that the predictor equation finds an edge-centered solution vector at the half-time-step point using the beginning-of-step values of the fluxes evaluated at zone centers. The corrector step uses the edge-centered fluxes derived from the half-time-step unknowns to make the usual conservative update of the zone-centered unknowns.

The two-step Lax–Wendroff method is second order and the dissipation is quite small, $O(\Delta x^2 \Delta t^2)$ (Richtmyer and Morton, 1967). Sod (1978) shows the numerical results for this and a number of other methods on his now well-known shock tube problem. While the first-order method of Godunov *et al.* (1962) smears the shock and contact discontinuity slightly while avoiding ringing, the second order Lax–Wendroff method has fairly sharp jumps combined with a substantial amount of ringing. The ringing has grown to be considered unacceptable, and this was a principal motivation for developing higher order Godunov schemes in which the ringing could be minimized by monotonicity constraints and nonlinear filters.

The ENO approach is another path toward this goal which avoids the use of a Riemann solver. The conservation laws are written in a form in which space has been discretized but time is continuous:

$$\frac{dU_j}{dt} = \frac{1}{\Delta x}\left(\hat{F}_{j-1/2} - \hat{F}_{j+1/2}\right), \tag{3.34}$$

in which the fluxes $\hat{F}_{j+1/2}$ come from the application of an appropriate operator to the U_j. This operator will act on values of U in perhaps several zones surrounding the edge $j+1/2$ in order to obtain $\hat{F}_{j+1/2}$. The next essential part of the ENO concept is the use of Runge–Kutta methods for the integration of (3.34). The predictor–corrector method is a kind of second order Runge–Kutta. The two-step Lax–Wendroff method departs from a second order Runge–Kutta solution of (3.34) by using a spatially staggered predictor step. One of the goals in the ENO and WENO schemes is to avoid the introduction of oscillations – ringing – hence the names of the methods. One aspect of this is the use of total variation diminishing (TVD) or total variation bounded (TVB) time integrators. TVD means essentially that the spatial fluctuation in the solution at the end of the time step cannot be larger than it was at the beginning of the time step. TVB means that, while the fluctuation may grow, it will always be less than some fixed bound. Some Runge–Kutta integration schemes are TVD and some are not. A popular choice, referred to as RK3, is this third order method:

$$U^{(1)} = U^n + \Delta t\, L[U^n], \tag{3.35}$$

$$U^{(2)} = \frac{3}{4}U^n + \frac{1}{4}U^{(1)} + \frac{1}{4}\Delta t\, L[U^{(1)}], \tag{3.36}$$

$$U^{n+1} = \frac{1}{3}U^n + \frac{2}{3}U^{(2)} + \frac{2}{3}\Delta t\, L[U^{(2)}], \tag{3.37}$$

in which $L[U]$ stands for the operator that yields the right-hand side of (3.34).

The heart of the ENO and WENO schemes comes in choosing the spatial difference operator. To begin with, this is an interpolation problem. Given data, in this case the $F_j^n = F(U_n^j)$, that represent the average values in the spatial zones

$[x_{j-1/2}, x_{j+1/2}]$, for a certain span of zones, find the polynomial of suitable order that fits these data, and thereby derive the edge-centered values $F^n_{j+1/2}$. Then $L[U^n]$ follows as in (3.34). For uniform spatial zoning and for stencils (the collection of zone-centers that are used for the interpolation) that consist of some number of contiguous zones, the formulae are easily written down: the single-zone stencils:

$$F_{j+1/2} = F_j, \qquad F_{j+1/2} = F_{j+1}; \tag{3.38}$$

the two-zone stencils:

$$\begin{aligned} F_{j+1/2} &= \frac{1}{2}\left(-F_{j-1} + 3F_j\right), \quad F_{j+1/2} = \frac{1}{2}\left(F_j + F_{j+1}\right), \\ F_{j+1/2} &= \frac{1}{2}\left(3F_{j+1} - F_{j+2}\right); \end{aligned} \tag{3.39}$$

and the three-zone stencils:

$$\begin{aligned} F_{j+1/2} &= \frac{1}{6}\left(2F_{j-2} - 7F_{j-1} + 11F_j\right), \\ F_{j+1/2} &= \frac{1}{6}\left(-F_{j-1} + 5F_j + 2F_{j+1}\right), \\ F_{j+1/2} &= \frac{1}{6}\left(2F_j + 5F_{j+1} - F_{j+2}\right), \\ F_{j+1/2} &= \frac{1}{6}\left(11F_{j+1} - 7F_{j+2} + 2F_{j+3}\right). \end{aligned} \tag{3.40}$$

The order of the interpolation polynomial is one less than the number of zones in the stencil in each case; the result is an approximation to $L[U]$ that is accurate to an order equal to the number of zones.

Next we discuss selecting among the stencils. As we know, stability of the method is aided by using preferentially data on the upwind side in space for the calculation of the flux. For a scalar equation the wind blows in the direction indicated by $\partial_U F(U)$: in the $+x$ direction if $\partial_U F(U) > 0$, in the $-x$ direction if $\partial_U F(U) < 0$. We thus discard one stencil from each list of candidates, the last one in each group for $\partial_U F(U) > 0$ and the first one for $\partial_U F(U) < 0$. If $F(U)$ is not a monotone function then it must be split into parts that are:

$$F(U) = F^+(U) + F^-(U), \tag{3.41}$$

with $\partial_U F^+(U) > 0$ and $\partial_U F^-(U) < 0$; there may be several ways to do this. The value of $\hat{F}$ is then calculated by summing the values of $\hat{F}^+$, calculated on stencils biased toward $-x$, with $\hat{F}^-$ calculated on stencils biased toward $+x$. For systems of conservations laws there is not just one derivative $\partial_U F(U)$ to examine, but the signs of the characteristic speeds, which are the eigenvalues of the Jacobian matrix. The eigenvectors of the Jacobian are formed into a group with positive characteristic speed and a group with negative characteristic speed. Projection operators for the positive and negative subspaces are formed from these, and the final $\hat{F}$ is built by combining the upwind stencils for each group using these operators. This is not the only method for splitting $\hat{F}$ into $\hat{F}^+$ and $\hat{F}^-$. A simpler alternative that seems to work well is based on the Lax–Friedrich method (Lax, 1954), for which

$$F^{\pm}(U) = \frac{1}{2}\left(F(U) \pm \alpha U\right) \tag{3.42}$$

with $\alpha = \max |\partial_U F(U)|$, taken over the full sampled range of U.

Now we come to the ENO/WENO concept. Consider the three-zone stencils, of which three survive after dropping the downwind-most of the original four. A figure of merit is constructed locally for each stencil by combining the squares of estimates of the first and second spatial differences of the Fs. The preferred stencil is the one for which this figure of merit is least. Near a shock or other discontinuity in the flow, this stencil will be the one that samples the properties downwind of the jump the least. For the ENO method the result, for this flux, is taken to be the one from the preferred stencil and the others are discarded. For the WENO method the flux is given by a weighted sum of the flux from the surviving stencils, with weights inversely related to the figure of merit mentioned. These weights have the effect, in smooth regions, of canceling some of the truncation error in the individual stencils. Thus each of the three-zone stencils gives a result that is third order accurate, the weighted sum is fifth order accurate. For an r-zone stencil, each individual is rth order accurate and the weighted sum is $(2r - 1)$th order accurate. The detailed expressions for the weights and the error analysis are found in Jiang and Shu (1996). This is the scheme referred to as WENO5. Balsara and Shu (2000) have given a ninth order WENO scheme that uses five different five-zone stencils. Ninth order pertains to the spatial differencing; WENO9 still uses the third order RK3 Runge–Kutta time integration scheme.

The test results using WENO5 in Jiang and Shu (1996) on Sod's problem and some of Woodward's problems (Woodward and Colella, 1984) look quite acceptable. The characteristic projection method for flux splitting has the edge in accuracy over the Lax–Friedrichs splitting. Shi, Zhang, and Shu (2003) show WENO5 and WENO9 results for Woodward's double Mach reflection problem and a Rayleigh–Taylor instability problem that contain a profusion of small-scale features for the high order methods.

3.2.4 Adaptive mesh refinement (AMR)

An Eulerian hydrodynamics variation is adaptive mesh refinement, or AMR. The concept of AMR is to provide greater grid resolution in the parts of the spatial domain that most need it any given time. While AMR might be possible in a

Lagrangian framework, the need for this capability is less since the Lagrangian zones already follow the material and tend to be where they are needed automatically. This is not completely true, since extra resolution may be needed near shocks, for example, which move through the material. The second reason for not using Lagrangian AMR is that it may be exceedingly cumbersome. The unstructured-mesh Lagrangian codes would be the best Lagrangian AMR vehicle, but we will discuss that no further.

The seminal paper describing block-structured AMR, a method that applies refinement in rectangular subdomains (patches), is Berger and Colella (1989). Details of the implementation are given by Bell, Colella, and Glaz (1989) and Bell, Berger, Saltzman, and Welcome (1994). A very nice capsule description of the method has been given by Grauer and Germaschewski (2001):

> We start by integrating the equations under consideration on a single grid with fixed resolution. Then, we constantly verify if the resolution is still sufficient by some appropriate criterion. If it is not we mark all points where the resolution is insufficient. Next, we try to cover these underresolved points by a collection of rectangles (or boxes in 3D) as effectively as possible using methods from image processing. These new rectangles need to be correctly embedded in their parent grids, otherwise they have to be modified accordingly. When the rectangles are filled with data, they are integrated using a discretization length and time step divided by certain refinement factors. In addition, all the steps described above are performed recursively. Important is the communication among the grids. The necessary boundary data are obtained from neighboring grids if present and otherwise from the parent grids. Similar treatment is applied to newly generated grids. Data are obtained from the old grids of the same level where possible, otherwise from the parent grids. This allows for effectively following moving structures. Finally, after each time step the coarse grids are updated using data from finer grids.

The AMR mesh may be thought of as hierarchical; at the top level is a uniform fairly coarse mesh, called the level 1 mesh. Portions of this are refined to produce a level 2 mesh, and so on to a maximum depth of refinement, perhaps to the third level, perhaps more. The data in any zone at a certain level are superseded by the data at the next level down if this zone is in a patch that is refined. The refinement factor may be chosen, for example, to be 4. With a refinement factor of 2 a few more levels are needed. Whatever refinement factor is used spatially, in each of the coordinate directions, the same factor is used for temporal refinement. Thus a patch that is refined four times in space is subcycled in the time integration; four of the refined times steps are computed to bring the refined patch up to the same new time level as the coarser region in which it is embedded. This is done so that each level of refinement uses the same Courant ratio. Of course, this is also done recursively when there are several levels of refinement. The maximum depth of refinement is a critical parameter in the application. If this is set too small, then important features will not be resolved. If it is set too large, then some hydrodynamics problems will "refine themselves to death" because the criterion for accepting the current mesh may not be satisfied until the available storage is exhausted. The maximum refinement depth must be chosen by the skilled code user to meet the requirements of the problem at hand.

The criterion for refinement, like the maximum depth of refinement, must also be tuned to fit the problem, and for the same reason: insufficient refinement defeats the hoped-for accuracy, and excessive refinement stops the calculation. The general criteria, such as the Richardson's extrapolation estimate of Berger and Colella (1989) can be useful, but without adjustment can also lead to excessive refinement. Here again, the skill of the experienced code user is necessary.

Since AMR puts the mesh where it is needed to resolve features in the flow, the total number of zones, counting all the levels of refinement, is much less than for the uniform mesh that provides the same maximum resolution. That is, this is the case if the areas of refinement are relatively few and localized. There are problems, such as homogeneous hydrodynamic turbulence, for which refinement would be necessary throughout the entire problem, and for which AMR is therefore of little value. Problems of this kind benefit from PPM or the WENO methods.

One group of closely related AMR codes is described by Klein, Greenough, Howell and coworkers (Truelove *et al.*, 1997, 1998; Klein (1999); Klein *et al.*, 2003, 2004). They use block-structured AMR as described above, and employ the PLMDE Godunov method. Block-structured AMR is not the only approach. The RAGE code developed by Gittings and others (Gittings, 1992; Holmes *et al.*, 1999) uses cell-by-cell refinement. All these codes have 3-D as well as 2-D versions. Operator splitting of the spatial coordinates can as easily be applied to three dimensions as two.

3.3 ALE methods

The ALE methods will not be described in detail. The difference between the ALE equations and the Eulerian equations is slight, as we see from a comparison of (2.15)–(2.17) with (2.2), (2.3), and (2.9). The computational procedure in most of the ALE codes follows the outline of the staggered-mesh Eulerian method described earlier. One difference is that, since the mesh is distorted rather than orthogonal as is the Eulerian mesh, the computation of the pressure gradient follows the prescription used in the Lagrangian codes. The ALE modifications to the Eulerian equations are treated fully by Bowers and Wilson (1991). What is left unaddressed is the very important question of specifying the grid motion. The grid motion algorithms have the vital task of following the material as closely as possible while at the same time preventing excessive distortion of the grid. The recipes for doing this are many, derived from a great number of hydrodynamic simulations that failed and needed a new grid motion algorithm to succeed. Unfortunately this experience is poorly documented. Some inkling as to the algorithms can be derived from the user's manual for CALE, Tipton's popular 2-D ALE code (Tipton, 1991).

3.4 3-D methods

The discussion of 3-D codes will also be very brief. The 3-D AMR codes have been mentioned. The extension from two to three dimensions of orthogonal-mesh Eulerian codes is relatively simple. There are some notable computer science issues involved with maintaining the data structures for the hierarchical mesh. The individual patches lead to box objects that carry the information about the cells they contain and also about how they interface to the boxes in which they are embedded. There are many more such items of information for the 3-D box than in two dimensions.

Besides these two codes there is PPM, which has been described already, and which exists in a 3-D version. This is the code that won the 1999 Gordon Bell award for its teraflop performance on a problem of Richtmyer–Meshkov instability (Mirin *et al.*, 1999). Another is the FLASH code developed by the ASCI Center for Thermonuclear Flashes at the University of Chicago (Fryxell *et al.*, 2000). FLASH incorporates PPM as one of its hydrodynamic options, but can also use block-structured AMR.

There are no good examples of a strictly Lagrangian 3-D code. This is for the simple reason that Lagrangian codes fail to run past the point in time when the mesh tangles, producing the bowties mentioned above. In three dimensions the opportunities to tangle are much richer, and the Lagrangian codes simply do not run long enough to be useful. An exception must be made for structural dynamics codes such as DYNA3D (Hallquist, 1982); these deal with solid materials that have strength and that therefore can resist the tangling to which fluids are susceptible.

There are 3-D ALE codes. We will mention two: the ARES code (Bazan, 1998) and the KULL code (Rathkopf *et al.*, 2000), both developed at Lawrence Livermore National Laboratory as part of the US Department of Energy's Advanced Simulation and Computing effort, which in fact supports all the 3-D codes mentioned above as well. ARES and KULL have many similarities, but ARES has a structured mesh that is almost entirely built of hexahedra, i.e., deformed cubes. The ARES mesh can have a limited number of points of irregularity where the connectivity is different from that for hexahedra. The MESH in KULL is unstructured, so in principle it is built of arbitrary polyhedra, but in practice KULL most often uses a hexahedral mesh like ARES's.

3.5 Other methods – SPH and spectral methods

Two other computational methods for hydrodynamics are current that avoid, either altogether or largely, the use of finite-difference or finite-element equations on a spatial mesh. These are smoothed particle hydrodynamics (SPH) and the spectral method. There is no mesh at all in SPH, unless self-gravity or magnetohydrodynamics are to be included. In the spectral method the mesh that is used is in Fourier (or Chebyshev) space, and normal space is used only for evaluating nonlinear products.

The SPH method was introduced by Lucy (1977) and Gingold and Monaghan (1977). Good descriptions of it are found in Monaghan (1992), Hernquist and Katz (1989), and Rasio (2000). At first blush SPH is a Monte Carlo method, since it uses a sampled set of discrete particles to represent the continuous fields of density and velocity. But the initial particles are not selected randomly, and the time advancement is completely deterministic. SPH is best thought of as a meshless Lagrangian method.

The following description of SPH is based largely on Rasio (2000). Each of the N particles in the simulation carries its particular values of mass m, velocity $\mathbf{u}$, internal energy e, and for some methods also the density ρ. The particle has an "uncertainty cloud" that is described by the function $W(\mathbf{r}-\mathbf{r}_i, h_i)$, where $\mathbf{r}_i$ is the current location of the particle and h_i is a parameter that sets its spatial extent. If the density is not an explicit attribute of the particles it can be computed from

$$\rho_i = \sum_j m_j W(\mathbf{r}_i - \mathbf{r}_j, h_j). \tag{3.43}$$

The kernel function W might be a Gaussian, or this spline function:

$$W(r,h) = \frac{1}{\pi h^3} \begin{cases} 1 - \frac{3}{2}\left(\frac{r}{h}\right)^2 + \frac{3}{4}\left(\frac{r}{h}\right)^3, & 0 \le \frac{r}{h} \le 1, \\ \frac{1}{4}\left[2 - \frac{r}{h}\right]^3, & 1 \le \frac{r}{h} \le 2, \\ 0, & \text{otherwise}. \end{cases} \tag{3.44}$$

The scale length h must be chosen to be small compared with the size of features in the solution, but large enough to encompass several of the neighbors of each particle. These are the same criteria one would use to select a zone size in a mesh-based simulation.

The evolution equations for $\mathbf{r}_i$, $\mathbf{u}_i$, and e_i are the following, for the simplest implementation:

$$\frac{d\mathbf{r}_i}{dt} = \mathbf{u}_i, \tag{3.45}$$

$$\frac{d\mathbf{u}_i}{dt} = -\sum_j m_j \left(\frac{p_i}{\rho_i^2} + \frac{p_j}{\rho_j^2} \right) \nabla_i W_{ij}, \tag{3.46}$$

$$\frac{de_i}{dt} = \frac{1}{2} \sum_j m_j \left(\frac{p_i}{\rho_i^2} + \frac{p_j}{\rho_j^2} \right) (\mathbf{u}_i - \mathbf{u}_j) \cdot \nabla_i W_{ij}, \tag{3.47}$$

where $W_{ij} = W_{ji} = W(\mathbf{r}_i - \mathbf{r}_j, h)$. (If h differs for different particles, a symmetrization procedure is required.) An important aspect of the summations over particles that appear in the dynamical equations is that, because the kernel W has a finite range, there is a very limited number of nonzero terms in the sum. The sum expression for ρ_i above introduces problems at free boundaries, and the procedure of carrying ρ explicitly on each particle may be preferred, using the following form for the equation of continuity:

$$\frac{d\rho_i}{dt} = \sum_j m_j (\mathbf{u}_i - \mathbf{u}_j) \cdot \nabla_i W_{ij}, \tag{3.48}$$

The pressure p_i is to be computed from the equation of state in terms of ρ_i and e_i.

These are the essential equations of the method. The further details involve including viscosity, constructing a dynamical equation for h_i, and other refinements. The TREESPH code described by Hernquist and Katz (1989) adds the physics of self-gravitation using the tree method. That is, a given particle exerts a force on its close neighbors that is calculated from the inverse-square law, but for remote particles the force is given by the gradient of a smoothed-out potential. A hierarchy of meshes is laid on the problem so that the cells in which the given particle lies, at all the levels of the hierarchy, form one branch of a tree. For each cell of the hierarchy a few of its multipole moments are computed. These multipole moments provide the force on the given particle from the more remote parts of the problem. If N is the total number of particles, then the number of close-neighbor interactions is $O(N)$, and, remarkably, the cost of the computation of the hierarchical multipole moments and their use to provide the remote contribution to the force is only $O(N \log N)$.

The time integration is commonly chosen to be the second order leapfrog method familiar from the staggered-mesh Lagrangian codes, *cf.*, (3.1), (3.2). There is a Courant-like limit to the time step for stability of order $0.5h/a$. The spatial accuracy is first order, i.e., the error is $O(h^2)$, and so the method is first order accurate overall. The SPH result converges to the continuum solution if $h \to 0$ but also the number of particles N_N within the volume $4\pi h^3/3$ tends to infinity, which means, of course, that $N \to \infty$. Monaghan (1992) describes a number of tests to which SPH has been subjected, including standard gas dynamics problems such as Sod's shock tube and Woodward and Colella's flow over a step; the results are acceptable, with the errors that would be expected for a first order code. Monaghan also provides an impressive list of astrophysical applications of SPH, ranging from tidal disruption in binary star systems, through asteroid impacts on the earth and molecular cloud fragmentation to large-scale structure in the universe.

Spectral methods, or more precisely pseudo-spectral methods, have a significant presence in the world of turbulence simulations and Navier–Stokes calculations. A standard text on spectral methods in fluid dynamics is Canuto, Hussaini, Quarteroni, and Zang (1988), and another reference is Boyd (2001). The major feature of the methods is the expansion of the unknowns ρ, $\mathbf{u}$, and so forth in terms of global basis functions such as trigonometric functions (Fourier series) or orthogonal polynomials (Chebyshev series). The spatial derivatives of the expanded functions are easily evaluated using the derivatives of the basis functions, which are available analytically. The technical distinction between "spectral" and "pseudo-spectral" methods is in whether the expansion should approximate the continuous solution in the sense of integral projections (the Galerkin concept) – the spectral method – or should be accurate at a certain set of mesh points (collocation points) – the pseudo-spectral method. The equations of fluid dynamics are nonlinear, of course, and terms such as $\mathbf{u} \cdot \nabla\mathbf{u}$ and $\nabla p/\rho$ in the Euler equation are only conveniently evaluated by combining $\mathbf{u}$ and $\nabla\mathbf{u}$ or ∇p and ρ at the collocation points, but relying on Fourier or Chebyshev space for the derivative calculations.

An ideal fluid dynamic application for the pseudo-spectral method is direct numerical simulation (DNS) of homogeneous turbulence at low Mach number using the incompressible Navier–Stokes equation

$$\frac{\partial \mathbf{u}}{\partial t} + \mathbf{u} \cdot \nabla\mathbf{u} = -\frac{1}{\rho}\nabla p + \nu\nabla^2\mathbf{u}, \tag{3.49}$$

in which the density ρ and the kinematic viscosity ν are constants, and for which the pressure is to be found from the incompressibility constraint,

$$\nabla \cdot \mathbf{u} = 0. \tag{3.50}$$

Equation (3.49) can be written like this:

$$\frac{\partial \mathbf{u}}{\partial t} = \mathbf{u} \times \boldsymbol{\omega} - \nabla\Pi + \nu\nabla^2\mathbf{u}, \tag{3.51}$$

in which $\boldsymbol{\omega} = \nabla \times \mathbf{u}$ is the vorticity and Π is the Bernoulli "constant" $\Pi = p/\rho + u^2/2$. The divergence of (3.51) leads to this Poisson equation for Π:

$$\nabla^2 \Pi = \boldsymbol{\nabla} \cdot (\mathbf{u} \times \boldsymbol{\omega}), \tag{3.52}$$

so that, in principle, Π can be eliminated from (3.51).

Poisson's equation is readily solved in Fourier space. We define the transform of a variable X by $\tilde{X} = \mathcal{F}_{\mathbf{k}}[X] = \iiint d^3\mathbf{r}\ \exp(i\mathbf{k} \cdot \mathbf{r}) X(\mathbf{r})$. The transform of (3.52) gives

$$-k^2 \tilde{\Pi} = -i\mathbf{k} \cdot \mathbf{s}, \tag{3.53}$$

in which the quantity $\mathbf{s}$ is the transform of the nonlinear term $\mathbf{u} \times \boldsymbol{\omega}$. The transform of the vorticity itself is given by

$$\tilde{\boldsymbol{\omega}} = -i\mathbf{k} \times \tilde{\mathbf{u}}. \tag{3.54}$$

Using (3.53), $\tilde{\Pi}$ can be eliminated from the transform of (3.51), which leads to

$$\left(\frac{d}{dt} + \nu k^2\right) \tilde{\mathbf{u}} = \left(\mathsf{I} - \frac{\mathbf{kk}}{k^2}\right) \cdot \mathbf{s}. \tag{3.55}$$

The tensor $\mathsf{I} - \mathbf{kk}/k^2$ is the projection operator perpendicular to $\mathbf{k}$, so its effect is to make only the transverse (divergence-free) component of $\mathbf{u} \times \boldsymbol{\omega}$ appear in the acceleration equation.

The integration of (3.55) is done by first choosing a uniform mesh in $\mathbf{r}$ space, to which corresponds a similarly shaped mesh in $\mathbf{k}$ space. An integration method such as Runge–Kutta is applied to follow the evolution of $\tilde{\mathbf{u}}$ on this $\mathbf{k}$ mesh. The evaluation of $d\tilde{\mathbf{u}}/dt$ at each step in the integration requires first forming $\tilde{\boldsymbol{\omega}}$ from $\tilde{\mathbf{u}}$, applying fast Fourier transforms (FFTs) to each of these to give $\mathbf{u}$ and $\boldsymbol{\omega}$, forming the cross product, then another FFT to give $\mathbf{s}$, from which $d\tilde{\mathbf{u}}/dt$ follows. An additional FFT provides Π and hence p if desired for edit purposes.

A very large-scale example of this approach to a DNS problem may be found in Yokokawa *et al.*, (2002).

3.6 Summary

Numerical hydrodynamics is a very rich field. It is a garden with many beautiful flowers, and it is not easy to compare them. Some of the methods have been developed within a certain area of application, and these methods are well suited to their area. We think of SPH and its astrophysical applications, for which a very robust Lagrangian method is highly desirable even if it is somewhat more dissipative than competing methods, and if the difficulties of adding other types of physics such as MHD or radiation transport can be overcome. Spectral methods might be very cumbersome indeed if applied to problems with real material properties and complicated structures, but for the ultimate in accuracy, as in DNS, they are perfectly suited. The work-horse 1-D method, for everything from inertial fusion capsule implosions to pulsating stars and the big bang, is von Neumann–Richtmyer staggered-mesh Lagrangian hydrodynamics. In two and three dimensions, for the problems that do have real materials or complicated structures, there are hard choices to make. The ALE codes naturally put the resolution where it is needed in the spatial structure; their down side is a lack of robustness that is constantly being improved by more sophisticated grid-motion algorithms. The Eulerian codes are much more robust, but accuracy and freedom from numerical diffusion have traditionally been their weak points. The Godunov and ENO/WENO methods have definitely corrected that. If a sufficiently fine mesh is used to resolve all the features of interest, or with AMR to achieve that, Eulerian methods are quite attractive.

The Godunov vs ENO/WENO competition is in a state of flux at this time, and the best methods may not have been created yet. It is interesting to look at the comparisons in Liska and Wendroff (2003) and Rider, Greenough, and Kamm (2003), as well as the examples shown by Shi *et al.* (2003). Liska and Wendroff do not include PLMDE or any AMR code in the list of tested codes, but they do include a WENO3 code, with simplified flux splitting, and a WENO5 code with characteristic-based flux splitting, as well as PPM and others. The test suite consists of 17 Riemann problems, the Noh problem (Noh, 1987), a Rayleigh–Taylor bubble growth problem, an implosion problem and an explosion problem. As the authors point out, no one code is best on all the problems, although WENO5 and PPM are near the top on most of them. Interestingly, WENO5 crashed attempting to run the Noh cylindrical stagnation problem. On the Rayleigh–Taylor bubble problem the amount of fine-scale structure that develops on the edges of the bubble due to secondary Kelvin–Helmholtz instability was quite different in the various codes. The low order, rather dissipative codes showed quite smooth bubbles, while PPM had quite a large amount of such structure; WENO5 less so. This is also similar to the result in Holmes *et al.* (1999), where the AMR code RAGE was compared with PROMETHEUS, a code quite similar to PPM, and a front-tracking code FronTier (Grove *et al.*, 1993) on a Richtmyer–Meshkov instability growth problem rather similar to the Rayleigh–Taylor one. In this case RAGE and FronTier were relatively smooth and PROMETHEUS stood out for its amount of small-scale structure. RAGE also compares favorably with results of a Richtmyer–Meshkov instability experiment, as seen in Baltrusaitis *et al.* (1996). Shi *et al.* also show a Rayleigh–Taylor problem as computed by WENO5 and WENO9; the fine-scale structure in the highly-zoned WENO9 calculation is prodigious. The experiment described by Holmes *et al.* did not have sufficient spatial resolution to select between the codes on the basis of the fine structure, so this issue remains unsettled.

Rider *et al.* (2003) focus on comparisons of WENO5 with several PLMDE and PPM variants. The test problems were the Sod shock tube, the Woodward and Colella colliding blast wave and the Shu and Osher (1989) steepening wave train. The high order accuracy of WENO5 did not result in an appreciably lower error for those problems featuring a strong shock. Neither was WENO5 much worse than the others; the error measures for the different methods spanned a surprisingly small range. An interesting comparison can be made if the accuracy results are normalized by the computing time required. That is, the codes are compared on the basis of cost to obtain a specified accuracy. Since for a given mesh WENO requires about three times the floating point operations of a Godunov method like PLMDE, and the accuracy is not very different for shock problems, the efficiency of WENO5 compares poorly with the PLM and PPM methods. It should be noted that this statement is specific to shock problems; in flows that are everywhere smooth the high order WENO performance is very good. Rider *et al.* also include in their paper a new method that adds some extremum-preserving features derived from WENO to the basic PPM scheme; that is, the slope limiting that is a normal part of PPM is relaxed somewhat to avoid clipping the peaks. This method turned out to perform exceptionally on the test problems, and to be efficient as well. Time will tell whether this new variation will fulfill its early promise.

It does seem that the high-accuracy Eulerian methods are continuing to develop and to represent a very attractive alternative to ALE methods. As the scale of computing systems expands, the ability of Eulerian codes to resolve all the relevant spatial structures will improve, and when that happens the scales will tip strongly in the Eulerian direction.

4
Description of radiation

We turn now to the subject of radiation transport. As much as possible, the present goal is to demystify this subject. Photons are just particles, like the others that make up our systems; they just happen to go faster and farther, and are therefore often of special importance in carrying energy and momentum from one place to another. In kinetic theory we introduce the phase-space distribution function for the atoms, develop the theory of the Boltzmann transport equation, and come up with some satisfactory approximate methods for solving it. Radiation transport is exactly the same; the transport equation is about the same, and the approximate methods are about the same as well. The difference is that the subject of radiation transport was elaborated by different people than was kinetic theory, using an entirely different notation, and we have that difference with us today. In the last two or three decades yet another community has joined the discussion of radiation transport, and these are the nuclear engineers, who have evolved a collection of methods for describing neutron transport, methods that are useful for photons as well as neutrons. The present discussion will not attempt to show, Rashomon-like, the same physical concepts from the varied points of view of several disciplines. We will stick with one, mainly the astrophysical notation found, for example, in Mihalas and Mihalas (1984). The elementary definitions of the radiation field quantities are found in many astrophysics books. One good treatment is Mihalas's *Stellar Atmospheres* (1978), and this is also found in Mihalas and Mihalas (1984). The coordinate-free equations are presented by Pomraning (1973) and by Cox and Giuli in the first volume of *Principles of Stellar Structure* (1968).

4.1 Intensity; flux; energy density; stress tensor

Astrophysical radiation transport begins with the intensity I_ν as the fundamental concept instead of the phase space distribution function; the content is equivalent, as we will see. Historically, this might be due to the fact that you can discuss the intensity entirely in the wave picture, and never make reference to those pesky quanta. With sufficient effort it is possible to derive the transport equation directly from Maxwell's equations of electromagnetism and avoid quantum theory entirely. In the present day this seems like a silly thing to do, and certainly more trouble than it is worth. As soon as one has to deal with matter–radiation interactions the quantum picture quickly becomes indispensable. So our present discussion will begin with the classical intensity, I_ν, but we will notice right away that it has an interpretation in terms of a distribution function for photons.

So let us define the intensity. It is a function of three spatial coordinates, two angle coordinates, the radiation frequency, and time, seven coordinates in all. (Polarization of the radiation is described with an additional coordinate, but for the moment we will lump the polarizations together.) The angle coordinates specify in which direction the radiation is going, the spatial coordinates specify the location of the intensity measurement. The frequency coordinate says that we measure the intensity in a small spectral band around one frequency in the spectrum. The intensity is in power units, energy per unit time, but is also expressed per unit area, per unit solid angle, and per unit frequency bandwidth.

The per-unit-area and per-unit-solid-angle attributes work in this way: We make an ideal apparatus, a pinhole camera, to measure intensity by setting up a screen at the place where we want to know the intensity that is opaque except for an aperture that has an area A_1. This is illustrated in Figure 4.1. This screen is oriented perpendicular to the direction we want to measure. The aperture has a shutter, and we will open the shutter at time t and leave it open for a duration Δt. On the downstream side of the screen we arrange for the matter to be cleared away, so the radiation streams freely through. At a suitable distance away from the screen we set up a second screen, which is used as a detector. The second screen has to be at a distance from the first screen that is much larger than the size of the aperture. At a location on the second screen exactly in line with the ray through the aperture that is going in the desired direction, there is a sensitive detector area of size A_2. This is a clever detector that responds only to the frequency bandwidth $\Delta\nu$ around frequency ν. The size of A_2 as viewed from A_1 has to be large compared with the diffraction angle for the aperture radius. The definition of intensity I_ν is that the energy collected by this detector is

$$\Delta E = I_\nu \frac{A_1 A_2}{r^2} \Delta t \Delta \nu. \tag{4.1}$$

The combination A_2/r^2 is the solid angle $d\Omega$ subtended by A_2 at the aperture, so we say the intensity is the energy crossing a unit area at a given point per unit time per unit frequency and per unit solid angle in the direction of interest. In this construction it is obvious that we can increase the distance r by a factor f,

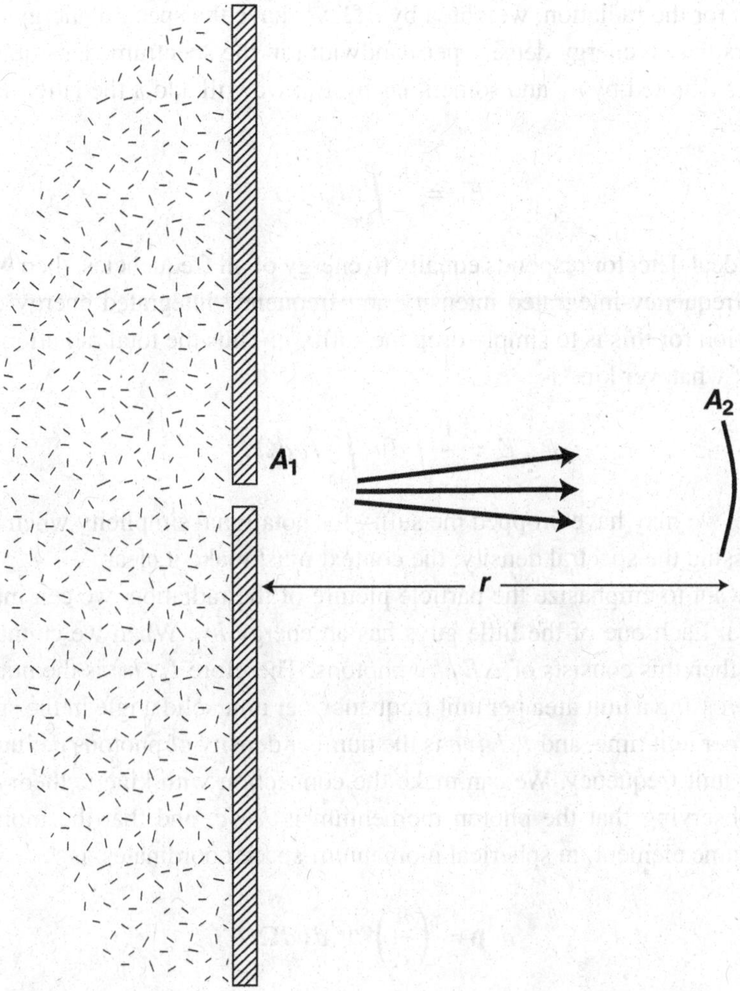

Fig. 4.1 Illustration of the ideal apparatus that serves to define the specific intensity.

and enlarge both of the lateral dimensions of the detector by the same factor thus increasing A_2 by a factor f^2, without changing $d\Omega$, and therefore the intensity is really a property of the radiation field at the place where we put the aperture.

The photons that comprise our bundle of radiant energy that will be registered by the detector travel a distance $c\Delta t$ during the time the shutter is open. Thus at any given time they occupy a cylindrical volume $c\Delta t$ long with a cross section A_1, and therefore a total volume $cA_1\Delta t$. Dividing ΔE by the volume tells us that there is a contribution to the radiation energy per unit volume of $(I_\nu/c)(A_2/r^2)\Delta\nu$ from the solid angle A_2/r^2 and the frequency band $\Delta\nu$. Thus I_ν/c is the radiation energy density per unit solid angle per unit frequency. If this quantity is integrated over all directions for the radiation, weighted by $d\Omega$, we have the spectral energy density, the total radiation energy density per bandwidth in the spectrum. This quantity is sometimes denoted by u_ν and sometimes by E_ν; we will adopt the latter notation, so we have

$$E_\nu = \frac{1}{c}\int_{4\pi} I_\nu d\Omega. \tag{4.2}$$

If our ideal detector responds equally to energy of all frequencies, then we measure the frequency-integrated intensity and frequency-integrated energy density. One notation for this is to simply drop the suffix ν. Thus the total radiation energy density of whatever kind is

$$E = \frac{1}{c}\int d\nu \int_{4\pi} I_\nu \, d\Omega. \tag{4.3}$$

Of course, we may have dropped the suffix for notational simplicity when we still are discussing the spectral density; the context must make it clear.

If we want to emphasize the particle picture of the radiation, we can introduce the quanta. Each one of the little guys has an energy $h\nu$. When we count an energy ΔE then this consists of $\Delta E/h\nu$ photons. Therefore $I_\nu/h\nu$ is the number of photons crossing a unit area per unit frequency per unit solid angle in the specified direction per unit time, and $I_\nu/h\nu c$ is the number density of photons per unit solid angle per unit frequency. We can make the connection with kinetic theory closer still by observing that the photon momentum is $h\nu/c$, and that the momentum space volume element, in spherical momentum-space coordinates, is

$$d^3\mathbf{p} = \left(\frac{h}{c}\right)^3 \nu^2 d\nu d\Omega. \tag{4.4}$$

That means that if we divide $I_\nu/h\nu c$ by $h^3\nu^2/c^3$ we get the phase space density:

$$f = \frac{dN}{dV d^3\mathbf{p}} = \frac{I_\nu}{h^4\nu^3/c^2}, \tag{4.5}$$

which is the usual unknown in the Boltzmann transport equation.

Along with the intensity, we will often work with its angle moments. The first of these has been introduced already, the energy density:

$$E_\nu = \frac{1}{c}\int_{4\pi} I_\nu d\Omega. \tag{4.6}$$

The next moment is the vector flux

$$\mathbf{F}_\nu = \int_{4\pi} \mathbf{n} I_\nu d\Omega. \tag{4.7}$$

If we consider an oriented area element represented by the vector $\mathbf{A}$, and form $\mathbf{F} \cdot \mathbf{A}$, then we see that it is the sum of contributions like $\mathbf{n} \cdot \mathbf{A} I_\nu d\Omega$, which gives the signed flow of energy from one side of the element to the other, positive if going toward $\mathbf{A}$, and negative for going toward $-\mathbf{A}$. In other words, $\mathbf{F}_\nu$ is a proper energy flux for radiation with the frequency ν per unit bandwidth. Notice that the flux moment does not contain a factor c as does the energy density. The ratio of $\mathbf{F}_\nu$ to E_ν is the "fluid velocity" of the radiation considered by itself as a fluid. From the integrals it is evident that its magnitude must be no larger than c. As a concept this radiation fluid velocity has not been found to be especially useful, although the parameter $\mathbf{F}/(cE)$ does enter some approximation schemes.

The third moment is defined by

$$\mathsf{P}_\nu = \frac{1}{c} \int_{4\pi} \mathbf{n}\mathbf{n} I_\nu d\Omega. \tag{4.8}$$

By appealing to our knowledge of kinetic theory we can recognize this as the pressure tensor, i.e., the pressure per unit bandwidth due to the radiation of frequency ν, to be exact. We have to be careful to remember that lower case p is the material pressure, and capital P refers to radiation. Notice that the factor $1/c$ that was present in the definition of energy density and absent in the definition of flux is present again in the definition of the radiation pressure.

In much of the astrophysical radiative transfer literature the moments are defined as averages over solid angle, i.e., are divided by 4π, and the factors of c are omitted. Thus the quantities J_ν, $\mathbf{H}_\nu$ and K_ν are defined by

$$J_\nu = \frac{1}{4\pi} \int_{4\pi} I_\nu d\Omega, \tag{4.9}$$

$$\mathbf{H}_\nu = \frac{1}{4\pi} \int_{4\pi} \mathbf{n} I_\nu d\Omega, \tag{4.10}$$

and

$$\mathsf{K}_\nu = \frac{1}{4\pi} \int_{4\pi} \mathbf{n}\mathbf{n} I_\nu d\Omega. \tag{4.11}$$

These definitions remove some of the 4πs and cs from radiative transfer theory, at the expense of introducing those factors into the radiation-hydrodynamic equations. We will stick with E, $\mathbf{F}$, and P.

The pressure tensor P_ν is represented by a 3×3 matrix. The trace of that matrix is defined as the sum of its diagonal elements. If we perform that operation on (4.8), the effect is to replace the factor $\mathbf{nn}$ with $\mathbf{n} \cdot \mathbf{n}$ inside the integral. But $\mathbf{n} \cdot \mathbf{n}$ is 1 since $\mathbf{n}$ is a unit vector, and therefore

$$\mathrm{Tr}(\mathsf{P}_\nu) = E_\nu. \tag{4.12}$$

Deep inside an opaque material all directions look the same, since the material far enough away from any given observation point to have different temperature, density, and so on is hidden from view by the opacity of the intervening material. Thus the radiation field tends to be isotropic, being the same in every direction. This makes the pressure tensor a scalar tensor, i.e., one which is a factor times the unit matrix. The reasoning goes like this: the diagonal elements should all be the same since we can turn x into y and vice-versa by making a suitable rotation of 90 degrees about the z axis, and this rotation does nothing to the isotropic intensity. Likewise for x and z and y and z. The off-diagonal elements should vanish, since otherwise a reflection like $x \to -x$ would flip the signs of P_{xy} and P_{xz}; but this reflection leaves the intensity unchanged and therefore the tensor elements should stay the same. When the pressure tensor is a scalar tensor, say

$$\mathsf{P}_\nu = P_\nu \mathsf{I}, \tag{4.13}$$

then the trace is

$$\mathrm{Tr}(\mathsf{P}_\nu) = 3P_\nu. \tag{4.14}$$

Since the trace is E_ν according to (4.12), the diagonal elements are all equal to $E_\nu/3$:

$$\mathsf{P}_\nu = \frac{1}{3}E_\nu \mathsf{I}. \tag{4.15}$$

(The unit matrix is denoted by I not I to avoid confusing it with the intensity.) This relation for pressure and energy density is exactly that for a $\gamma = 4/3$ ideal gas. Radiation *is* a $\gamma = 4/3$ ideal gas if the mean free paths are short enough that transport effects are small. This works very well inside stars.

For a final remark on the intensity concept, let us return to the topic of wave-particle duality and the coarse-graining that was slipped in earlier when we said, "take a distance large enough to avoid the diffraction effects." There will not be such a distance unless the system is quite large compared with the wavelength. And if we have chosen a truly tiny bandwidth $\Delta\nu$, then the wave train will be at least $1/\Delta\nu$ in duration, since otherwise the light pulse has sidebands outside the chosen bandwidth. It can easily happen that $1/\Delta\nu$ is longer than the shutter time Δt we selected. In other words, the Fourier relations between localization in coordinate space and spreading in Fourier space, and between location in time and spreading in frequency, will make the classical definition of the intensity given here nonsense unless the system is very large compared with the wavelength and the times of interest are much longer than the wave period. Radiative transfer is a geometrical optics concept that makes no allowance for wave optics.

4.2 The transport equation; absorptivity, emissivity

The idea of the transport equation is very simple: the intensity does not change as a bundle of radiation travels along. If we think again of that cylindrical bundle of radiation that contains a total energy $I_\nu A_1 \Delta\nu \Delta t d\Omega$, we see that at a later time $t + \tau$ the same bundle of radiation is located at a different place, but occupies the same bandwidth $\Delta\nu$ and fills the same solid angle $d\Omega$; furthermore it will take the same time Δt to pass by the new location. In other words, the intensity at the displaced location at the later time is also I_ν, unless, that is, radiation has been gained or lost by the bundle through interaction with the matter it had to pass through.

We need a notation for the direction of the radiation propagation. We will use the unit vector **n** for that. The vector **n** can be defined by two angles, such as colatitude and longitude. We might say

$$\mathbf{n} = \sin\theta\cos\phi\mathbf{e}_x + \sin\theta\sin\phi\mathbf{e}_y + \cos\theta\mathbf{e}_z, \tag{4.16}$$

with θ for the colatitude and ϕ for the longitude. In 3-D space the vectors **n** trace out a unit sphere; the element of area on that sphere is $d\Omega$. In terms of the angle coordinates, it is

$$d\Omega = \sin\theta\, d\theta d\phi. \tag{4.17}$$

The angle θ goes from 0 to π, and ϕ goes from 0 to 2π. It is easily checked that the total solid angle, the surface area of the unit sphere, is 4π. There is a conceptual pitfall in using these coordinates θ and ϕ for angle space. We might confuse them for the angles that are part of the spherical coordinate set in real space! In fact, they are completely independent. One set is for momentum space and the other is for configuration space or real space. Some authors use Θ and Φ for the momentum coordinates and θ and ϕ for the real space coordinates, hoping to avoid confusion that way.

The other new notation we need is the name of the position vector, i.e., the vector (x, y, z) in real space. We will use **r** for that. The boldface matters here; vector **r** is (x, y, z) while scalar r is the distance between two points. Returning to the bundle of radiation, we now have the notation to say that I_ν does not change as the bundle moves over the time τ:

$$I_\nu(\mathbf{r} + \mathbf{n}c\tau, \mathbf{n}, t + \tau) = I_\nu(\mathbf{r}, \mathbf{n}, t). \tag{4.18}$$

Next we Taylor-expand the left-hand side around the point **r** and time t, and discard the terms of order τ^2 or higher. Subtract the term $I_\nu(\mathbf{r}, \mathbf{n}, t)$, divide by $c\tau$, and we have it:

$$\frac{1}{c}\frac{\partial I_\nu}{\partial t} + \mathbf{n}\cdot\nabla I_\nu = 0. \tag{4.19}$$

(The operator ∇ is the spatial gradient operator, and it will keep that meaning for us hereafter.) This is the radiation transport equation, minus the source and sink terms, which come next. If this resembles the Boltzmann transport equation, that is not an accident. The latter equation might have extra terms involving the momentum derivatives of the phase space density, which are proportional to the forces acting on the particles. There are no forces acting on our photons, so those terms are missing here.

Next we consider first the absorption and then the emission. The empirical law is that radiation impinging on a thin slab of matter is attenuated by a small fraction. This is a definite fraction that does not depend on the intensity, so if the intensity doubles, so does the amount of energy removed from the beam. The fraction is also proportional to the thickness of the slab, if this is not too large. So the intensity change is

$$\Delta I_\nu = -k_\nu c\tau I_\nu \tag{4.20}$$

if none of the absorbed energy is replaced. The proportionality coefficient k_ν is the absorptivity, or absorption coefficient.[1] If this model of pure attenuation is applied to a thicker slab (thickness L), the result is

$$I_\nu = I_\nu^0 \exp(-k_\nu L), \tag{4.21}$$

which is called Beer's law after an application in atmospheric physics. In the atomic model of radiation–matter action, we would say that there is a probability per unit length k_ν that a photon will interact with the matter. For a thin slab $c\tau$ thick, there is a probability $1 - k_\nu c\tau$ of no interaction, so the photon makes it through to the other side, and a probability $k_\nu c\tau$ of having an interaction, in which case it is gone from this beam. The mean loss to I_ν is then $k_\nu c\tau I_\nu$, as before.

Optical depth, a well-loved concept in astrophysics, is defined to be the exponent in Beer's law: $k_\nu L$, or $\int k_\nu dl$ in general. The usual notation for optical depth is τ, perhaps with a subscript ν, and we will adopt this in later sections when we are no longer using τ for the lag time in discussing the transport equation.

The emissivity determines the amount of energy added by a thin slab to a beam of radiation that is passing through it. Normally this does not depend on what the

[1] There is a more precise terminology that is used when the process of scattering, in which the radiation is modified and redirected by its interaction with the matter, is distinguished from true absorption, in which the radiation is actually removed. Then the coefficient of absorption gives the probability of the second process alone, while the combined probability of both processes is called the *extinction* coefficient.

intensity of the beam is, but it does depend on the thickness of the slab:

$$\Delta I_\nu = +j_\nu c\tau. \tag{4.22}$$

The emissivity or emission coefficient j_ν and the absorptivity k_ν are not in the same units. Absorptivity has the dimensions of inverse length, while emissivity has the dimensions of I_ν divided by length, i.e., energy per unit volume per unit bandwidth per unit solid angle per unit time.

The consideration of what the absorptivity and emissivity actually are will be taken up in Chapter 8; this is a topic in atomic physics. We should mention here that there are processes that are nonlinear in the intensity, for example, multiphoton absorption, for which the energy loss is proportional to two or more factors of the intensity for different frequencies and directions. This problem is so specialized that it can be treated in its own context should the need arise. One important process that modifies this discussion is stimulated emission. This will be treated at more length later, but the end result is that j_ν contains a part that is proportional to the intensity in the beam to which j_ν is contributing, i.e., to I_ν. This is exactly accounted for by making a subtraction from k_ν ("negative absorption"). In thermodynamic equilibrium the subtracted piece is $\exp(-h\nu/kT)$ times the original value, and therefore the subtraction can be performed at the outset. We will talk more about scattering in Chapter 8 as well. It is normally a linear process, i.e., scattering contributes something to k_ν, and the photons thus removed are returned at other angles and frequencies (perhaps), and therefore appear as a term in j_ν involving an integration of I_ν over angle and frequency. Stimulated emission raises its head here as well, making the out-scattering and in-scattering terms quadratic in the intensity. For the present, we will suppose that k_ν has been corrected for stimulated emission, and that j_ν has lumped into it any of the more complex processes. We proceed to put the loss and gain terms into the intensity budget, Taylor expand and cancel as before, and end up with the radiation transport equation in the most general form we need right now:

$$\frac{1}{c}\frac{\partial I_\nu}{\partial t} + \mathbf{n}\cdot\nabla I_\nu = j_\nu - k_\nu I_\nu. \tag{4.23}$$

If we prefer to call out the scattering processes explicitly, then the equation takes this form (see Section 12.3)

$$\begin{aligned}\frac{1}{c}\frac{\partial I_\nu}{\partial t} + \mathbf{n}\cdot\nabla I_\nu &= j_\nu - k_\nu I_\nu \\ &\quad + \sigma_\nu \int_0^\infty d\nu' \int_{4\pi} d\Omega' \left\{ -R(\nu'\mathbf{n}', \nu\mathbf{n})\frac{\nu}{\nu'} I_\nu \left[1 + \frac{c^2 I_{\nu'}}{2h\nu'^3}\right] \right. \\ &\quad \left. + R(\nu\mathbf{n}, \nu'\mathbf{n}') I_{\nu'} \left[1 + \frac{c^2 I_\nu}{2h\nu^3}\right] \right\},\end{aligned} \tag{4.24}$$

in which j_ν and k_ν now pertain only to absorptive processes, σ_ν is the scattering coefficient and the R function is the (normalized) scattering redistribution function in frequency and angle. As we will discuss in Sections 8.3 and 12.3, the frequency shift in scattering is often negligible and the cross product terms due to stimulated scattering cancel out; furthermore, isotropic scattering can be a good approximation. In this case the transport equation with scattering is simpler:

$$\frac{1}{c}\frac{\partial I_\nu}{\partial t} + \mathbf{n} \cdot \nabla I_\nu = j_\nu + \sigma_\nu J_\nu - (k_\nu + \sigma_\nu) I_\nu. \tag{4.25}$$

4.3 Radiation moment equations

Earlier we discussed the first three angle moments of the intensity, namely the energy density, the flux, and the pressure tensor. If we chose to, we could add tensors of higher rank to this set, the moments of rank 3, 4, ... We can also take moments of the transport equation by integrating over angles after multiplying by 1, **n**, **nn**, ... Sad to say, each moment of the transport equation introduces the next higher moment of the intensity, so the set of moment equations up through a given order is always one equation short of having as many equations as there are unknowns. The system of equations must be closed by using an *ad hoc* relation that gives the highest moment, say, as an expression involving lower moments. The kinetic theory of gases and plasma physics are richer in closure theories than is radiation transport; in the latter case no closure theory beyond the pressure tensor has had any currency. Our discussion here will therefore be limited to the first two moments of the transport equation. In the kinetic theory of gases these are precisely the moments that lead to Euler's equations after the pressure tensor is approximated.

We will start taking integrals of the transport equation (4.23), but two points must be mentioned first. The transport equation is written above in a coordinate-free notation. Its simplicity in that case is somewhat deceptive. If a curvilinear system of spatial coordinates is used then it is not correct to think of **n** as a constant vector during spatial transport. Instead we have to imagine that the vector **n** is moved along remaining parallel to itself in a physical sense, which means that its components along the three coordinate directions are changing as it goes. This would take us deeply into Riemannian geometry, which cannot be undertaken in this book. This is avoided if we always refer to a Cartesian system of fixed-space coordinates, for then the components of **n** truly are constant.

The second point is that isotropic k_ν and j_ν is by no means the only possibility. The absorptivity can depend on direction if the absorbers are preferentially oriented a certain way, as is the case with ice crystals in cirrus clouds, for example, since the crystals tend to float with their flat sides horizontal. Another example is given by a highly magnetized plasma, for which the normal processes of atomic absorption are highly modified by the Zeeman effect and depend on the photon direction with respect to the magnetic field. Anisotropy of the emissivity is more common, since scattering is in most cases different for different angles between the incoming and outgoing photons; since the intensity itself may be quite anisotropic, the scattered intensity is anisotropic too. An additional reason for anisotropy will be discussed below, which is that absorption and emission that occur isotropically in the rest frame of the material fluid are not isotropic in the fixed frame owing to the Doppler and aberration effects. Having said this, we will proceed to ignore these anisotropies for now, but you were warned.

As a first step to taking angle moments we take advantage of the constancy of **n** and take it inside the spatial derivative in the transport equation:

$$\frac{1}{c}\frac{\partial I_\nu}{\partial t} + \nabla \cdot (\mathbf{n} I_\nu) = j_\nu - k_\nu I_\nu. \tag{4.26}$$

Now we multiply successively by 1 and **n**, and integrate over angles. The factor **n** passes through the divergence by the same argument we just gave. We immediately obtain

$$\frac{\partial E_\nu}{\partial t} + \nabla \cdot \mathbf{F}_\nu = 4\pi j_\nu - k_\nu c E_\nu, \tag{4.27}$$

$$\frac{1}{c}\frac{\partial \mathbf{F}_\nu}{\partial t} + c\nabla \cdot \mathsf{P}_\nu = -k_\nu \mathbf{F}_\nu. \tag{4.28}$$

Since E_ν, $\mathbf{F}_\nu$, and P_ν are a proper scalar, vector, and tensor, respectively (not in space-time as we discuss later, but just in space), we are free to use curvilinear coordinates for these moment equations, which are written in coordinate-free form.

Rather than commenting on the meaning of these equations we pass on to their frequency-integrated form:

$$\frac{\partial E}{\partial t} + \nabla \cdot \mathbf{F} = \int d\nu \, (4\pi j_\nu - k_\nu c E_\nu), \tag{4.29}$$

$$\frac{1}{c}\frac{\partial \mathbf{F}}{\partial t} + c\nabla \cdot \mathsf{P} = -\int d\nu \, k_\nu \mathbf{F}_\nu. \tag{4.30}$$

For a nonrelativistic gas the first two moment equations express mass conservation and momentum conservation. The second of the radiation moment equations is indeed an expression of conservation of radiation momentum. But photons have no rest mass, and the first equation here is not about the conservation of that. Rather, it is about the conservation of relative mass, which is to say, of energy. The terms on the left-hand side of the first equation have the normal conservation law form for an energy density and an energy flux. The integral on the right-hand side represents the rates of gain and loss of radiation energy per unit volume. Emissivity is an energy gain for the radiation; it would be a loss for material energy. Likewise the absorptivity term is a loss of radiation energy; it would be a gain for the material.

Turning to the second equation, the first thing we need to do is divide by c:

$$\frac{1}{c^2}\frac{\partial \mathbf{F}}{\partial t} + \nabla \cdot \mathsf{P} = -\int d\nu\, k_\nu c \frac{\mathbf{F}_\nu}{c^2}. \tag{4.31}$$

Now we read off the terms. The radiation momentum density is $\mathbf{F}/c^2$ and the radiation momentum flux (pressure) is P. On the right-hand side we can regard $k_\nu c$ as the absorption probability per unit time and $\mathbf{F}_\nu/c^2$ as the spectral momentum density, so the integral is the momentum lost per unit time by the radiation and transferred to the matter. There is an important caveat with (4.31): the neglected anisotropy of absorption and emission introduces an important correction to this term if the material is not at rest; this is treated in Section 6.3.

4.4 Diffusion approximation

The diffusion approximation is by far the most important approximate treatment of radiation transport; it pertains to the limit in which radiation is treated as an ideal fluid with small corrections. The approximation becomes accurate when the photon mean free paths are small compared with other length scales, i.e., when $k_\nu L \gg 1$, where L is a typical length. The diffusion approximation is found to be so much simpler than solving the full transport equation that every effort is made to adapt it to problems where $k_\nu L < 1$, for which it is not expected to be accurate. In this case the goal is somewhat different: it is understood that the results will not be precise, but they may well be qualitatively correct, and the error, perhaps 20–30%, may be tolerable.

The following development of the diffusion approximation is motivated by the discussion in Cox and Giuli (1968), Section 6.3, and a similar discussion in Schwarzschild's book *Structure and Evolution of the Stars* (1958).

The diffusion approximation is an expansion in a small parameter – the mean free path or $1/k_\nu$ – truncated after the first two terms. We begin by rearranging the transport equation in this way:

$$I_\nu = \frac{j_\nu}{k_\nu} - \frac{1}{k_\nu}\left(\frac{1}{c}\frac{\partial I_\nu}{\partial t} + \mathbf{n} \cdot \nabla I_\nu\right). \tag{4.32}$$

Since we suppose that k_ν is large the second expression on the right-hand side should be a small correction to the first term. So we use this equation as a basis for obtaining I_ν by successive approximations. The first approximation is

$$I_\nu^0 = \frac{j_\nu}{k_\nu}. \tag{4.33}$$

This is a local balance approximation which says that the radiation is in equilibrium with its sources. If the source and sink terms are those for matter at a temperature T, then I_ν^0 should be the thermodynamic equilibrium radiation field, which is the Planck function $B_\nu(T)$. The next step of the successive approximations is to put I_ν^0 in for I_ν on the right-hand side of (4.32), which leads to

$$I_\nu^1 = \frac{j_\nu}{k_\nu} - \frac{1}{k_\nu}\left(\frac{1}{c}\frac{\partial j_\nu/k_\nu}{\partial t} + \mathbf{n}\cdot\nabla\frac{j_\nu}{k_\nu}\right). \tag{4.34}$$

For the standard diffusion approximation we stop here. We could include more terms, but there is no good reason to do so. If k_ν is indeed large, then the next term would be too small to care about. If k_ν is not large, then we anticipate that including more terms will make the answer worse rather than better. The infinite series is really an asymptotic expansion, and including one term after another will make the answer better for a while, then it will start to diverge. Thus we may as well stop after the second term. You will notice that this series contains no reference whatsoever to the possible existence of a nearby boundary that may have a large effect on the intensity; perhaps there is vacuum just a short distance away, and therefore some of the radiation that would have come from that direction if the medium had gone on indefinitely is actually missing. A helpful way of looking at this is that the presence of a boundary modifies the solution for a certain distance into the interior (see Figure 4.2); this region of boundary influence is the boundary layer. The diffusion expansion will never give the right answer in the boundary layer; if we are lucky the error is tolerable. Farther into the interior of the problem is a region where the boundary influence is not felt. It is here that the diffusion approximation is valid.

At first sight we would say that the thickness of a boundary layer should be one or two mean free paths, since that is the distance a photon is likely to penetrate before being absorbed. Indeed, in many cases this is a good estimate. But we must mention here that scattering complicates this discussion considerably. Scattering is described with an albedo, which is the probability that the photon survives one interaction with the matter. If the albedo is very close to unity, then a photon will survive a large number of scatterings before finally being destroyed. In this case the thickness of the boundary layer becomes the distance the photon can move in a random walk with that many steps; this can be many mean free paths, depending on the albedo. Thus with scattering the boundary layers are thicker and encompass a large part of what we might have thought was the interior of the problem. In

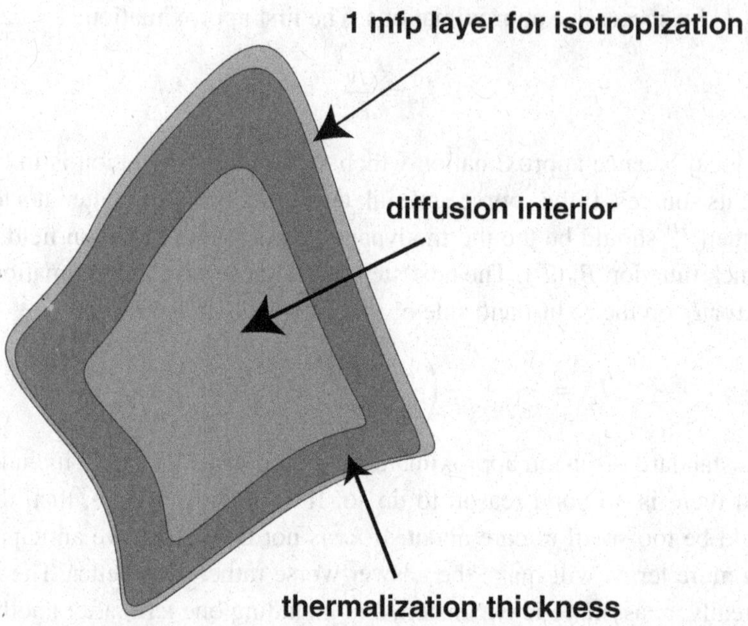

Fig. 4.2 Illustration of a generic radiative transfer problem showing the three qualitatively different regions: the isotropization layer within about one mean free path (mfp) of the outside boundary; the thermalization layer, perhaps many mean free paths thick, interior to which the radiation field is well approximated as Planckian; and the diffusion interior, this region of near equilibrium. Only in the innermost region is diffusion a good approximation.

the limit of unit albedo the problem is all boundary layer. There is no interior at all in this case, just as there is none for problems in potential theory – Laplace's equation does not allow an interior region free from the influence of the boundary. Fortunately for us, the scattering albedo in the large majority of practical problems is either small, or in any case not too close to unity, so the boundary layers are only a small number of mean free paths thick.

We turn again to the diffusion approximation (4.34) and consider what the radiation moments become in this case. For the energy density:

$$E_\nu = \frac{4\pi}{c}\frac{j_\nu}{k_\nu} - \frac{4\pi}{k_\nu c^2}\frac{\partial}{\partial t}\left(\frac{j_\nu}{k_\nu}\right). \tag{4.35}$$

The gradient term cancels in E_ν since $\mathbf{n}$ is odd and its angle average vanishes. The integrals for $\mathbf{F}_\nu$ and P_ν are facilitated by noticing that the angle average of the tensor $\mathbf{nn}$ is $\mathsf{I}/3$. The $\mathbf{n}\cdot\nabla()$ term can be changed to $\nabla\cdot(\mathbf{n}())$, and then when another factor of $\mathbf{n}$ is put in, for obtaining the flux, the term becomes $\nabla\cdot(\mathbf{nn}())$. As noted, averaging over angles changes $\mathbf{nn}$ into $\mathsf{I}/3$, and we can use the fact that the divergence of a scalar tensor is the gradient of one of the diagonal elements. In the flux integral this particular term is the only one that survives:

$$\mathbf{F}_\nu = -\frac{4\pi}{3k_\nu}\nabla\frac{j_\nu}{k_\nu}. \tag{4.36}$$

For the pressure moment the odd term goes away and the even terms survive. All the even terms lead to the angle average of **nn**, and so

$$\mathsf{P}_\nu = \frac{1}{3}\mathsf{I}\left[\frac{4\pi}{c}\frac{j_\nu}{k_\nu} - \frac{4\pi}{k_\nu c^2}\frac{\partial}{\partial t}\left(\frac{j_\nu}{k_\nu}\right)\right]. \tag{4.37}$$

This result is crucial: the first two terms in the diffusion approximation lead to an isotropic pressure tensor. The intensity itself is not isotropic because of the first order gradient term, but to this order there is no anisotropic correction to I that is even in **n**. This would appear in the next term in the asymptotic series, of order $1/k_\nu^2$. Thus the diffusion approximation leads to Eddington's approximation

$$\mathsf{P}_\nu = \frac{1}{3}E_\nu\mathsf{I}. \tag{4.38}$$

It is important to distinguish Eddington's approximation from the diffusion approximation. The diffusion approximation is stronger, i.e., a more severe approximation. Eddington's approximation follows from the diffusion approximation, as we have seen, but the reverse is not true. In the example discussed above, where it was noted that scattering with an albedo very close to unity results in thick boundary layers in which the radiation field is strongly influenced by the boundary, it would be found that Eddington's approximation would be valid in most of the boundary layer except the one or two means free paths next to the boundary. Scattering quickly produces isotropy of the intensity, even though the mean intensity might be far from the diffusion value.

We will leave the diffusion approximation aside for now and look at the implications of Eddington's approximation for the radiation moment equations. We substitute (4.38) into (4.28) and find

$$\frac{1}{c}\frac{\partial \mathbf{F}_\nu}{\partial t} + \frac{c}{3}\nabla E_\nu = -k_\nu \mathbf{F}_\nu. \tag{4.39}$$

Equations (4.27) and (4.39) form a closed set for the moments E_ν and $\mathbf{F}_\nu$, so these can be solved, possibly in conjunction with the hydrodynamic equations. The radiation equations themselves form a hyperbolic system. If we drop the j_ν and k_ν terms temporarily (not a good idea in general, since Eddington's approximation is based on $k_\nu L \gg 1$!) we can get a single equation for E_ν by combining the time derivative of (4.27) with the divergence of (4.39),

$$\frac{\partial^2 E_\nu}{\partial t^2} - \frac{c^2}{3}\nabla^2 E_\nu = 0. \tag{4.40}$$

This is the wave equation for a wave speed of $c/\sqrt{3}$. It is the wrong vacuum solution, of course. In reality, taking the j_ν and k_ν terms into account, the $c/\sqrt{3}$ waves can never be observed unless the equations are applied to a case for which Eddington's approximation is not valid. Before the wave can have propagated one wavelength it will have been absorbed, since $k_\nu \times$ wavelength should be large. The $\partial \mathbf{F}/\partial t$ term serves to keep the information propagation bounded; no signal will propagate faster than $c/\sqrt{3}$. But if this limit is being exercised, the solution is probably wrong.

As an alternative to the hyperbolic system we consider modifying the second moment equation (4.39) by discarding the $\partial \mathbf{F}_\nu/\partial t$ term to get

$$\frac{c}{3}\nabla E_\nu = -k_\nu \mathbf{F}_\nu. \tag{4.41}$$

By doing this we have lost the finite propagation speed, and we have also lost the radiation momentum density; when radiation imparts some momentum to the matter using this picture, there is no compensation in radiation momentum, so there is an error in the total momentum budget. The radiation equations are now substantially simpler. The flux need not be kept as a separate variable, but it can be eliminated between (4.27) and (4.41), which take the combined form

$$\frac{\partial E_\nu}{\partial t} - \nabla \cdot \left(\frac{c}{3k_\nu} \nabla E_\nu \right) = 4\pi j_\nu - k_\nu c E_\nu. \tag{4.42}$$

This is an equation of parabolic type, like the equation for the diffusion of heat. If the absorption and emission terms are dropped temporarily, then the terms that are left correspond to a radiation wave that spreads according to $x \sim \sqrt{ct/(3k_\nu)}$, which is plausible. Even so, the propagation of radiation from a pulse at $t = 0$ can be faster than the speed of light at early time when $k_\nu ct < 1$, and the transport in reality is not diffusive. This is related to the fact that there is no limit to the flux given by (4.41). *Ad hoc* methods for preventing numerical calculations from giving unphysical results on this account when they adopt an equation like (4.42) are referred to as flux limiting. These are discussed in Section 11.5. For additional reading on the merits of the Eddington closure of the radiation moment equations see Mihalas and Mihalas (1984) and also the literature that has developed on flux limiting (see Section 11.5), especially Pomraning (1982).

4.5 Coupling terms in Euler's equations

The topic of the coupling of radiation and matter concerns the source-sink terms in the conservation equations. In this discussion we generally follow Mihalas and Mihalas (1984), Section 94, with some parts from Castor (1972). We have already discussed how the right-hand side of the transport equation, i.e., $j_\nu - k_\nu I_\nu$, is the energy gained by the radiation field at the expense of the matter per unit volume per unit time per unit bandwidth and per unit solid angle. This is quite general, even when there are anisotropies, Doppler shifts, and so forth. Therefore the quantities g^0 and $\mathbf{g}$ defined by

$$g^0 = \int d\nu \int_{4\pi} d\Omega\,(j_\nu - k_\nu I_\nu) \tag{4.43}$$

and

$$\mathbf{g} = \frac{1}{c}\int d\nu \int_{4\pi} d\Omega\,\mathbf{n}(j_\nu - k_\nu I_\nu) \tag{4.44}$$

are the correct energy and momentum exchange rates. We should put the negatives of these on the right-hand side of the material total energy and momentum equations:

$$\frac{\partial}{\partial t}\left(\rho e + \frac{1}{2}\rho u^2\right) + \nabla\cdot\left(\rho \mathbf{u} h + \frac{1}{2}\rho \mathbf{u} u^2\right) = -g^0, \tag{4.45}$$

$$\frac{\partial \rho \mathbf{u}}{\partial t} + \nabla\cdot(\rho \mathbf{u}\mathbf{u}) + \nabla p = -\mathbf{g}. \tag{4.46}$$

We have left out the nonradiative body force and heat deposition terms for simplicity. Summing the frequency-integrated radiation moment equations and the material equations gives the overall conservation laws

$$\frac{\partial}{\partial t}\left(\rho e + E + \frac{1}{2}\rho u^2\right) + \nabla\cdot\left(\rho \mathbf{u} h + \frac{1}{2}\rho \mathbf{u} u^2 + \mathbf{F}\right) = 0, \tag{4.47}$$

$$\frac{\partial}{\partial t}\left(\rho \mathbf{u} + \frac{\mathbf{F}}{c^2}\right) + \nabla\cdot(\rho \mathbf{u}\mathbf{u} + \mathsf{P}) + \nabla p = 0. \tag{4.48}$$

Here again we see that we have to choose between keeping the awkward radiation momentum density term or not having exact momentum conservation.

One remark should be made about the radiation–matter energy exchange. We often see treatments that use $-\nabla\cdot\mathbf{F}$ for the matter heating rate. That is incorrect, since it fails to account for the rate of change of the radiation energy density. In fact, $-\nabla\cdot\mathbf{F}$ is a contribution to the rate of increase of the density of matter energy plus radiation energy, as shown by (4.47). The correct matter heating rate due to the radiation is $-g^0$. In a steady state $\nabla\cdot\mathbf{F}$ and g^0 are equal, as shown by (4.29) and (4.43).

5

Steady-state transfer

In this chapter we will review some of the standard results of radiative transport/transfer theory. A large part of the theoretical development has been done for the steady-state case and for slab geometry. Within these approximations the results can be made quite refined in the sense of avoiding approximations to the angle dependence of the radiation field such as diffusion or Eddington's approximation. These results lead to concepts that are helpful in understanding more complicated cases.

5.1 Formal solution in three dimensions

Since the transport equation (4.23) says in effect that

$$\frac{dI}{ds} = j - kI \tag{5.1}$$

along the ray path, where j and k are the local in space and time emissivity and absorptivity, the solution is just

$$I = I_{\mathrm{B}} \exp\left(-\int_{s_{\mathrm{B}}}^{s} ds'\, k(s')\right) + \int_{s_{\mathrm{B}}}^{s} ds' \exp\left(-\int_{s'}^{s} ds''\, k(s'')\right) j(s'); \tag{5.2}$$

see, for example, Pomraning (1973). The notations ν and $\mathbf{n}$ have been omitted here, since these parameters are constant in this fixed-frame picture. The suffix "B" denotes a point on the spatial boundary of the problem, or a point at the initial time, where the boundary or initial data prescribe I_{B}. The values of j and k inside the integrals over s' or s'' are at the displaced location and retarded time given, for example, by $\mathbf{r}' = \mathbf{r} - \mathbf{n}(s - s')$ and $t' = t - (s - s')/c$. Thus a more notationally complete but harder to read version of the same formal solution is

$$\begin{aligned} I(\mathbf{r}, t) &= I(\mathbf{r} - (s - s_B)\mathbf{n}, t - (s - s_B)/c) \\ &\quad \times \exp\left[- \int_{s_B}^{s} ds'\, k(\mathbf{r} - (s - s')\mathbf{n}, t - (s - s')/c) \right] \\ &\quad + \int_{s_B}^{s} ds' \exp\left[- \int_{s'}^{s} ds''\, k(\mathbf{r} - (s - s'')\mathbf{n}, t - (s - s'')/c) \right] \\ &\quad \times j(\mathbf{r} - (s - s')\mathbf{n}, t - (s - s')/c). \end{aligned} \tag{5.3}$$

This form of the transport equation is not too useful as it stands, but after making the simplifications of time independence and slab symmetry we will get a form that lends itself to the theoretical elaborations that were promised. Equation (5.3) as it stands corresponds well with the Monte Carlo approach to radiation transport that will be touched on later.

5.2 Time-independent slab geometry

We assume now that our problem is in a steady state and that there is translational symmetry in x and y, so the z coordinate is the only nontrivial one. This also results in the problem being axially symmetric about the z axis. Radiative transfer in this kind of geometry – slab geometry – has an extensive literature, including the older books by Chandrasekhar (1960) and Kourganoff (1963), and more general stellar atmospheres texts like Mihalas (1978). The notation in these works is standard, and has been followed here. As a result of the symmetry the intensity can depend only on the component of $\mathbf{n}$ in the z direction. In astrophysics it is traditional to include a minus sign here, so a positive direction cosine will refer to propagation toward $-z$. The reason is that z is thought of as a coordinate measured from the outside inward toward the center of a star, but a positive direction cosine is used for the radiation the external observer views. So we take

$$n_z \equiv -\mu, \tag{5.4}$$

and use μ as our label for angles.

We also introduce the optical depth variable τ at this point. There is a powerful justification for this, which is that a variety of transfer problems which differ from each other just in how the opacity is distributed in the z coordinate become the same problem when viewed in τ space. The notational problem of carrying factors of $k(z)$ vanishes. At the end it is very simple to transform back to physical space. The variable τ is reckoned from the outside of the star inward, in the same sense as z, but in the direction of negative μ. Thus

$$\tau = \int_{-\infty}^{z} dz' \, k(z'). \tag{5.5}$$

The assumption here is that the star (cloud,...) has an "outside" beyond which there is vacuum, and the zero point of τ is located there. If the star trails off gradually to infinite radius, it is assumed that the optical depth integral converges in that limit. If this is not the right picture, if instead a "wall" of some kind is encountered on the outside, then this must be accounted for in the boundary condition. We mention again that all quantities are in fact frequency-dependent, which will not be shown explicitly unless it is necessary to do so.

In τ space the equation of transfer, which is what we call transport in the steady-state case, becomes

$$\mu \frac{dI}{d\tau} = I - \frac{j}{k}. \tag{5.6}$$

The mapping to τ space allows us to largely ignore the spatial variation of $k(z)$; we make this complete by introducing the *source function* S defined by

$$S \equiv \frac{j}{k}. \tag{5.7}$$

The disadvantage of having to remember the meaning of yet another symbol is compensated by the greater simplicity of the equations, besides which it has been found by old hands of radiative transfer that the source function concept is actually an aid to understanding.

In thermodynamic equilibrium the radiation field is in equilibrium with its sources, and therefore the right-hand side of (5.6) must be zero. Also in this case the intensity is the Planck function, and therefore the source function becomes the Planck function in thermodynamic equilibrium. Since j and k are properties of the matter – scattering aside, which we discuss next – the source function is the Planck function whenever the state of the matter is the same as if it actually were in thermodynamic equilibrium, whether the intensity I is the Planck function or not. This is the concept of LTE.

Scattering complicates this picture. The atomic basis of scattering will be discussed later, but here we can simply say that a scattering process is one in which a photon that is removed by the interaction is returned to the radiation field *instantly* instead of causing an excitation of the atoms. The emissivity due to this process depends directly on the actual radiation field rather than on the state of the matter, and therefore the corresponding source function need not be the Planck function even when the matter is in LTE.

The emissivity due to scattering can often be treated as isotropic, and for pure scattering every photon removed from the radiation field by a scattering process comes back at some other angle. The consequence of these two conditions is that the scattering source function is given by

$$S = \frac{1}{2}\int_{-1}^{1} d\mu\, I(\mu) = J = \frac{cE}{4\pi}. \tag{5.8}$$

One comment that must be made here concerns the frequencies. Energy is conserved overall in a scattering process, and we are considering processes that leave the scatterer (electron or atom) in the same state after the event as before. The only way that energy can be lost by the photon (or conceivably gained) is in the recoil kinetic energy of the scatterer. This effect is exceedingly small except for x-rays and gamma-rays being scattered by electrons, and for Raman scattering. For electron scattering processes in the IR, optical, and UV it is a very good assumption that there is no frequency change in the scattering. That means that the monochromatic source function in (5.8) depends on I, J or E at exactly the same frequency. The relaxation of this assumption for Compton scattering of x-rays by electrons will be discussed below.

When the absorptivity and emissivity include both scattering processes and what is called "true" absorption, i.e., everything that is *not* scattering, then the emissivity in the LTE case can be expressed as

$$j = kS = k_a B + k_s J, \tag{5.9}$$

where k_a is the absorptivity for "true" absorption and k_s is the absorptivity for scattering. Thus the total source function is

$$S = \frac{k_a}{k_a + k_s} B + \frac{k_s}{k_a + k_s} J = \frac{k_a}{k_a + k_s} B + \frac{k_s}{k_a + k_s}\frac{1}{2}\int_{-1}^{1} d\mu\, I(\mu). \tag{5.10}$$

The ratio $k_s/(k_a + k_s)$ might be called the *single-scattering albedo*, although sometimes that term is reserved for things like scattering of light by dirty water droplets. If we denote the albedo by ϖ then we can write

$$S = (1 - \varpi)B + \frac{\varpi}{2}\int_{-1}^{1} d\mu\, I(\mu). \tag{5.11}$$

The formal integral (5.3) turns into the following form with our simplified geometry:

$$I = \begin{cases} I_{\mathrm{B}} \exp(\tau/\mu) + \displaystyle\int_0^{\tau} \frac{d\tau'}{|\mu|} \exp\left[(\tau - \tau')/\mu\right] S(\tau') & \mu < 0 \\ \displaystyle\int_{\tau}^{\infty} \frac{d\tau'}{\mu} \exp[(\tau - \tau')/\mu] S(\tau') & \mu > 0 \end{cases}. \tag{5.12}$$

If the source function is already known, e.g., it is the Planck function, the job is done at this point. Otherwise, as for instance in the case that S is given by a relation like (5.11), we proceed further. Since S depends in such cases on B, which is known, and on J, which is unknown, we will want the expression for J that can be derived from (5.12), and while we are at it, we will get the flux and pressure moments as well. The integrals over μ of the exponential functions like $\exp(-x/\mu)$ are defined in terms of the special functions called exponential integral functions. The generic exponential integral function is $E_n(x)$ defined by

$$E_n(x) = \int_1^\infty \frac{e^{-xy}}{y^n}\, dy. \tag{5.13}$$

The properties of the exponential-integral functions are described at length in Abramowitz and Stegun (1964). Another form that is useful here is

$$E_n(x) = \int_0^1 t^{n-2} e^{-x/t}\, dt. \tag{5.14}$$

These functions obey the recursion relations

$$E_n'(x) = -E_{n-1}(x), \tag{5.15}$$

and

$$E_n(x) = \frac{1}{n-1}\left[e^{-x} - xE_{n-1}(x)\right]. \tag{5.16}$$

It is sometimes useful to know that the E_n functions are related to the incomplete gamma function.

Forming the moments of $I(\mu)$ given by (5.12) leads to

$$J(\tau) = \frac{cE}{4\pi} = \frac{1}{2}E_2(\tau)I_{\mathrm{B}} + \frac{1}{2}\int_0^\infty d\tau'\, E_1(|\tau' - \tau|)S(\tau'), \tag{5.17}$$

$$H(\tau) = \frac{F}{4\pi} = -\frac{1}{2}E_3(\tau)I_{\mathrm{B}} + \frac{1}{2}\int_0^\infty d\tau'\, \mathrm{sgn}(\tau' - \tau)E_2(|\tau' - \tau|)S(\tau'), \tag{5.18}$$

and

$$K(\tau) = \frac{cP}{4\pi} = \frac{1}{2}E_4(\tau)I_{\mathrm{B}} + \frac{1}{2}\int_0^\infty d\tau'\, E_3(|\tau' - \tau|)S(\tau'), \tag{5.19}$$

which also identically obey the moment equations

$$\frac{dH}{d\tau} = J - S \tag{5.20}$$

and

$$\frac{dK}{d\tau} = H. \tag{5.21}$$

All the functions $E_n(x)$ behave at large x as $\exp(-x)/x$. Therefore when τ is significantly larger than 1, say $\tau > 4$, the boundary terms in J, H, and K are negligible, and the lower limits of the integrals in (5.17)–(5.19) can be extended to $-\infty$ without affecting the answer. Thus it is as if the medium were infinitely extended. These are the hallmarks of the interior part of the problem. If S is slowly varying there, as it should be, then the integrals in the expressions for J and K can be approximated by taking S outside the integral and evaluating it at τ. Since also

$$\int_{-\infty}^{\infty} E_n(|x|)\, dx = \frac{2}{n}, \tag{5.22}$$

the conclusion is that $K \approx J/3$, which is Eddington's approximation. In other words, the radiation becomes isotropic once $\tau \gg 1$.

Another insight into the behavior of the transfer equation is obtained by taking S out of the integral in (5.17) whether τ is large or not. The resulting relation

$$\begin{aligned} J &\approx \frac{1}{2}E_2(\tau)I_\mathrm{B} + S\frac{1}{2}\int_0^{\infty} E_1(|\tau' - \tau|)\, d\tau' \\ &= \frac{1}{2}E_2(\tau)I_\mathrm{B} + \left[1 - \frac{1}{2}E_2(\tau)\right] S \end{aligned} \tag{5.23}$$

is called the *escape-probability* approximation. The expression

$$p_\mathrm{esc} = \frac{1}{2}E_2(\tau) \tag{5.24}$$

is called the escape probability, or the two-sided single-flight escape probability, to be more precise. We can regard this as the average of the one-sided escape probability going toward larger τ, namely zero, and the one-sided escape probability going toward smaller τ, which is $E_2(\tau)$. The E_2 function goes between the limits of 1 and 0 as τ goes from 0 to ∞, so it is reasonable to think of it as a probability. We will return to escape probabilities in Sections 6.8 and 11.8.

5.3 Milne's equation

Some of the standard problems in radiative transfer theory involve solving for J in the case of a semi-infinite atmosphere with no externally-incident radiation, when there is a relation of the kind in (5.11) between S and J. Thus $I_\mathrm{B} = 0$ and there is a vacuum boundary at $\tau = 0$ but the problem extends to infinity in the positive τ

direction. The simplest of these is the homogeneous Milne problem:

$$S = J = \frac{1}{2}\int_0^{\infty} S(\tau')E_1(|\tau' - \tau|)\, d\tau. \tag{5.25}$$

This arises in conservative scattering, i.e., when there is no absorption at all. It also is a model for an atmosphere in radiative energy balance, for which the absorption and emission rates are just equal, with a frequency-independent (gray) opacity that permits integrating all the radiation quantities over frequency. This problem is described at length by Chandrasekhar (1960) and Kourganoff (1963), and the latter reference gives the exact solution obtained using the Wiener–Hopf method.

Sometimes we need a notation for the kernel in (5.25), and for this we use $K_1(\tau)$:

$$K_1(\tau) = \frac{1}{2}E_1(|\tau|). \tag{5.26}$$

If the medium were infinite the lower limit of integration in (5.25) would be $-\infty$ not 0, and then we see that the equation would have a displacement kernel, i.e., be of convolution type. We would want to know the Fourier transform of the kernel. We can readily calculate that from the definition of the E_1 function, as follows:

$$\begin{aligned}\tilde{K}_1(k) &= \frac{1}{2}\int_{-\infty}^{\infty} d\tau\, \exp(-ik\tau)\int_0^1 \frac{du}{u}\exp(-|\tau|/u)\\ &= \Re\left[\int_0^1 \frac{du}{u}\left(\frac{1}{u} + ik\right)^{-1}\right]\\ &= \int_0^1 \frac{du}{1 + k^2u^2} = \frac{\tan^{-1} k}{k}.\end{aligned} \tag{5.27}$$

Equation (5.25) is a Fredholm equation of the second kind on a half-space. This is the kind of equation for which the Wiener–Hopf method was designed, and the solution has been obtained in terms of definite integrals. Since $J = S$, (5.20) says that the total flux H is constant, in which case (5.21) can be integrated to give

$$K = H(\tau + \text{constant}). \tag{5.28}$$

The Eddington approximation should become accurate at large τ, so the solution for J can be written

$$J = 3H(\tau + q(\tau)), \tag{5.29}$$

where $q(\tau)$, called the Hopf function, tends to a constant at large τ.

The constant value of the flux $F = 4\pi H$ defines the effective temperature of a star according to

$$F = \sigma T_{\text{eff}}^4. \tag{5.30}$$

Furthermore, the frequency-integrated mean intensity J and the source function S are both the same, in the gray atmosphere problem,[1] as the frequency-integrated Planck function B, which is

$$B = \frac{\sigma}{\pi} T^4. \tag{5.31}$$

Therefore (5.29) gives the temperature distribution in the gray atmosphere

$$T^4 = \frac{3}{4} T_{\text{eff}}^4 (\tau + q(\tau)). \tag{5.32}$$

The Hopf function is the main result of Milne's problem. As we see, in Eddington's approximation it is replaced by a constant. The value of the constant is derived from the relation imposed at $\tau = 0$, of the kind

$$\frac{H(0)}{J(0)} = \frac{1}{3q(0)} = \frac{\int_0^1 \mu I(0, \mu)\, d\mu}{\int_0^1 I(0, \mu)\, d\mu} = \langle \mu \rangle. \tag{5.33}$$

Different estimates of $\langle \mu \rangle$ have been proposed, each with its corresponding value of $q(\tau)$. If I is approximately constant on the range $[0, 1]$ then $\langle \mu \rangle$ turns out to be $1/2$, and so the constant value of $q(\tau)$ should be $2/3$. A quadrature for the integrals based on two-point Gaussian quadrature (abscissae at $\pm 1/\sqrt{3}$) leads to $\langle \mu \rangle = 1/\sqrt{3}$ and to $q(\tau) = 1/\sqrt{3}$. Neither estimate of $q(\tau)$ is perfect. The exact function is shown in Figure 5.1. The value of $q(\tau)$ for $\tau \to \infty$ is $0.710\,446\,09\ldots$. As it happens, the exact value of $q(0)$ is $1/\sqrt{3} = 0.577\,35\ldots$, but the estimate $q(\tau) = 2/3$ is closer to the exact solution in the average. The exact function is very nearly constant for $\tau > 4$ as expected based on the earlier discussion. In fact, $q(\tau)$ is within 1% of q_∞ for $\tau > 1.3$, and within 0.1% of q_∞ for $\tau > 3$.

[1] When the opacity is *gray*, i.e., independent of frequency, the radiative equilibrium condition becomes $J = B$, and therefore $S = B$ in that case.

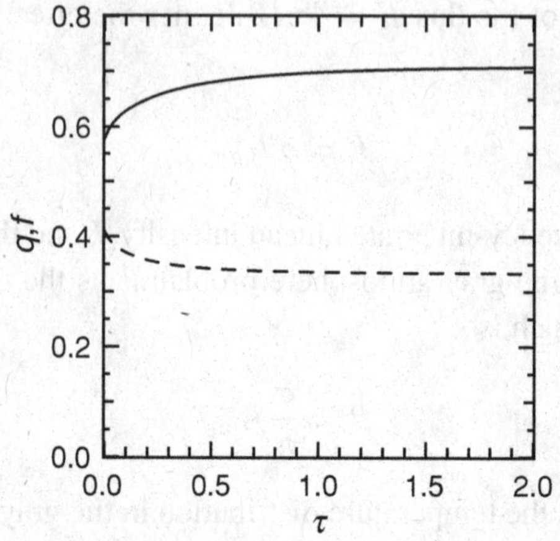

Fig. 5.1 Solid curve: exact Hopf function $q(\tau)$; dashed curve: corresponding Eddington factor, vs optical depth τ.

5.4 Eddington factor

The ratio of the zz component of the radiation pressure to the energy density, i.e., of K to J, defines the *Eddington factor*:

$$f \equiv \frac{P_{zz}}{E} = \frac{K}{J}. \tag{5.34}$$

It is a variable, not a constant, unless the Eddington approximation is being made, in which case it has the value $1/3$. If the Eddington factor is somehow already known, then the system of the first two moment equations is closed, just as in the Eddington approximation. Furthermore, if f is known then probably the mean cosine $f_H \equiv \langle\mu\rangle$ in (5.33) is also known. This allows us to write down the answer to the Milne problem immediately:

$$fJ = H\left[\tau + \frac{f(0)}{f_H}\right]. \tag{5.35}$$

This is the idea behind variable-Eddington-factor approximation methods, about which we will say more later. For the Milne problem there is a simple relation between the Eddington factor and the Hopf function, which is

$$3f(\tau) = \frac{\tau + q_\infty}{\tau + q(\tau)}. \tag{5.36}$$

This result for f based on the exact Hopf function is also shown in Figure 5.1.

5.5 Milne's second equation – thermalization

The next standard problem is the inhomogeneous version of the first. The model in this case is that there is a specified Planck function that may depend on position in the atmosphere and a scattering albedo that is smaller than unity, $\varpi < 1$. The objective is to solve for S in this case. The integral equation is

$$S(\tau) = (1 - \varpi)B(\tau) + \frac{\varpi}{2}\int_0^\infty E_1(|\tau' - \tau|)S(\tau'), \tag{5.37}$$

a Fredholm equation of the first kind. The solution in terms of integrals has been derived using another application of the Wiener–Hopf method, and also, by Sobolev (1963), using "elementary" methods, which is to say, without using analytic function theory. The general solution is too complicated to derive here. It is given in the form of integrals for the resolvent function $R(\tau, \tau')$ in terms of which the solution to (5.37) is

$$S(\tau) = (1 - \varpi)B(\tau) + \varpi\int_0^\infty R(\tau, \tau')B(\tau'). \tag{5.38}$$

The second Milne equation and the theory of the resolvent are discussed at length in Sobolev, and very similar material pertaining to spectral line transport is found in Ivanov (1973). The Wiener–Hopf theory is based on the factorization of a function $T(z)$ related to the Fourier transform of the kernel:

$$T(z) \equiv 1 - \varpi\tilde{K}_1(i/z) = 1 - \varpi z \coth^{-1} z, \tag{5.39}$$

where the last equality comes from (5.27). The nontrivial part is to factor $T(z)$ into parts that are analytic and nonvanishing in the left and right half-planes,

$$T(z) = \frac{1}{H(z)H(-z)}, \tag{5.40}$$

where the H function, introduced by Chandrasekhar,[2] is analytic and nonvanishing in $\Re z \geq 0$. Finding $H(z)$ can be done formally, by applying Cauchy's integral formula to $\log[T(z)]$, or computationally using the method of discrete ordinates or a numerical inversion of an integral equation satisfied by H. The result for the resolvent function is given in terms of its double Laplace transform, i.e., transformed with respect to both τ and τ':

$$\tilde{\tilde{R}}(p, q) = \frac{1 - \varpi}{\varpi}\frac{H(1/p)H(1/q) - 1}{p + q}. \tag{5.41}$$

[2] Also attributed to V. A. Ambartsumian.

The integral formulae by which $R(\tau, \tau')$ can be obtained from its double transform are developed in the radiative transfer literature, e.g., Sobolev (1963) and Ivanov (1973).

The physical interpretation of (5.38) is the following. The first term in S is the contribution from the local source, i.e., the photons emitted for the first time at this location. The second term is the intensity at the location τ of the photons first emitted at another location τ' by the primary source, which then travel from τ' to τ in any number of flights with scattering events between successive flights. Now imagine we time-reverse this picture. A collection of photons at τ will contain some that are destroyed on the spot by absorption, and others that travel from τ to τ' in any number of flights before being destroyed at that location. This final destruction of a photon after perhaps a large number of scatterings is the process called thermalization of the radiation. The meaning of $R(\tau, \tau')$ is therefore the conditional probability distribution of thermalization positions τ' given that a photon originates at τ. It is not too hard to show that $R(\tau, \tau') = R(\tau', \tau)$, and therefore R is also the conditional probability of thermalization at τ for photons created at τ'.

When the albedo is small, R is nearly the same as $E_1(|\tau' - \tau|)/2$, in other words, it is sharply peaked at $\tau' = \tau$. As the albedo approaches 1 the distribution becomes broader and broader. The typical extent of R in $|\tau' - \tau|$ is referred to as the *thermalization length*, and it tends to infinity as $\varpi \to 1$.

This progression in R as the albedo approaches unity is easiest to illustrate for the infinite medium, for which all the integrals are taken over the range $-\infty$ to ∞. The integral equation is then a convolution integral and its solution is easily found using Fourier transforms; the inverse transformation to obtain $R(\tau' - \tau)$ can be done by the method of residues. Figure 5.2 shows how R progresses as the albedo

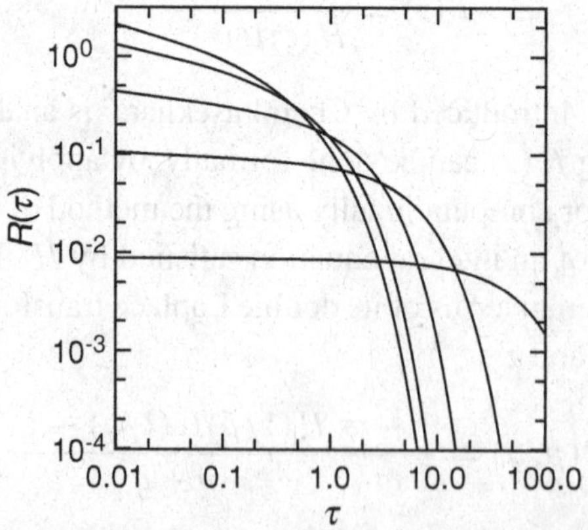

Fig. 5.2 Resolvent function for the infinite medium vs optical depth τ. The curves are ordered at the left, top to bottom, as $\epsilon = 1, 0.5, 0.1, 0.01, 0.0001$.

is raised from zero to 0.9999. When $\epsilon \equiv 1 - \varpi$ is small the entire contribution to R comes from a single pole in $\tilde{R}(k)$. This pole is located at $k = i\kappa_0$, where κ_0 is the root of

$$\frac{\tanh^{-1}\kappa_0}{\kappa_0} = \frac{1}{\varpi}, \tag{5.42}$$

which also means that $z = 1/\kappa_0$ is a zero of $T(z)$. This root appears in various guises in asymptotic diffusion theory. For small ϵ the root is approximately

$$\kappa_0 \approx \sqrt{3\epsilon}. \tag{5.43}$$

This is identical to what would be obtained using the Eddington approximation, and in fact the whole solution $R(\tau)$ tends to the Eddington approximation result when ϵ is small. In the half-space case only the long-range part of R is well approximated by Eddington, and the short-range part shows some effects of angle-dependent transfer.

The half-width of $R(\tau)$ at the $1/e$ point, in the small-ϵ case, is just

$$L = \frac{1}{\sqrt{3\epsilon}}, \tag{5.44}$$

and this is as good a definition of the thermalization length as any. This expression has a simple interpretation. Since ϵ is the probability that a photon will be absorbed on any single material interaction, then the mean number of times it will survive scattering is $1/\epsilon - 1$, and the mean number of total flights is $1/\epsilon$. The rms displacement in τ per flight is $1/\sqrt{3}$. Thus in a random walk with $1/\epsilon$ steps the net displacement in τ will be $1/\sqrt{3\epsilon}$.

In summary, the results for the inhomogeneous Milne problem are the following. The solution for the source function depends both locally and nonlocally on the primary source, which is the Planck function in this case. The nonlocal term is expressed by a resolvent function. When the albedo is close to unity the resolvent includes the effect of a large number of scatterings, and the typical extent of the resolvent in optical depth space is what you would expect for a random walk with a large number of steps. This extent varies in proportion to $1/\sqrt{\epsilon}$, where ϵ is the destruction probability per scattering, 1− the scattering albedo. In a large but finite volume filled with scatterers and with a smoothly distributed primary source of photons, the source function and the mean intensity will be depressed from the local equilibrium value for all points that are within a thermalization depth of the boundary. This may be quite a thick layer. But except within one or two mean free paths of the boundary the radiation will be nearly isotropic, albeit perturbed in magnitude owing to the presence of the boundary.

5.6 The Feautrier or even-parity equation

A simple manipulation of the equation of transfer in slab geometry turns out to be very useful in numerical solution methods. This is to define the even and odd combinations of the intensity $I(\mu)$ and $I(-\mu)$. For $0 \leq \mu \leq 1$ those two variables are replaced by the combinations

$$j(\mu) = \frac{1}{2}[I(\mu) + I(-\mu)], \qquad h(\mu) = \frac{1}{2}[I(\mu) - I(-\mu)]. \tag{5.45}$$

The use of j for the even-parity combination should not be confused with the emissivity function j_ν; the notation here is meant to suggest that $j(\mu)$ is like an angle-dependent J and $h(\mu)$ is like an angle-dependent H. In fact, they are related in this way

$$J = \int_0^1 d\mu\, j(\mu), \qquad H = \int_0^1 d\mu\, \mu\, h(\mu). \tag{5.46}$$

We now introduce the important assumption that the absorption coefficient and the source function are isotropic. Taking the even and odd combinations of (5.6) for $\pm\mu$, given this assumption, leads to the following:

$$\mu \frac{dh}{d\tau} = j - S, \qquad \mu \frac{dj}{d\tau} = h. \tag{5.47}$$

Now it is obvious that h can easily be eliminated to produce this second order equation:

$$\mu^2 \frac{d^2 j}{d\tau^2} = j - S. \tag{5.48}$$

Equation (5.48) is the heart of Feautrier's method (Feautrier, 1964). Sometimes this is called the second order form of the transfer equation while (5.6) is called the first order form. The great virtue of this equation is that it lends itself to this simple and accurate finite-difference form:

$$\mu_k^2 \frac{j_{i-1,k} - 2j_{i,k} + j_{i+1,k}}{\Delta\tau^2} = j_{i,k} - S_i, \tag{5.49}$$

for a uniformly-space grid in τ, in which i is the index of τ points, and k is the index for μ values, which all lie in the range $0 \to 1$. There is a simple extension of this to non-equally-spaced τ values. This discretization is second order accurate, and the solution $j_{i,k}$ is guaranteed to be positive for all i and k if the source function values S_i are all positive. As it turns out, these two properties are difficult to obtain simultaneously with finite-difference forms of the first order (5.6). The large majority of the work in astrophysical radiative transfer for slab geometry since 1964 has used Feautrier's equation (5.48). Many of the exceptions have been when fluid motion effects are included, which cause the absorption coefficient to depend on direction, or with nonisotropic scattering or angle-dependent frequency redistribution in lines, for which the source function is angle-dependent.

5.7 Eddington–Barbier relation

We now address the question: what intensity does an external observer see when he looks into an atmosphere with a certain temperature distribution? The rigorous result requires solving the general Milne equation for the source function, then doing an additional integration to get the emergent intensity. But a useful semi-quantitative formula can be found by supposing that the Planck function varies slowly with optical depth; this is the Eddington–Barbier relation.

We consider the case without scattering first. The emergent intensity is

$$I(0,\mu) = \int_0^\infty \frac{d\tau}{\mu} \exp(-\tau/\mu) B(\tau). \tag{5.50}$$

Now suppose we adopt a linear approximation for $B(\tau)$, namely

$$B(\tau) \approx a + b\tau. \tag{5.51}$$

Then it is easy to see that

$$I(0,\mu) \approx a + b\mu = B(\mu). \tag{5.52}$$

The intensity seen by the observer is the Planck function one mean free path into the atmosphere as measured along the ray. This is the basic form of the Eddington–Barbier relation. Another relation gives the value of the total flux leaving the atmosphere at the surface, which we get by multiplying the intensity by $2\pi\mu$ and integrating over μ:

$$F(0) \approx \pi\left(a + \frac{2}{3}b\right) = \pi B\left(\frac{2}{3}\right), \tag{5.53}$$

in other words, the flux is π times the Planck function at a location two-thirds of a mean free path into the atmosphere from the outside. Not coincidentally, the value of the temperature at optical depth 2/3 is just the effective temperature, if (5.32) is used and $q(\tau) \approx 2/3$.

The Eddington–Barbier relation is due to Barbier (1943), and a good discussion is given by Kourganoff (1963), Section 18. The two forms of the Eddington–Barbier relation are found to be very useful in understanding qualitatively what the spectrum should look like for atmospheres with complicated temperature structure, like the sun, when the opacity is very different at different frequencies.

When scattering is included the results are more complicated. The Planck function should be replaced by the source function $S(\tau)$ in (5.50), from which we see that the emergent intensity apart from a factor μ is the Laplace transform of the source function at $p = 1/\mu$. We can get this for the case that the Planck function is given by (5.51) from the double transform of the resolvent (5.41), by a suitable limiting procedure $q \to 0$. What we get is

$$\begin{aligned} I(0,\mu) &= \sqrt{\epsilon} H(\mu) \left[a + b \left(\mu + \frac{1-\epsilon}{2\sqrt{\epsilon}} \alpha_1 \right) \right] \\ &= \sqrt{\epsilon} H(\mu) B \left(\mu + \frac{1-\epsilon}{2\sqrt{\epsilon}} \alpha_1 \right). \end{aligned} \tag{5.54}$$

Here $H(\mu)$ is the Chandrasekhar H function introduced earlier, which depends on $\epsilon = 1 - \varpi$ in addition to μ, and α_1 is its first moment:

$$\alpha_1 = \int_0^1 d\mu \, \mu H(\mu). \tag{5.55}$$

The function $H(\mu)$ is identically 1 when there is no scattering, $\epsilon = 1$, and in that case we see that (5.54) reduces to (5.52). The scattering-dominated case, with an albedo that approaches 1 so ϵ is small, is more interesting. The H function tends to a limit, the function for conservative scattering, which is roughly equal to $1 + \sqrt{3}\mu$. The exact value of α_1 for this limiting H function is $2/\sqrt{3}$. We see that the emergent intensity depends in this limit on the Planck function at a large value of τ, which is approximately the thermalization depth $1/\sqrt{3\epsilon}$ since the added term μ is negligible. Thus for $\epsilon \to 0$ the emergent intensity is

$$I(0,\mu) \approx \sqrt{\epsilon} H(\mu) B \left(\frac{1}{\sqrt{3\epsilon}} \right). \tag{5.56}$$

We also note that the emergent intensity is a lot *smaller* than the Planck function at the thermalization depth. We can also obtain an expression for the emergent flux by doing the integration over $2\pi \mu d\mu$ as before. We note that the integral of H becomes α_1, for which we substitute $2/\sqrt{3}$. We find

$$F(0) \approx 4\pi \sqrt{\frac{\epsilon}{3}} B \left(\frac{1}{\sqrt{3\epsilon}} \right). \tag{5.57}$$

There is a cartoon-level way of understanding (5.56), which is the following. While in reality the primary source term ϵB acts throughout the whole problem, including in the thermalization layer, the cartoon version is to suppose that in the whole of this layer, down to the thermalization depth, there is only conservative scattering, but from the thermalization depth on down the radiation field is thermalized, i.e., is exactly Planckian. The radiation that emerges at the surface is therefore the Planckian emission at the thermalization depth, but attenuated by the effect of multiple scatterings between there and the surface, which reduce its intensity by the diffuse transmission factor for that layer. For a thick scattering layer we know that the fraction of radiation that is reflected is almost 100% and the transmitted fraction is small, of order the reciprocal of the optical depth. For our problem the optical thickness of the layer is the thermalization depth, $\approx 1/\sqrt{3\epsilon}$, and so the diffuse transmission fraction is about $\sqrt{3\epsilon}$. Except for the numerical factors, this gives (5.56).

6
The comoving-frame picture

There is just one good reason for viewing the radiation field in the comoving frame of the fluid, and developing equations based on this picture, and it is an important one. It is that owing to the Doppler and aberration effects it is only in the comoving frame that the emissivity and absorptivity have the values specified by atomic physics. When the photorecombination process results in the emission of photons, only in the comoving frame will this emission be isotropic. Only in the comoving frame does the photoionization edge appear in the absorptivity at the same frequency for every angle, and is that frequency the one in the tables of atomic absorption energies. Isotropic emission and absorption in the comoving frame mean that the equilibrium intensity, (4.33), will be isotropic, and that therefore the flux will vanish. As mentioned before, the flux F in the fixed frame does not vanish no matter how opaque the medium may be. The plan of this chapter is to present the transformation relations for the various radiation quantities for going from the comoving frame to the fixed frame (or vice-versa), and to follow through some of the implications of these relations. The development will be carried out only to order u/c, not because the relativistic treatment is especially hard, but because we have no use for this when we are using Newtonian mechanics for the fluid equations. We note that all the complexities of comoving-frame transport arise from the space and time variation of the fluid velocity. Indeed, if the velocity is a uniform constant then the comoving frame is an inertial frame and all the earlier simple relations apply in it, just as in any other inertial frame.

This material is admittedly complicated. The reader is encouraged to find the other references, including the key paper of L. H. Thomas (1930) and the report by Fraser (1966). This topic is Chapter 7 in Mihalas and Mihalas (1984), and the present discussion also draws on Castor (1972).

6.1 The Doppler and aberration transformations

We have to dip into the regime of relativistic kinematics for a little while. We will talk about the position four-vector $(x^\mu) = (t, \mathbf{r})$. The Greek index μ runs from 0 to 3, with the 0 value designating the time component and 1, 2, and 3 designating the space components. If a particle moves from $(t, \mathbf{r})$ to $(t + dt, \mathbf{r} + d\mathbf{r})$ in the (relative) time dt, then the amount of proper time elapsed in its own frame is ds given by

$$ds^2 = dt^2 - \frac{d\mathbf{r}^2}{c^2}. \tag{6.1}$$

The particle's four-velocity is

$$(dx^\mu/ds) = \frac{dt}{ds}(1, \mathbf{v}) = (\gamma, \gamma\mathbf{v}), \tag{6.2}$$

where $\gamma = 1/\sqrt{1 - v^2/c^2}$ (not to be confused with the ratio of specific heats). Its four-momentum is

$$(p^\mu) = \left(m\frac{dx^\mu}{ds}\right) = (m\gamma, m\gamma\mathbf{v}). \tag{6.3}$$

We will use m only for the rest masses of particles. The time component of the four-momentum is E/c^2, where E is the relativistic energy of the particle. For photons, whose speed is c, $\mathbf{v}$ is $c\mathbf{n}$, but γ tends to infinity. However, the rest mass also vanishes, and $m\gamma$ tends to a finite limit such that E is $h\nu$, i.e., $m\gamma = h\nu/c^2$. Thus the photon four-momentum is

$$(p^\mu) = \frac{h\nu}{c^2}(1, \mathbf{n}c). \tag{6.4}$$

We consider the Lorentz transformation. We want the transformation from one set of coordinates $(x^\mu_{(0)}) = (t_0, \mathbf{r}_0)$ to the $(t, \mathbf{r})$ set such that a point with a fixed $\mathbf{r}_0$ is moving with the velocity $\mathbf{u}$ in the $(t, \mathbf{r})$ coordinates. We quote the result:

$$x^\mu = A^\mu_\lambda x^\lambda_{(0)} \tag{6.5}$$

(using the summation convention), where (A^μ_λ) is the 4×4 matrix given by

$$(A^\mu_\lambda) = \begin{pmatrix} \gamma_u & \gamma_u\mathbf{u}^\mathrm{T}/c^2 \\ \gamma_u\mathbf{u} & \mathsf{I} + (\gamma_u - 1)\mathbf{u}\mathbf{u}^\mathrm{T}/u^2 \end{pmatrix}. \tag{6.6}$$

The velocity $\mathbf{u}$ is represented by a column vector here, and its transpose is the row vector $\mathbf{u}^T$. The scalar γ_u is $1/\sqrt{1 - u^2/c^2}$. The 3×3 matrix in the lower right-hand corner is arranged to leave unchanged a vector it multiplies that is perpendicular to $\mathbf{u}$, and to multiply by γ_u one that is parallel to $\mathbf{u}$.

The very point of the Lorentz transformation is to leave the proper time element ds unchanged in the transformation, so the transformation relation for a four-velocity is the same as for the coordinates,

$$\frac{dx^\mu}{ds} = A^\mu_\lambda \frac{dx^\lambda_{(0)}}{ds}. \tag{6.7}$$

Therefore the same is true for the four-momentum:

$$p^\mu = A^\mu_\lambda p^\lambda_{(0)}. \tag{6.8}$$

Now we can apply this to photons and get the Doppler and aberration relations:

$$\nu = \nu_0 \gamma_u \left(1 + \frac{\mathbf{n}_{(0)} \cdot \mathbf{u}}{c}\right), \tag{6.9}$$

$$\mathbf{n} = \frac{\gamma_u \mathbf{u}/c + \mathbf{n}_{(0)} + (\gamma_u - 1)(\mathbf{n}_{(0)} \cdot \mathbf{u})\mathbf{u}/u^2}{\gamma_u (1 + \mathbf{n}_{(0)} \cdot \mathbf{u}/c)}. \tag{6.10}$$

For small values of the relative velocity $\mathbf{u}$ of the two frames these relations simplify considerably. In that case we can neglect all the u^2/c^2 terms, which means that γ_u can be replaced by 1. Therefore

$$\nu = \nu_{(0)} \left(1 + \frac{\mathbf{n}_{(0)} \cdot \mathbf{u}}{c}\right), \tag{6.11}$$

$$\mathbf{n} = \frac{\mathbf{n}_{(0)} + \mathbf{u}/c}{1 + \mathbf{n}_{(0)} \cdot \mathbf{u}/c}. \tag{6.12}$$

These relations will be adequate for most (all?) of the subsequent discussion.

6.2 Transforming I, k, j

Phase-space density functions for relativistic gases require a little more of our relativistic kinematics. The reason is that the momentum space volume element $d^3\mathbf{p}$ is not a Lorentz invariant. There are good discussions of this in Synge (1957), and in Mihalas and Mihalas (1984) . We will give the quick version here. All of the possible four-momentum values p^μ that are allowed for a given kind of particle do not fill up space-time, since they must all be consistent with the value of the proper mass, and therefore

$$(p^0)^2 - \mathbf{p}^2/c^2 = m^2. \tag{6.13}$$

Thus the allowed four-momentum values form a hypersurface (mass shell) in space-time. An infinitesimal piece of the mass shell is an oriented three-volume in space-time, just as a patch on an ordinary surface in space is an oriented 2-D element. Oriented means that it is associated with a particular vector perpendicular to it, namely the surface normal at that point. The oriented three-volume element for an infinitesimal piece of the mass shell is a good four-vector that points along the hypersurface normal for the mass shell at the momentum value in question. What direction is that? The answer is that it is along p^μ itself. The reason is that p^μ is a four-vector of constant length, as indicated by (6.13), and therefore the allowed displacements dp^μ consistent with staying on the mass shell are perpendicular to p^μ (in the sense of the Minkowski metric). The surface normal is the one vector perpendicular to all the allowed displacements in the surface, therefore it is along p^μ.

Thus the oriented three-volume element for the mass shell is one good four-vector, and p^μ is another good four-vector, and we now know that they are parallel. The constant of proportionality between them is therefore a Lorentz invariant, and this is the invariant mass shell volume element. In a particular frame, the time-like component of the oriented three-volume is just $d^3\mathbf{p}$ if we choose to use p^1, p^2, and p^3 as the coordinates on the hypersurface. Since the time-like component of p^μ is proportional to the relativistic energy E, this combination is a Lorentz invariant:

$$\frac{d^3\mathbf{p}}{E}. \tag{6.14}$$

For photons, the momentum-space volume element is proportional to $\nu^2 d\nu d\Omega$ (see (4.4)), and E is proportional to ν, so the invariant volume element is $\nu d\nu d\Omega$.

What about the ordinary spatial volume element $d^3\mathbf{r}$? This is not Lorentz invariant either. In a particular coordinate frame, three-space is a slice through space-time at a constant value of the appropriate time variable; it is another hypersurface. The hypersurface normal is in the direction of the four-velocity corresponding to the velocity of that frame. Let us call that four-velocity U^μ. In this one frame U^μ has the components $(1, 0, 0, 0)$. We conclude that the oriented 3-volume element for this time slice is $d^3\mathbf{r}U^\mu$. We next make use of the fact that dot products of four-vectors are Lorentz invariant. Since p^μ is one four-vector and $d^3\mathbf{r}U^\mu$ is another, we conclude that $Ed^3\mathbf{r}$ is invariant. Here E is the energy of one particular photon in the frame for which $d^3\mathbf{r}$ is the correct 3-space volume element.

Notice that we have to divide the momentum space volume element by E to get an invariant and we have to multiply the coordinate volume element by E to get an invariant. That means that the phase-space volume element $d^3\mathbf{r}d^3\mathbf{p}$ is an invariant by itself. We recall from the discussion above that the phase-space density of photons is

$$\frac{I_\nu}{h^4\nu^3/c^2}. \tag{6.15}$$

If we change the units here to measure the number per volume h^3 in phase space, and divide by 2 to find the average number per mode of polarization, we get

$$\mathcal{I}_\nu = \frac{I_\nu}{2h\nu^3/c^2} \tag{6.16}$$

for the Lorentz-invariant intensity. This is exactly the quantity referred to as the number of photons per mode. This is the central quantity in the general covariant formulation of radiation transport by Lindquist (1966). Since $\mathcal{I}_\nu$ is invariant, the rule for transforming the intensity when going between different frames is just

$$I_\nu = \left(\frac{\nu}{\nu_0}\right)^3 I_\nu^{(0)}. \tag{6.17}$$

We can get the rules for transforming the absorptivity and emissivity by considering two time slices separated by dt, with matching three-volumes $d^3\mathbf{r}$ on each, in addition to a momentum volume $d^3\mathbf{p}$. The number of photons added to $d^3\mathbf{r}$ during dt that lie in $d^3\mathbf{p}$ is equal to

$$\Delta N = \frac{j_\nu}{h\nu} d^3\mathbf{r}dtd\nu d\Omega = \frac{j_\nu}{h^4\nu^3/c^3} d^3\mathbf{r}dtd^3\mathbf{p}. \tag{6.18}$$

The product $d^3\mathbf{r}dt$ by itself is the four-volume element, and it is invariant. Thus $d^3\mathbf{r}dtd^3\mathbf{p}$ is not invariant, but $d^3\mathbf{r}dtd^3\mathbf{p}/E$ is, in view of the earlier result. So we rearrange (6.18) as

$$\Delta N = \frac{j_\nu}{h^3\nu^2/c^3} \frac{d^3\mathbf{r}dtd^3\mathbf{p}}{h\nu} \tag{6.19}$$

and since ΔN should be invariant, we conclude that

$$e_\nu = \frac{c}{2} \frac{j_\nu}{\nu^2} \tag{6.20}$$

is Lorentz invariant. The same reasoning applies to $k_\nu I_\nu$, therefore

$$a_\nu = \frac{h}{c} \nu k_\nu \tag{6.21}$$

is invariant. (The constant factors included in these definitions are for the purpose of making (6.24) consistent with our prior notation.) This gives the transformation relations

$$j_\nu = \left(\frac{\nu}{\nu_0}\right)^2 j_\nu^{(0)}, \tag{6.22}$$

$$k_\nu = \frac{\nu_0}{\nu} k_\nu^{(0)}. \tag{6.23}$$

More discussion of the covariant absorption and emission is found in Linquist (1966).

As a final note to the business of Lorentz transformations of the intensity, we quote the covariant form of the transport equation for Cartesian coordinate systems (see Lindquist (1966)):

$$p^{\mu} \mathcal{I}_{\nu,\mu} = \mathcal{e}_{\nu} - \mathcal{a}_{\nu} \mathcal{I}_{\nu}. \tag{6.24}$$

The comma subscript here indicates that the following subscripts denote the coordinates by which this quantity is differentiated. Thus $f_{,\mu}$ means $\partial f/\partial x^{\mu}$. Putting in the appropriate powers of ν shows that this is identical to the equation we usually use, (4.23).

6.3 Transforming E, F, and P

The moments of the radiation field integrated over frequency have a natural physical interpretation as the parts of the stress-energy tensor. The tensor is defined by

$$T^{\lambda\mu} = \int \frac{d^3\mathbf{p}}{E} p^{\lambda} p^{\mu} \mathcal{I}. \tag{6.25}$$

This is a good contravariant second rank tensor since it is built from the product of two contravariant four-vectors multiplied by scalars. We evaluate the integral using the relations we just derived and find, after discarding some irrelevant factors of h and c,

$$(T^{\lambda\mu}) = c \int \int_{4\pi} \begin{pmatrix} 1 \\ \mathbf{n}c \end{pmatrix} (1\ \mathbf{n}^{\mathrm{T}}c) \left(\frac{\nu}{c}\right)^2 \frac{I_\nu}{\nu^3} \frac{\nu^2 d\nu d\Omega}{\nu} \tag{6.26}$$

$$= \int d\nu \int_{4\pi} d\Omega\, I_\nu \begin{pmatrix} 1/c & \mathbf{n}^{\mathrm{T}} \\ \mathbf{n} & \mathbf{n}\mathbf{n}^{\mathrm{T}}c \end{pmatrix} \tag{6.27}$$

$$= \begin{pmatrix} E & \mathbf{F}^{\mathrm{T}} \\ \mathbf{F} & c^2\mathsf{P} \end{pmatrix}. \tag{6.28}$$

Thus the time–time part of $T^{\lambda\mu}$ is the energy density, the time-space part is the flux vector, and the space–space part is the pressure tensor, apart from factors of c for unit conversion.

We are familiar with stress-energy tensors like this in other fields. The electromagnetic stress-energy tensor is of this kind (see Panofsky and Philips (1962)). Its time–time part is the electromagnetic energy density

$$\frac{1}{8\pi}(\mathcal{E}^2 + \mathcal{H}^2), \tag{6.29}$$

the space–time part is the Poynting vector, and the space–space part is the Maxwell stress tensor. As a matter of fact, the radiation stress-energy tensor is the electromagnetic stress-energy tensor after coarse-graining.

Since $T^{\lambda\mu}$ is a good tensor, we can use the Lorentz transformation matrix A to obtain its components in the fixed frame from those in the comoving frame. The rule is to multiply the T matrix by one factor of A from the left and by one factor of A from the right. We carry this out for the small $\mathbf{u}$ case and discard terms of order u^2/c^2 and higher. The result is

$$\begin{pmatrix} E & \mathbf{F}^{\mathrm{T}} \\ \mathbf{F} & c^2\mathsf{P} \end{pmatrix} = \begin{pmatrix} E_0 + \dfrac{2\mathbf{u}}{c^2}\cdot\mathbf{F}_0 & (\mathbf{F}_0 + \mathbf{u}E_0 + \mathbf{u}\cdot\mathsf{P}_0)^{\mathrm{T}} \\ \mathbf{F}_0 + \mathbf{u}E_0 + \mathbf{u}\cdot\mathsf{P}_0 & c^2\mathsf{P}_0 + \mathbf{F}_0\mathbf{u}^T + \mathbf{u}\mathbf{F}_0^{\mathrm{T}} \end{pmatrix}. \tag{6.30}$$

The transformation relation for the flux is the one that was used earlier in discussing the total energy equation for matter and radiation.

The quantities cE, $\mathbf{F}$, and $c\mathsf{P}$ are potentially all the same order of magnitude. In diffusion regions the flux will be smaller than the other two, however. So the velocity corrections to the moments are formally of order u/c, but in a diffusion region the relative magnitude of the correction to the flux may be much larger than that, while the relative corrections to the energy density and pressure are even smaller than u/c. In other words, the correction to the flux is the important one to keep in mind.

The next interesting thing that we can do with the stress-energy tensor is to form its divergence,

$$T^{\lambda\mu}{}_{,\lambda} = g^\mu. \tag{6.31}$$

The four-vector g^μ is made up of the energy and momentum source rates for the radiation field. We can find an expression for g^μ by multiplying the invariant transport equation (6.24) by p^μ and then integrating with $d^3\mathbf{p}/E$. This leads to

$$(T^{\lambda\mu}{}_{,\lambda}) = \int\!\!\int_{4\pi} \begin{pmatrix} 1 \\ \mathbf{n}c \end{pmatrix} (j_\nu - k_\nu I_\nu)d\nu d\Omega, \tag{6.32}$$

and therefore

$$g^0 = \int\!\!\int_{4\pi} (j_\nu - k_\nu I_\nu)d\nu d\Omega \tag{6.33}$$

and

$$\mathbf{g} = (g^i) = \int\!\!\int_{4\pi} \mathbf{n}c(j_\nu - k_\nu I_\nu)d\nu d\Omega, \tag{6.34}$$

exactly as given earlier, in (4.43) and (4.44). Writing out the components of the tensor divergence in the fixed frame and substituting these results for g^μ into (6.31) recovers (4.27) and (4.28) discussed earlier. What is new is the realization that g^μ is a four-vector, and that therefore it can be evaluated by going to the comoving frame where the atomic properties are known, and then transforming back to the fixed frame; see Mihalas and Mihalas (1984), Section 91. The Lorentz transformation to first order in u/c is

$$g^0 = g^0_{(0)} + \frac{\mathbf{u}}{c^2} \cdot \mathbf{g}_{(0)}, \tag{6.35}$$

$$\mathbf{g} = \mathbf{g}_{(0)} + \mathbf{u} g^0_{(0)}. \tag{6.36}$$

The second term on the right-hand side of the $\mathbf{g}$ equation is another of those pesky ones that do not have an analog in nonrelativistic mechanics. It comes about because the addition of energy increases the relativistic mass density and therefore also the relativistic momentum density. When we treat the material fluid nonrelativistically there is no similar term in the material momentum equation to compensate this one. Fortunately it is small and we can discard it or hide it somewhere. The velocity term in the g^0 equation does have a nonrelativistic meaning. It is the rate of doing work by the force exerted on the radiation by the matter.

The components of g^μ in the fluid frame are easily found from the atomic properties of the matter, without the need for Doppler and aberration transformations. Furthermore, we can almost always assume isotropy of the absorptivity and emissivity in the fluid frame. With that assumption we obtain

$$g^0_{(0)} = \int d\nu \left(4\pi j^{(0)}_\nu - k^{(0)}_\nu c E^{(0)}_\nu\right), \tag{6.37}$$

$$\mathbf{g}_{(0)} = -c \int d\nu\, k^{(0)}_\nu \mathbf{F}^{(0)}_\nu. \tag{6.38}$$

These are the same expressions that we used for the right-hand sides of the fixed-frame moment equations earlier, only repeated here using the fluid-frame absorptivity and emissivity and the fluid-frame radiation moments. The important difference is that the earlier fixed-frame expressions are wrong if there is an appreciable velocity while the fluid-frame expressions are correct as long as the velocity is nonrelativistic. The components g^0 and $\mathbf{g}$ obtained by substituting (6.37) and (6.38) into (6.35) and (6.36) give the correct quantities to put on the right-hand sides of the fixed-frame moment equations.

We have not described transforming the monochromatic radiation moments from the comoving frame to the fixed frame. That is because the monochromatic moments are not space-time tensors, and therefore the Lorentz transformation rules do not apply. That is one mathematical reason. A second mathematical reason is that an angle moment at a constant value of the fixed-frame frequency is an integral over a different slice through photon momentum space than an angle moment at any constant fluid-frame frequency. In other words, there is no simple relation between the two kinds of moments. The relations that have been obtained involve substituting the transformation relations for the intensity and the direction vector into the integrals for the fixed-frame moments, then using Taylor expansions to first order of $I^{(0)}$ to perform the mappings $\nu_0 \to \nu$ and $\mathbf{n}_0 \to \mathbf{n}$. The results will not be given here because this Taylor expansion technique is a very poor idea. It makes the implicit assumption that $v/c \ll \Delta\nu/\nu$, where $\Delta\nu$ is the width of the narrowest feature in the spectrum. It is easy to think of important examples that violate this limit by a wide margin: supernovae and stellar winds, to name just two.

6.4 The comoving-frame transport equation

There are two basically different approaches to solving radiation transport problems involving time-dependent flows. The first is to solve partial differential equations for the radiation intensity or its moments as viewed in a fixed frame of reference. The second is to let the unknown be the radiation field in the comoving frame of the fluid. There are advantages and disadvantages of each approach. The fixed-frame method has the advantage of simplicity in the partial differential equation, and the disadvantage of complexity in the absorption and emission coefficients, and in the corresponding energy/momentum terms for the matter. For the treatment with comoving-frame radiation the advantages and disadvantages are exchanged. In this section we examine the comoving-frame view of the radiation.

There are also two approaches to deriving the comoving-frame transport equation. One is to substitute the transformation relations for the radiation variables into the fixed-frame transport equation and perform the necessary expansions (Buchler, 1979, 1983). The other is to treat the Lagrangian frame as a curvilinear coordinate system, work out the appropriate metric tensor, and write the transport equation using a curvilinear generalization of (6.24) (see Castor (1972) and Mihalas and Mihalas (1984), Section 95). In spherical symmetry the second approach is in fact somewhat simpler than the first, and is an aid to understanding the problem. This technique fails to generalize to higher-dimensional cases, however. The reason is interesting: the curvilinear coordinate system is very awkward to deal with unless the metric tensor is diagonal, so that locally it agrees with a Lorentz transformation from the fixed space. This imposes certain conditions on the mapping from fixed space to comoving space, and it turns out that these are impossible to satisfy if the fluid velocity is rotational. Rather than discuss this further, we will turn immediately to the technique of transforming the fixed-frame transport equation.

The objective is very simple: Substitute the relations (6.11), (6.12), (6.17), (6.22), and (6.23) into the transport equation (4.23) and expand, discarding all the terms that are $O(u^2/c^2)$ or higher. This result is obtained:

$$\left(1+\frac{\mathbf{n}_0}{c}\cdot\mathbf{u}\right)\left(\frac{1}{c}\frac{\partial I^{(0)}}{\partial t}+\frac{\mathbf{u}}{c}\cdot\nabla I^{(0)}\right)+\mathbf{n}_0\cdot\nabla I^{(0)}$$
$$-\frac{\nu_0}{c}\left(\frac{\mathbf{a}}{c}+\mathbf{n}_0\cdot\nabla\mathbf{u}\right)\cdot\nabla_{\nu_0\mathbf{n}_0}I^{(0)}$$
$$+\frac{3}{c}\left(\frac{\mathbf{n}_0\cdot\mathbf{a}}{c}+\mathbf{n}_0\cdot\nabla\mathbf{u}\cdot\mathbf{n}_0\right)I^{(0)}=j^{(0)}-k^{(0)}I^{(0)}. \tag{6.39}$$

It was first given in this form by Buchler (1983). For simplicity here and below, the subscripts indicating that all the frequency-dependent quantities in the comoving frame are evaluated at ν_0 have been omitted. There are two features of this equation on which we should comment first. The term $\mathbf{n}_0\cdot\mathbf{u}/c$ in the coefficient of the first term would go away if the transformation from the fixed frame to the comoving frame had really been a Lorentz transformation, since a spatial derivative at constant time in the moving frame is not the same as a spatial derivative at constant time in the fixed frame owing to the relativity of time. But we are not altering the time coordinate in our procedure, so this term is left over. The terms involving the acceleration, $\mathbf{a}$, arise from $\partial\mathbf{u}/\partial t$. The actual acceleration is $\partial\mathbf{u}/\partial t+\mathbf{u}\cdot\nabla\mathbf{u}$, of course, so we might wonder about the other piece. But the extra parts would bring in terms that are of order u^2/c^2, and all such terms have been discarded. It is our opinion that the acceleration terms, as well as the $\mathbf{n}_0\cdot\mathbf{u}/c$ part of the coefficient of the first term, should be discarded. The reasoning is as follows. In a fluid flow problem the time derivative and the flow derivative $\mathbf{u}\cdot\nabla$ are generally of the same order of magnitude. If the acceleration terms above are ordered in this way then all of them are seen to be of order u^2/c^2 compared with the dominant terms, and may be discarded. The same is true of the $\mathbf{n}_0\cdot\mathbf{u}/c$ term multiplying $\partial I^{(0)}/\partial t$.

We suggest that (6.39) be retained for those problems involving nonrelativistic velocities that evolve on a light-transit time scale, and we simplify this equation by dropping the subject terms for problems with the fluid-flow time scale. Carrying out the simplification leads to

$$\frac{1}{c}\frac{DI^{(0)}}{Dt}+\mathbf{n}_0\cdot\nabla I^{(0)}-\frac{\nu_0}{c}\mathbf{n}_0\cdot\nabla\mathbf{u}\cdot\nabla_{\nu_0\mathbf{n}_0}I^{(0)}$$
$$+\frac{3}{c}\mathbf{n}_0\cdot\nabla\mathbf{u}\cdot\mathbf{n}_0 I^{(0)}=j^{(0)}-k^{(0)}I^{(0)}. \tag{6.40}$$

The operator D/Dt that appears here is the Lagrangian time derivative discussed in Section 2.2, $D/Dt\equiv\partial/\partial t+\mathbf{n}\cdot\nabla$.

We have so far not commented on the gradient with respect to momentum components. The reason for this term is simple. A photon with a fixed momentum travels along its ray and, as the local fluid velocity changes, so do its momentum components when referred to the local comoving frame. Thus the transport operator must account for this change with a gradient term multiplied by the rate of change of the comoving momentum along the ray. We normally use spherical coordinates, i.e., ν_0 and $\mathbf{n}_0$, rather than Cartesian coordinates for momentum space. So let's separate the vector $\mathbf{n}_0 \cdot \nabla\mathbf{u}$ into its radial and angular components in momentum space:

$$\mathbf{n}_0 \cdot \nabla\mathbf{u} = (\mathbf{n}_0 \cdot \nabla\mathbf{u} \cdot \mathbf{n}_0)\mathbf{n}_0 + \mathbf{n}_0 \cdot \nabla\mathbf{u} \cdot (\mathsf{I} - \mathbf{n}_0\mathbf{n}_0). \tag{6.41}$$

When dotted with the momentum-space gradient, the first term picks up the radial derivative, i.e., $\partial/\partial\nu_0$, and the second one picks up $1/\nu_0$ times the angle gradient, the gradient on the unit sphere. We end up with a form like this:

$$\frac{1}{c}\frac{DI^{(0)}}{Dt} + \mathbf{n}_0 \cdot \nabla I^{(0)} - \frac{1}{c}\mathbf{n}_0 \cdot \nabla\mathbf{u} \cdot \mathbf{n}_0\nu_0\frac{\partial I^{(0)}}{\partial\nu_0} - \frac{1}{c}\mathbf{n}_0 \cdot \nabla\mathbf{u} \cdot (\mathsf{I} - \mathbf{n}_0\mathbf{n}_0) \cdot \nabla_{\mathbf{n}_0} I^{(0)}$$
$$+ \frac{3}{c}\mathbf{n}_0 \cdot \nabla\mathbf{u} \cdot \mathbf{n}_0 I^{(0)} = j^{(0)} - k^{(0)} I^{(0)}. \tag{6.42}$$

The projection operator term here looks more complicated than it really is. The frequency-derivative term here is the Doppler correction, and the angle-derivative term is the abberation correction.

We will give the monochromatic and frequency-integrated moment equations that follow from this comoving-frame transport equation as they were derived by Buchler (1983). The only fussy part of the derivation is an integration by parts of the angle-derivative term. The frequency-dependent moment equations based on (6.39) are found to be

$$\frac{\partial E_\nu}{\partial t} + \nabla \cdot (\mathbf{u}E_\nu) + \frac{1}{c^2}\frac{D(\mathbf{u} \cdot \mathbf{F}_\nu)}{Dt} + \nabla \cdot \mathbf{F}_\nu + \left(\mathsf{P}_\nu - \frac{\partial(\nu\mathsf{P}_\nu)}{\partial\nu}\right) : \nabla\mathbf{u}$$
$$+ \frac{1}{c^2}\left(\mathbf{F}_\nu - \frac{\partial(\nu\mathbf{F}_\nu)}{\partial\nu}\right) \cdot \mathbf{a} = 4\pi j_\nu - k_\nu c E_\nu \tag{6.43}$$

and

$$\frac{1}{c}\frac{\partial\mathbf{F}_\nu}{\partial t} + \frac{1}{c}\nabla \cdot (\mathbf{u}\mathbf{F}_\nu) + \frac{1}{c}\frac{D(\mathbf{u} \cdot \mathsf{P}_\nu)}{Dt} + c\nabla \cdot \mathsf{P}_\nu + \frac{\mathbf{a}}{c}E_\nu + \frac{1}{c}\mathbf{F}_\nu \cdot \nabla\mathbf{u}$$
$$- \frac{1}{c}\frac{\partial(\nu\mathsf{Q}_\nu)}{\partial\nu} : \nabla\mathbf{u} - \frac{1}{c}\frac{\partial(\nu\mathsf{P}_\nu)}{\partial\nu} \cdot \mathbf{a} = -k_\nu\mathbf{F}_\nu. \tag{6.44}$$

To avoid the dreadful profusion of superscripts and subscripts the designations $^{(0)}$ or $_{(0)}$ have been omitted from all the quantities. The object Q_ν is the symmetric third rank tensor

$$\mathsf{Q}_\nu = \int_{4\pi} d\Omega\, \mathbf{nnn} I_\nu . \tag{6.45}$$

The colon ":" operator indicates summing the product of the tensor on the left with the tensor on the right over two indices, viz.,

$$\mathsf{R}:\mathsf{S} \equiv R_{ij} S_{ij} . \tag{6.46}$$

The factors are symmetrical in their indices in such cases.

The frequency-integrated moment equations are

$$\frac{\partial E}{\partial t} + \nabla \cdot (\mathbf{u}E) + \frac{1}{c^2}\frac{D(\mathbf{u}\cdot\mathbf{F})}{Dt} + \nabla \cdot \mathbf{F} + \mathsf{P}:\nabla\mathbf{u} + \frac{\mathbf{a}}{c^2}\cdot\mathbf{F}$$
$$= \int d\nu\, (4\pi j_\nu - k_\nu c E_\nu) = g^0_{(0)} \tag{6.47}$$

and

$$\frac{1}{c}\frac{\partial \mathbf{F}}{\partial t} + \frac{1}{c}\nabla \cdot (\mathbf{uF}) + \frac{1}{c}\frac{D(\mathbf{u}\cdot\mathsf{P})}{Dt} + c\nabla \cdot \mathsf{P} + \frac{\mathbf{a}}{c}E + \frac{1}{c}\mathbf{F}\cdot\nabla\mathbf{u}$$
$$= -\int d\nu\, k_\nu \mathbf{F}_\nu = \frac{\mathbf{g}_{(0)}}{c}. \tag{6.48}$$

The term in third rank tensor Q vanishes in the frequency integration.

After considering the ordering of the terms in several different hydrodynamic regimes, Buchler (1983) suggests that certain of the terms can be dropped that would not be of relative magnitude u/c or larger in any regime. His suggestions are to drop all terms with c^2 in the denominator in (6.39) (thus reducing it to (6.40)), the terms with c^2 in the denominator in (6.43) and the terms in (6.44) with c in the denominator other than the $\partial\mathbf{F}_\nu/\partial t$ term. His suggested frequency-dependent moment equations are therefore

$$\frac{\partial E_\nu}{\partial t} + \nabla \cdot (\mathbf{u}E_\nu) + \nabla \cdot \mathbf{F}_\nu + \left(\mathsf{P}_\nu - \frac{\partial(\nu \mathsf{P}_\nu)}{\partial \nu}\right):\nabla\mathbf{u} = 4\pi j_\nu - k_\nu c E_\nu \tag{6.49}$$

and

$$\frac{1}{c}\frac{\partial \mathbf{F}_\nu}{\partial t} + \frac{1}{c}\nabla \cdot (\mathbf{uF}_\nu) + c\nabla \cdot \mathsf{P}_\nu = -k_\nu \mathbf{F}_\nu . \tag{6.50}$$

The zeroth moment equation (6.49) follows exactly from (6.40), but the first moment equation (6.50) does not. Terms of order $\mathbf{F}_\nu \nabla \cdot \mathbf{u}$ and $\mathsf{Q}_\nu:\nabla\mathbf{u}$ are neglected by Buchler in deriving (6.50). Buchler's simplified forms of the frequency-integrated equations are

$$\frac{\partial E}{\partial t} + \nabla \cdot (\mathbf{u}E) + \nabla \cdot \mathbf{F} + \mathsf{P}:\nabla\mathbf{u} = \int d\nu\, (4\pi j_\nu - k_\nu c E_\nu) = g^0_{(0)} \tag{6.51}$$

and

$$\frac{1}{c}\frac{\partial \mathbf{F}}{\partial t} + \frac{1}{c}\nabla \cdot (\mathbf{uF}) + c\nabla \cdot \mathsf{P} = -\int d\nu\, k_\nu \mathbf{F}_\nu = \frac{\mathbf{g}_{(0)}}{c}. \quad (6.52)$$

6.5 A common-sense summary

After spending page after page of messy algebra deriving all the u/c corrections to radiation transport, it is easy to lose sight of the fact that some of the corrections really *are* significant while others are less so. Here are some of the main points:

- Ignoring the Doppler effect entirely and using the fixed-frame transport equation with the absorptivity and emissivity appropriate for matter at rest gives the wrong answer for the radiation when the velocity is supersonic and spectral lines dominate the opacity, and it also makes a significant error in the energy coupling rate of radiation and matter.
- The coupling terms are correctly given by the usual relations (6.33), (6.34) but those values are in the fluid frame. The transformations (6.35) and (6.36) have to be used to get the correct coupling terms for the fixed-frame equations.
- The coupling term in the material *internal* energy equation is indeed the fluid-frame energy term $g^0_{(0)}$ from (6.37), but in the material *total* energy equation it is combined with the work done by the radiation force, which turns it into the fixed-frame energy term equation (6.35).
- The diffusion limit gives a Fick's law form ($\mathbf{F} \propto -\nabla E$) for the flux in the fluid frame, not the fixed frame. In the fixed frame the convective flux of radiation enthalpy is added on.
- *Confusion of the correct frames for the coupling terms and the Fick's law flux is more serious than the other moderate corrections the u/c terms give.*
- The velocity terms in the comoving-frame energy equation (6.49) matter, especially the Doppler-shift frequency-derivative term; the ones in the comoving-frame momentum equation are less important. The aberration (angle-derivative) term in the transport equation does not survive in the energy moment and does not matter in the flux moment, so this can be dropped with little consequence.

The remaining part of this chapter is concerned with the problem of solving the comoving-frame equation if the choice is made to take $I_\nu^{(0)}$ or its moments as the basic variable(s).

6.6 The comoving-frame equation as a boundary-value problem

From a mathematical point of view the comoving-frame transport equation, say (6.40), is a partial differential equation for one scalar dependent variable in seven independent variables, namely x, y, z, ν, the two angles that describe $\mathbf{n}$, and t.

By contrast the fixed-frame equation has only four independent variables, x, y, z, and t, and the three photon momentum coordinates enter only as parameters, not as differentiation variables. The seven-dimensional problem has not been attacked yet, but efforts have been made to solve the comoving-frame equation in cases of lower dimensionality.

To illustrate the ideas in this case we will consider a problem with one spatial dimension, and for which photon flight time can be neglected, which allows us to drop the time derivatives. Spherical geometry is the interesting case, but we will discuss slab geometry for simplicity. Let the nontrivial space coordinate be x. If we particularize the simplified transport equation (6.40) to this case, and also drop the angle-derivative term, we end up with

$$\left(n_x + \frac{u_x}{c}\right)\frac{\partial I}{\partial x} - \frac{\nu}{c}n_x^2\frac{du_x}{dx}\frac{\partial I}{\partial \nu} = j_\nu - k_\nu I. \tag{6.53}$$

The seven independent variables have been reduced to two by the symmetry assumptions and by neglecting aberration. It is now quite feasible to apply a numerical technique for solving a hyperbolic equation in two dimensions to (6.53).

The concept of characteristics applies to (6.53). In fact, the fixed-frame transport equation is already in characteristic form, so the characteristic equation for (6.53) is just the equation for the Doppler effect:

$$\left(n_x + \frac{u_x}{c}\right)dI = (j_\nu - k_\nu I)dx \qquad \text{on} \qquad \frac{d\nu}{dx} = -\frac{\nu}{cn_x + u_x}n_x^2\frac{du_x}{dx}. \tag{6.54}$$

The characteristic slope here can be simplified by dropping the u_x term in the denominator; the difference this makes is $O(u^2/c^2)$. The slope becomes

$$\frac{d\nu}{dx} \approx -\frac{\nu}{c}n_x\frac{du_x}{dx}. \tag{6.55}$$

We see that in (6.54) the direction in which the radiation flows is toward $+x$ if $n_x + u_x/c > 0$ and toward $-x$ if $n_x + u_x/c < 0$. In either case the sign of the change in $d\nu$ is the sign of $-du_x/dx$. That is, the comoving-frame frequency decreases along the ray if u_x increases, and increases if u_x decreases. Figure 6.1 illustrates three cases of characteristic morphology depending on whether du_x/dx is positive, negative or indefinite. The ordinate is $\log \nu$ in these figures, so the characteristics are parallel curves. Since u_x/c is most often relatively small the slopes of the characteristics are modest. This suggests solving the PDE as an initial-value problem with x as the time-like variable. If $n_x > 0$ we can sweep in the direction of increasing x, solving for one slice in the ν direction at each step. This procedure

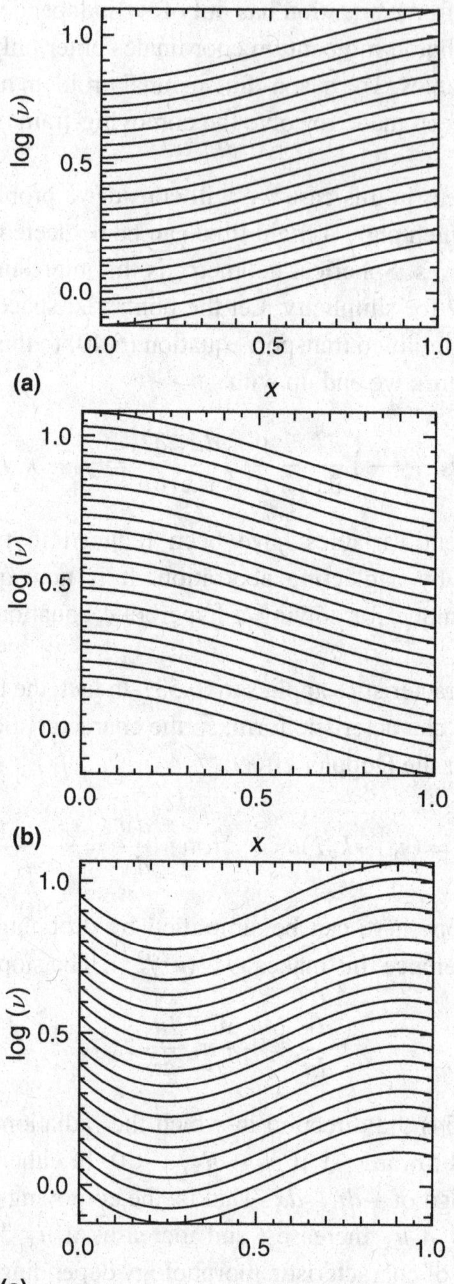

Fig. 6.1 Characteristics in ν vs x if $n_x u_x$: (a) decreases, (b) increases or (c) is nonmonotonic with x. Ordinate: log (ν) abscissa: **x**.

will be subject to a Courant-like condition

$$\frac{\Delta x |du_x/dx|}{c\Delta \log \nu} < 1 \tag{6.56}$$

unless implicit x-differencing is used, which requires solving a system of equations at each ν slice. Normally this would not be a problem, since $|u_x| \ll c$. However, in order to solve transfer problems involving spectral lines we may want to use very fine frequency meshes, and if the flow is supersonic, (6.56) may be violated by a large factor. We can then turn to implicit differencing, which ensures stability, but accuracy may still be badly compromised.

The alternative to a spatial sweep is a frequency sweep. The question is, in which direction? If u_x is monotonically increasing with x, then a sweep from high frequency to low works. If u_x is decreasing, a sweep from low frequency to high works. The direction of the sweep also agrees with the frequency boundary on which we can give "initial" values that make the problem well posed. This boundary must be an "inflow" boundary. In order to have a well-posed problem, we should specify one and only one condition on each characteristic, at a place where it enters the problem on either a spatial or a frequency boundary. With $n_x > 0$ and $du_x/dx > 0$ that will be either the lower boundary in x or the upper boundary in ν, depending on which one the characteristic intersects. If $du_x/dx < 0$ the lower frequency boundary is used instead. Then what about the nonmonotonic velocity case? The frequency sweep is simply not feasible in that case, and we have to fall back on the spatial sweep, and use implicit x-differencing if necessary. In the nonmonotonic case the same characteristic can easily intersect the same frequency boundary two or more times, which makes the problem badly posed.

This discussion has been based on the transport equation, i.e., the equation for $I_\nu^{(0)}$. Very similar considerations arise if the system of comoving-frame moment equations is solved instead. The two equations for $E_\nu^{(0)}$ and $F_\nu^{(0)}$ (in one dimension) can be rearranged by addition and subtraction to give two equations mathematically similar to the transport equation, one with $n_x > 0$ and one with $n_x < 0$, to which the preceding remarks apply.

The successful applications of the comoving-frame transfer or moment equations have been to cases with a monotonic velocity.

6.7 Diffusion in the comoving frame

We have mentioned several times now that the diffusion approximation should be applied in the comoving frame if the intent is to obtain a flux that tends to zero as the mean free path becomes small, i.e., one which obeys Fick's law. We will substantiate that claim by rederiving the diffusion results beginning with the comoving-frame transport equation. The present discussion is new, and is intended as a complement to Section 97 of Mihalas and Mihalas (1984).

We begin with (6.39), which we arrange in the form

$$I^{(0)} = \frac{j^{(0)}}{k^{(0)}} - \frac{1}{k^{(0)}}\left[\left(1+\frac{\mathbf{n}_0}{c}\cdot\mathbf{u}\right)\left(\frac{1}{c}\frac{\partial I^{(0)}}{\partial t} + \frac{\mathbf{u}}{c}\cdot\nabla I^{(0)}\right) + \mathbf{n}_0\cdot\nabla I^{(0)} - \frac{\nu_0}{c}\left(\frac{\mathbf{a}}{c}+\mathbf{n}_0\cdot\nabla\mathbf{u}\right)\cdot\nabla_{\nu_0\mathbf{n}_0} I^{(0)} + \frac{3}{c}\left(\frac{\mathbf{n}_0\cdot\mathbf{a}}{c}+\mathbf{n}_0\cdot\nabla\mathbf{u}\cdot\mathbf{n}_0\right) I^{(0)}\right]. \tag{6.57}$$

The zeroth approximation is that $I^{(0)}$ is $j^{(0)}/k^{(0)}$ which we identify with the thermodynamic equilibrium value, the Planck function,

$$\frac{j^{(0)}}{k^{(0)}} = B_\nu(T). \tag{6.58}$$

Making this replacement for $I^{(0)}$ in (6.57) leads to

$$I^{(0)} = B_\nu(T) - \frac{1}{k^{(0)}}\left[\left(1+\frac{\mathbf{n}_0}{c}\cdot\mathbf{u}\right)\left(\frac{1}{c}\frac{\partial B_\nu}{\partial t} + \frac{\mathbf{u}}{c}\cdot\nabla B_\nu\right) + \mathbf{n}_0\cdot\nabla B_\nu - \frac{\nu_0}{c}\left(\frac{\mathbf{a}}{c}+\mathbf{n}_0\cdot\nabla\mathbf{u}\right)\cdot\nabla_{\nu_0\mathbf{n}_0} B_\nu + \frac{3}{c}\left(\frac{\mathbf{n}_0\cdot\mathbf{a}}{c}+\mathbf{n}_0\cdot\nabla\mathbf{u}\cdot\mathbf{n}_0\right) B_\nu\right]. \tag{6.59}$$

Now the task is to expand (6.59) and keep terms of first order in the velocity. The Planck function is isotropic, but it does have spatial and temporal gradients. Its momentum-space gradient is

$$\nabla_{\nu_0\mathbf{n}_0} B_\nu = \frac{\partial B_\nu}{\partial \nu}\mathbf{n}_0. \tag{6.60}$$

What results is

$$I^{(0)} = B_\nu - \frac{1}{k^{(0)}}\left[\frac{dB_\nu}{dT}\left(\frac{1}{c}\frac{DT}{Dt} + \mathbf{n}_0\cdot\nabla T + \frac{\mathbf{n}_0\cdot\mathbf{u}}{c^2}\frac{DT}{Dt}\right) - \frac{1}{c}\left(\frac{\mathbf{a}\cdot\mathbf{n}_0}{c}+\mathbf{n}_0\cdot\nabla\mathbf{u}\cdot\mathbf{n}_0\right)\nu_0\frac{\partial B_\nu}{\partial \nu} + \frac{3}{c}\left(\frac{\mathbf{n}_0\cdot\mathbf{a}}{c}+\mathbf{n}_0\cdot\nabla\mathbf{u}\cdot\mathbf{n}_0\right) B_\nu\right]. \tag{6.61}$$

Integrating (6.61) over angles leads to the diffusion formula for the comoving-frame monochromatic energy density

$$E_\nu^{(0)} = \frac{4\pi B_\nu}{c} - \frac{4\pi}{k_\nu^{(0)}c^2}\left[\frac{dB_\nu}{dT}\frac{DT}{Dt} + \frac{\nabla\cdot\mathbf{u}}{3}\left(3B_\nu - \nu\frac{\partial B_\nu}{\partial \nu}\right)\right]. \tag{6.62}$$

By virtue of Wien's displacement law, which is

$$B_\nu(T) = \nu^3 \times \text{function}(\nu/T), \tag{6.63}$$

it follows that

$$3B_\nu - \nu\frac{\partial B_\nu}{\partial \nu} = T\frac{dB_\nu}{dT}. \tag{6.64}$$

Making this replacement in (6.62) leads to the simple form

$$E_\nu^{(0)} = \frac{4\pi B_\nu}{c} - \frac{4\pi}{k_\nu^{(0)}c^2}\frac{dB_\nu}{dT}\left[\frac{DT}{Dt} + \frac{\nabla\cdot\mathbf{u}}{3}T\right]. \tag{6.65}$$

A frequency integration of (6.65) leads to the following expression for the total comoving-frame energy density

$$E = \frac{4\pi B}{c}\left[1 - \frac{4}{k_R c}\left(\frac{1}{T}\frac{DT}{Dt} + \frac{\nabla\cdot\mathbf{u}}{3}\right)\right], \tag{6.66}$$

where the Rosseland mean has been put in to replace its defining integral,

$$\frac{1}{k_R} = \frac{\int d\nu\,(1/k_\nu^{(0)})dB_\nu/dT}{\int d\nu\, dB_\nu/dT}, \tag{6.67}$$

and where the notation $B = \int d\nu\, B_\nu$ has been used. In fact, $4\pi B/c$ is just the thermodynamic-equilibrium radiation energy density aT^4, and the fact that $dB/dT = 4B/T$ has been used in deriving (6.66). If the characteristic length scale for the problem is L and the characteristic time scale is τ, then the correction term in (6.66) is of order $\lambda_R/[c\min(\tau, L/u)]$, where λ_R denotes the Rosseland mean of the mean free path. So if the flow time L/u is longer than the characteristic time τ then the order of the correction is $\lambda_R/(c\tau)$, while if the flow time is shorter then the order is $(\lambda_R/L)(u/c)$. If the characteristic time is so short that it is comparable with the light-transit time c/L, then the order is just λ_R/L. When the flow time is shortest the size of the correction is the product of two small quantities, λ_R/L and u/c. This is why Mihalas and Mihalas (1984), Section 97, refer to this order of diffusion expansion as "second order diffusion."

The monochromatic flux derived from (6.61) is

$$\mathbf{F}_\nu^{(0)} = -\frac{4\pi}{3k_\nu^{(0)}}\frac{dB_\nu}{dT}\left(\nabla T + \frac{\mathbf{u}}{c^2}\frac{DT}{Dt} + \frac{\mathbf{a}}{c^2}T\right), \tag{6.68}$$

where we have again been able to use Wien's displacement law (6.64). We have the problem now of finding the physical explanations for the at-first-sight puzzling terms in the velocity and the acceleration. The velocity one is relatively easy. We need to recall that we are using the fixed-frame coordinates, and that in the Lorentz transformation to a locally comoving frame not only does the time derivative transform according to

$$\frac{\partial}{\partial t} \rightarrow \frac{D}{Dt} = \frac{\partial}{\partial t} + \mathbf{u} \cdot \nabla, \tag{6.69}$$

but the spatial derivative changes from one at constant t to one at constant comoving time t' according to the other Lorentz relation

$$\nabla \rightarrow \nabla' = \nabla + \frac{\mathbf{u}}{c^2} \frac{\partial}{\partial t}. \tag{6.70}$$

As we see, the $\mathbf{u}$ term in (6.68) is absorbed in changing ∇T to $\nabla' T$, apart from an error that is $O(u^2/c^2)$. In short, this velocity term is *real* and represents a relativistic effect on the diffusion flux. The acceleration term has a physical interpretation as well. The mean free path $1/k_\nu^{(0)}$ corresponds to a flight time $1/(k_\nu^{(0)} c)$ during which the local fluid velocity has increased by an amount $\mathbf{a}/(k_\nu^{(0)} c)$. The incremental boost by this velocity change makes a contribution to the flux of $-\mathbf{a}/(k_\nu^{(0)} c)$ times the sum of E and P, which gives the term in question. To put it another way, if the flux is zero at the time when the photon flight begins, then by the end of the flight the fluid has accelerated away from the rest frame of that radiation, which produces a flux in the new fluid rest frame in the direction opposite to the acceleration that is proportional to the acceleration times the flight time.

The frequency-integrated flux is simply

$$\mathbf{F}^{(0)} = -\frac{4\pi}{3k_R} \frac{dB}{dT} \left(\nabla' T + \frac{\mathbf{a}}{c^2} T \right), \tag{6.71}$$

which, since $B = acT^4/(4\pi)$, can be written

$$\mathbf{F}^{(0)} = -K_R \left(\nabla' T + \frac{\mathbf{a}}{c^2} T \right), \tag{6.72}$$

in terms of the Rosseland-mean radiative conductivity

$$K_R = \frac{4\pi}{3k_R} \frac{dB}{dT} = \frac{4acT^3}{3k_R}. \tag{6.73}$$

The gradient operator here is the Lorentz-corrected one from (6.70).

Before writing the result for the pressure tensor we have to obtain the fully symmetric angle average of the product of four **n**s. In other words, we want to evaluate

$$\frac{1}{4\pi} \int_{4\pi} d\Omega \, n_i n_j n_k n_l. \tag{6.74}$$

First we note that at least two of the four indices must be equal, since there are four indices and only three possible values they can take. If one pair out of the four are equal, say $i = j$, then it must be true that the other pair are equal as well, $k = l$, or otherwise the integral vanishes. Thus one contribution to the integral is proportional to $\delta_{ij}\delta_{kl}$. By choosing the first pair in the three possible ways and adding the results we get the required fully-symmetric form

$$\delta_{ij}\delta_{kl} + \delta_{ik}\delta_{jl} + \delta_{il}\delta_{jk}. \tag{6.75}$$

The integral must be proportional to this. We get the proportionality factor by taking the case $i = j = k = l = 1$, for which the angle average of n_x^4 is seen to be 1/5 by doing the integrations in that case, and for which the sum of delta functions is 3. Thus

$$\frac{1}{4\pi}\int_{4\pi} d\Omega\, n_i n_j n_k n_l = \frac{1}{15}\left(\delta_{ij}\delta_{kl} + \delta_{ik}\delta_{jl} + \delta_{il}\delta_{jk}\right). \tag{6.76}$$

Now we can do the integrations to find the monochromatic pressure tensor:

$$\mathsf{P}_\nu^{(0)} = \left(\frac{4\pi B_\nu}{c} - \frac{4\pi}{k_\nu^{(0)}c^2}\frac{dB_\nu}{dT}\frac{DT}{Dt}\right)\frac{\mathsf{I}}{3} - \frac{4\pi}{15k_\nu^{(0)}c^2}T\frac{dB_\nu}{dT}\left[\nabla\mathbf{u} + (\nabla\mathbf{u})^{\mathrm{T}} + \nabla\cdot\mathbf{u}\,\mathsf{I}\right] \tag{6.77}$$

and the frequency-integrated form

$$\mathsf{P}^{(0)} = \frac{1}{3}aT^4\left(1 - \frac{4}{k_R c}\frac{1}{T}\frac{DT}{Dt}\right)\mathsf{I} - \frac{4aT^4}{15k_R c}\left[\nabla\mathbf{u} + (\nabla\mathbf{u})^T + \nabla\cdot\mathbf{u}\,\mathsf{I}\right]. \tag{6.78}$$

The form of this pressure tensor implies that radiation in the diffusion limit contributes normal and bulk viscosity coefficients (cf., (2.19))

$$\mu_R = \frac{4aT^4}{15k_R c} \tag{6.79}$$

and

$$\zeta_R = \frac{5}{3}\mu_R = \frac{4aT^4}{9k_R c} \tag{6.80}$$

to the mixed fluid. The bulk viscosity in this picture is not at all close to zero, which is a cause of consternation in view of some very general relativistic results. Mihalas and Mihalas (1984) discuss this subtle point and its explanation by Weinberg in terms of a slight renormalization of the definition of temperature.

The order of magnitude of the shear stress contribution to P compared with the isotropic pressure is $O(\lambda_R u/(Lc))$ in the notation used above. In other words, isotropy of P is a very good approximation in a diffusion regime owing to the smallness of both λ_R/L and u/c.

We want to compare the size of the radiative viscosity with the gas-kinetic viscosity of the matter, and also estimate the size of a Reynolds number based on radiative viscosity. For gas-kinetic viscosity we use an estimate $\mu = \rho\lambda_g v_{th}$, where λ_g is a gas-kinetic mean free path, and v_{th} is the mean thermal speed of a gas particle. In plasmas the electron and ion contributions must be summed, but it turns out that the ion contribution dominates by a factor $\approx \sqrt{m_H/m}$, where m_H is the hydrogen-atom mass. Thus the v_{th} that is appropriate is comparable with the sound speed. An estimate of $\rho\lambda_g$ is m_H/σ_C, where $\sigma_C \approx (e^2/kT)^2$ is the Coulomb scattering cross section. Working out the ratio of μ_R to μ_g gives this estimate

$$\frac{\mu_R}{\mu_g} \approx \frac{\sigma_C}{\sigma_R}\frac{aT^4}{p_g}\frac{c_s}{c}, \tag{6.81}$$

where σ_R is the Rosseland mean photo cross section, p_g is the gas pressure and c_s is the (gas) speed of sound. Radiative viscosity is only relevant in a diffusion region, which means inside a star as opposed to high in its atmosphere.[1] As a result the radiation pressure may be a few times larger than the gas pressure, but not orders of magnitude larger. The Coulomb cross section can be orders of magnitude larger than the photo cross section, however. For example, deep in a stellar envelope, say where kT is about 100 eV, σ_C is around 10^{-17} cm^2 while the Rosseland mean photo cross section is only 10^{-24}–10^{-23}. Thus the ratio of the cross sections can be of order 10^6. The ratio of the sound speed to the speed of light is the small factor, about 5×10^{-4} at 100 eV. The effect of the large cross section ratio more than makes up for the speed ratio, and we see that radiative viscosity dominates gas viscosity unless the radiation pressure is much smaller than the gas pressure.

The estimate of the Reynolds number based on radiative viscosity is

$$\mathrm{R_e} = \frac{\rho v L}{\mu_R} \approx k_R L \frac{p_g}{aT^4}\frac{v}{c_s}\frac{c}{c_s}. \tag{6.82}$$

We see that $\mathrm{R_e} \gg 1$ is quite likely whenever diffusion is valid, since then the optical depth $k_R L$ is large, the ratio of gas pressure to radiation pressure is not *very* small, the Mach number v/c_s is not too small, and c/c_s is large, perhaps $O(10^3)$.

6.8 Sobolev approximation

In a number of astrophysical environments, such as active galactic nuclei, molecular clouds, and some circumstellar outflows, as well as in laboratory plasmas that may be generated by high power lasers or magnetic pinches, a flow speed that is highly supersonic is combined with spectral line transport. This is the regime that Sobolev's approximation addresses. The original work on the high-velocity-gradient approximation is that of Sobolev (1960). The present discussion follows

[1] The momentum effects of radiation may be large for $\tau \ll 1$, but viscosity is not a useful way of treating them.

Castor (1970, 1974b). A special topic is needed in this case because when the flow velocity is supersonic the common approximation of forgetting about the Doppler shifts in calculating the opacity is quite wrong. An alternative approximation, of fully accounting for the fluid velocity but ignoring some spatial gradients, is better suited to these problems. This approximation, called after the first to study it, offers a relatively easy way to analyze a non-LTE problem with a myriad of lines and level populations to be computed. It has also led to a simple but useful way of approximating the force on the stellar material due to the large number of lines, and some simple but realistic stellar wind models (see Section 12.5).

The environment to which we apply the Sobolev approximation is a fluid flow with a velocity field $\mathbf{u}$ and a corresponding rate-of-strain tensor $\nabla\mathbf{u}$. A restriction that has to be imposed at the outset is that the principal strains all have the same sign in all parts of the flow. In other words, either the whole flow is expanding or it is contracting, but nowhere is there stretch in one direction and compression in another. (This case, which introduces nonlocal coupling, is treated by Rybicki and Hummer (1978).) We suppose that the velocity is not so terribly large, perhaps at most a few percent of c, so most of the $\mathbf{u}/c$ corrections to transport are relatively unimportant. The radiation–enthalpy–advection correction is probably important, but that is not the one we are going to talk about here. We will focus on line transport, and on the role the Doppler frequency-derivative term plays in line transport when the velocity is supersonic. If we let our imagination work on a typical ray cutting through the flow, and suppose that we follow a photon as it tracks along the ray, we see that the photon changes its local comoving-frame frequency as it goes, and the whole range it scans is typical of the flow velocity. Since the flow is supersonic, that range is larger than a typical line width. For some of the wind models, and for supernovae, the flow is actually around Mach 100, which means that not only does the range of comoving-frame frequencies cover one line, it can span the spacing between lines, and so the photon may even visit the frequencies of several lines as it moves along a single path.

We quote the comoving-frame transport equation again:

$$\frac{1}{c}\frac{DI_\nu}{Dt} + \mathbf{n}\cdot\nabla I_\nu - \frac{1}{c}\mathbf{n}\cdot\nabla\mathbf{u}\cdot\mathbf{n}\,\nu\frac{\partial I_\nu}{\partial\nu} - \frac{1}{c}\mathbf{n}\cdot\nabla\mathbf{u}\cdot(\mathsf{I}-\mathbf{nn})\cdot\nabla_{\mathbf{n}}\,I_\nu$$
$$+\frac{3}{c}\mathbf{n}\cdot\nabla\mathbf{u}\cdot\mathbf{n}\,I_\nu = j_\nu - k_\nu I_\nu, \tag{6.83}$$

where from now on we will understand that the comoving frame is used without explicitly adding superscripts to indicate this. In particular, ν will be the comoving-frame frequency.

As we just mentioned, the $1/c$ terms are relatively unimportant with the exception of the frequency-derivative term that expresses the Doppler effect. So we will first of all discard the time-derivative, angle-derivative, and dilation terms and keep only the spatial transport and the frequency-derivative terms on the left-hand side. The dimensional estimate of the ratio of these is

$$\frac{\text{frequency derivative}}{\text{spatial transport}} = \frac{u}{c}\frac{\nu}{\Delta\nu}, \tag{6.84}$$

where $\Delta\nu$ is the scale of the variations with frequency; we have assumed that the spatial scale lengths for the velocity and the comoving intensity are the same order. Thus in the flows being discussed, for which the velocity is larger than the line width converted to velocity units, the spatial transport term is dominated by the frequency-derivative term. Dropping the smaller spatial term leads to the *Sobolev equation*

$$-\frac{1}{c}\mathbf{n}\cdot\nabla\mathbf{u}\cdot\mathbf{n}\,\nu\frac{\partial I_\nu}{\partial\nu} = j_\nu - k_\nu I_\nu. \tag{6.85}$$

This can be integrated immediately, but first we want to discuss what we should integrate over. We will handle individual lines singly using (6.85), which means we select a frequency band that contains that line, with the endpoints located in stretches of continuous spectrum just outside the line. Depending on the strain tensor $\nabla\mathbf{u}$, the comoving photon frequency either decreases as the photon travels its path (expansion) or increases (compression). We pick an initial value for the intensity on the high-frequency side in the first case and on the low-frequency side in the second case. We call that value I_c, where the $_c$ stands for "continuum". We assume that the opacity and emissivity are due to the line alone, because the line most often dominates the total opacity at frequencies within the line. The addition of a background continuum adds other terms that Hummer and Rybicki (1985) discuss. The expressions for absorptivity and emissivity of a line will be discussed in detail later, see (9.4) and (9.5). We write them here as

$$k_\nu = k_L\phi(\nu), \tag{6.86}$$

$$j_\nu = k_L S_L\phi(\nu). \tag{6.87}$$

The quantity k_L depends on the upper and lower level populations and the oscillator strength of the line:

$$k_L = \frac{\pi e^2}{mc}(gf)_{\ell u}\left(\frac{N_\ell}{g_\ell} - \frac{N_u}{g_u}\right) = \frac{h\nu}{4\pi}\left(N_\ell B_{\ell u} - N_u B_{u\ell}\right), \tag{6.88}$$

and S_L is the line source function which is either the Planck function or the result of a non-LTE kinetics model, as in the case of resonance line scattering. The line profile function $\phi(\nu)$ is normalized over frequency,

$$\int d\nu\, \phi(\nu) = 1, \tag{6.89}$$

and is given very often by the Doppler broadening formula, a Gaussian. That is what we will assume here. The Gaussian profile provides sharp edges to the line bandwidth, and avoids concerns about the effect of an extended tail of the profile function that complicates or even invalidates the integration over frequency.

We need a variable for the indefinite integral of the profile function. We let this be y defined by

$$y = \begin{cases} \displaystyle\int_{\nu}^{\infty} d\nu' \phi(\nu') & \mathbf{n} \cdot \nabla\mathbf{u} \cdot \mathbf{n} > 0 \\ \displaystyle\int_{-\infty}^{\nu} d\nu' \phi(\nu') & \mathbf{n} \cdot \nabla\mathbf{u} \cdot \mathbf{n} < 0 \end{cases}, \tag{6.90}$$

i.e., the beginning point for y is on the side of the line where the initial value I_c is specified. The variation of y is from $y = 0$ on the incoming side of the line to $y = 1$ on the outgoing side. When integrating (6.85) over frequency we neglect the variation of the factor ν that multiplies the frequency derivative and replace it with the line-center frequency ν_0. Using the variable y in place of ν gives this result for the differential equation:

$$\frac{\nu_0 |\mathbf{n} \cdot \nabla\mathbf{u} \cdot \mathbf{n}|}{c} \frac{dI(y)}{dy} = k_L [S_L - I(y)], \tag{6.91}$$

which is immediately integrated using the boundary condition $I = I_c$ at $y = 0$:

$$I(y) = I_c \exp[-\tau(\mathbf{n}) y] + S_L\{1 - \exp[-\tau(\mathbf{n}) y]\}, \tag{6.92}$$

where $\tau(\mathbf{n})$, a direction-dependent quantity, is the *Sobolev optical depth*, defined by

$$\tau(\mathbf{n}) = \frac{k_L c}{\nu_0 |\mathbf{n} \cdot \nabla\mathbf{u} \cdot \mathbf{n}|}. \tag{6.93}$$

This differs from the usual optical depth that increases from one spatial side of the problem to the other in that this is a strictly local variable. It is a measure of the optical thickness of the resonance zone along a given ray where a particular photon might be absorbed by the line. This resonance zone is quite limited in extent, because the velocity gradient is large, when the Sobolev approximation is appropriate.

We find the quantity $\bar{J}$ to which the photoabsorption rate is proportional by integrating $I(y)$ from $y = 0$ to $y = 1$, which is equivalent to first multiplying by $\phi(\nu)$ then integrating over ν, and then forming the average over direction. The first integration gives

$$\bar{I}(\mathbf{n}) = I_c(\mathbf{n})\beta(\mathbf{n}) + S_L[1 - \beta(\mathbf{n})], \tag{6.94}$$

in which $\beta(\mathbf{n})$ is what we shall call the *angle-dependent escape probability*,

$$\beta(\mathbf{n}) = \frac{1 - \exp[-\tau(\mathbf{n})]}{\tau(\mathbf{n})}. \tag{6.95}$$

The angle average gives

$$\bar{J} = \frac{1}{4\pi}\int_{4\pi} d\Omega\, \beta(\mathbf{n}) I_c(\mathbf{n}) + (1 - \beta) S_L, \tag{6.96}$$

and β, the *Sobolev escape probability*, is the angle average of $\beta(\mathbf{n})$:

$$\beta = \frac{1}{4\pi}\int_{4\pi} d\Omega\, \beta(\mathbf{n}) = \frac{1}{4\pi}\int_{4\pi} d\Omega\, \frac{1 - \exp[-\tau(\mathbf{n})]}{\tau(\mathbf{n})}. \tag{6.97}$$

We have allowed for the direction dependence of the continuum intensity $I_c(\mathbf{n})$.

The Sobolev escape probability result (6.96) was given in 3-D form by Rybicki and Hummer (1983), and earlier for spherical geometry by Castor (1970). Equation (6.96) has the same structure as the static escape probability approximation (5.23) and also the accelerated lambda iteration (ALI) *ansatz* (11.142), to be discussed later. In all the escape probability approximations, including ALI which uses such an approximation as the acceleration operator, $\bar{J}$ is replaced by a linear expression in terms of the local source function. Among these methods, the Sobolev approximation is unique in that the approximation is actually accurate in the circumstances for which it is intended – high Mach number flows. It has been applied extensively to the study of stellar winds and supernovae as part of non-LTE modeling of the spectra using multilevel model atoms.

A further application of the Sobolev approximation is to the calculation of the body force associated with the absorption of radiation by the lines. We recall that the body force (per unit mass) is given by (see (6.38))

$$\mathbf{g}_R = \frac{1}{\rho c}\int_0^\infty d\nu\, k_\nu \mathbf{F}_\nu. \tag{6.98}$$

The contribution to $\mathbf{g}_R$ from a single line treated in the Sobolev approximation can be evaluated by using the Sobolev expression for the intensity and forming the integral over frequency before integrating over direction. What results is

$$\mathbf{g}_R = \frac{k_L}{\rho c}\int_{4\pi} d\Omega\, \mathbf{n}\{I_c(\mathbf{n})\beta(\mathbf{n}) + S_L[1 - \beta(\mathbf{n})]\}. \tag{6.99}$$

We observe that $\beta(\mathbf{n})$ is an even function of angle, and that therefore the contribution of the local emissions to the flux, and thus to the net force, vanishes. This becomes a somewhat subtle point when we examine the diffusion-like corrections to the Sobolev approximation, because a moderate gradient in the source function produces another force contribution in the opposite direction. The size of the latter contribution compared with the Sobolev formula is proportional to the ratio v_{th}/u and is therefore small in a hypersonic flow. We drop the S_L term in the force and obtain

$$\mathbf{g}_R = \frac{k_L}{\rho c} \int_{4\pi} d\Omega \, \mathbf{n} I_c(\mathbf{n}) \beta(\mathbf{n}). \tag{6.100}$$

The Sobolev approximation has therefore given a formula for the force that requires no further calculations if $I_c(\mathbf{n})$ is just the free-streaming radiation of a stellar photosphere, for instance.

The applications of Sobolev theory have so far been almost exclusively to spherically symmetric problems. We recall that the strain tensor in spherical coordinates looks like

$$\begin{pmatrix} du/dr & 0 & 0 \\ 0 & u/r & 0 \\ 0 & 0 & u/r \end{pmatrix}, \tag{6.101}$$

and that therefore the strain rate projected on the ray direction is

$$\mathbf{n} \cdot \nabla \mathbf{u} \cdot \mathbf{n} = \mu^2 \frac{du}{dr} + (1 - \mu^2) \frac{u}{r}. \tag{6.102}$$

We introduce an auxiliary quantity σ, *not* a cross section, by

$$\sigma = \frac{r}{u} \frac{du}{dr} - 1, \tag{6.103}$$

so we can write the projected strain rate as

$$\mathbf{n} \cdot \nabla \mathbf{u} \cdot \mathbf{n} = \frac{u}{r}(1 + \sigma \mu^2). \tag{6.104}$$

This means that the directional Sobolev optical depth is

$$\tau(\mathbf{n}) = \frac{\tau_0}{1 + \sigma \mu^2}, \tag{6.105}$$

in which the angle-independent optical depth τ_0 is

$$\tau_0 = \frac{k_L c r}{\nu_0 u}. \tag{6.106}$$

The formula for the escape probability then becomes

$$\beta = \int_0^1 d\mu \, \frac{1 + \sigma \mu^2}{\tau_0} \left[1 - \exp\left(-\frac{\tau_0}{1 + \sigma \mu^2} \right) \right]. \tag{6.107}$$

For the useful case that the continuum intensity is uniform within a cone $\mu > \mu_c$, which arises when the continuum radiation is supposed to come from a well-defined photosphere with no limb darkening, the integral for the I_c contribution to $\bar{J}$ can be expressed in terms of β:

$$\frac{1}{4\pi}\int_{4\pi} d\Omega\, \beta(\mathbf{n}) I_c(\mathbf{n}) = \beta_c I_c \tag{6.108}$$

with

$$\beta_c(\tau_0, \sigma) = \frac{1}{2}[\beta(\tau_0, \sigma) - \mu_c \beta(\tau_0, \sigma\mu_c^2)]. \tag{6.109}$$

The behavior of $\beta(\tau_0, \sigma)$ is simple. When the optical depth is zero the escape probability is unity. For large optical depth the integral (6.107) becomes

$$\beta \sim \frac{1 + \sigma/3}{\tau_0}. \tag{6.110}$$

The variation as the inverse of the optical depth is reminiscent of line scattering with complete redistribution over a Doppler profile, although the meaning of optical depth is different in the two cases. The static atmosphere gray escape probability decays exponentially at large optical depth, so that is quite unlike either case of line transfer.

The integral for $\beta(\tau_0, \sigma)$ cannot be evaluated in terms of elementary functions or the common special functions, but some excellent numerical approximations are available. Results for the 3-D case are given by Rybicki and Hummer (1983) as a single integral involving complete elliptic integrals of the first kind. Rybicki (private communication, August, 1978) suggested a very useful method of approximating $\beta(\tau_0, \sigma)$. It is based on a rational approximation to $[1 - \exp(-x)]/x$:

$$\frac{1 - \exp(-x)}{x} \approx \frac{P_{n-1}(x)}{Q_n(x)}, \tag{6.111}$$

in which P_{n-1} is a polynomial of the $(n-1)$th degree and Q_n is a polynomial of the nth degree. The zeroes of Q_n are complex in general. A partial-fraction expansion of P_{n-1}/Q_n leads to

$$\frac{1 - \exp(-x)}{x} \approx \sum_{i=1}^{n} \frac{r_i}{x - z_i}. \tag{6.112}$$

If this approximation is inserted into (6.107), the integration over μ can be done analytically, with the result

$$\beta(\tau_0, \sigma) \approx -\sum_{i=1}^{n} \frac{r_i}{z_i}\left[1 + \frac{\tau_0}{2t_i\sigma z_i}\log\left(\frac{t_i - 1}{t_i + 1}\right)\right], \tag{6.113}$$

in which

$$t_i = \sqrt{\frac{1}{\sigma}\left(\frac{\tau_0}{z_i} - 1\right)}. \tag{6.114}$$

The complex square root and logarithm functions are needed in these expressions. The $n = 2$ approximation of this kind, constrained to be accurate at $x = 0$ and $x \to \infty$, is

$$\frac{1 - \exp(-x)}{x} = \frac{1 + c_1 x}{1 + c_2 x + c_1 x^2}[1 + \epsilon(x)], \tag{6.115}$$

with $|\epsilon(x)| < 1.61 \times 10^{-2}$. The coefficients are $c_1 = 0.422\,26$ and $c_2 = 0.820\,47$. There is one complex-conjugate pair of roots z, given by

$$z = -0.971\,515 \mp 1.193\,464i, \qquad \frac{r}{z} = -0.5 \pm 0.011\,933\,91i. \tag{6.116}$$

Since the roots and residues are complex conjugates, it is sufficient to calculate just one of the terms in (6.113) and keep twice the real part. The approximation for $\beta(\tau_0, \sigma)$ has the same global relative accuracy as the rational approximation for $[1 - \exp(-x)]/x$, namely 1.61%. If more accuracy is needed, the next good approximation, for $n = 4$, may be used. The accuracy in that case is 1.5×10^{-4}, and there are two complex-conjugate pairs of roots and residues, $z = -1.915\,394 \pm 1.201\,751i$ with $r/z = -0.492\,975 \pm 0.216\,820i$, and $z = -0.048\,093 \pm 3.655\,564i$ with $r/z = -0.007\,025 \pm 0.050\,338i$.

Returning to the general discussion, we want to evaluate the radiation force. The reduction of the formula for $\mathbf{g}_R$ in spherical geometry gives

$$g_R = \frac{2\pi \nu_0}{\rho c^2}\frac{du}{dr}\int_{-1}^{1} \mu d\mu\, I_c(\mu)\frac{1 + \sigma\mu^2}{1 + \sigma}\left[1 - \exp\left(-\frac{\tau_0}{1 + \sigma\mu^2}\right)\right]. \tag{6.117}$$

When the continuum flux is confined to a narrow cone about the radial direction it is a good approximation to evaluate $1 + \sigma\mu^2$ as $1 + \sigma$ inside the integral, which then depends on

$$\tau_{\rm rad} = \frac{\tau_0}{1 + \sigma} = \frac{k_L c}{\nu_0 (du/dr)}. \tag{6.118}$$

In this radial-beaming approximation g_R is a function of the radial velocity gradient. We see that optically thin lines contribute an amount $(k_L F_{\nu_0}/c)$ that depends on the opacity and the continuum flux, but not on the velocity gradient. Optically thick lines contribute an amount that depends on the flux and the velocity gradient, but not on the opacity. The physical interpretation of this latter result is important. It says that a band $\Delta\nu$ of the continuous spectrum, which contains a momentum flux $\Delta\nu F_c/c$, gives its momentum to the matching spherical shell in the expanding envelope. The column thickness of that shell is $\rho\Delta r = \rho c\Delta\nu/(\nu du/dr)$, and dividing the momentum by the mass thickness gives the body force $(\nu F_c/\rho c^2)du/dr$.

6.9 Expansion opacity

Expansion opacity is a concept that grows out of Sobolev escape probability theory. It was introduced with that name by Karp, Lasher, Chan, and Salpeter (1977). The concept treats the situation in which the spectrum is filled with a forest of lines; there is a fluid velocity giving a significant strain-rate tensor $\nabla\mathbf{u}$ as discussed in the previous section; and the lines are spaced in $\log\nu$ by an amount that is comparable to or smaller than u/c. When these conditions occur, a photon may fly along until it hits resonance with one of the lines. If the strain rate in the medium corresponds to expansion, i.e., the eigenvalues of the strain-rate tensor are positive, then the photon's fluid-frame frequency steadily decreases as it goes along its path. When it hits resonance it is absorbed or scattered with a probability $1 - \exp[-\tau(\mathbf{n})]$, where $\tau(\mathbf{n})$ is the Sobolev optical depth discussed earlier, see (6.93). If this does occur, then the photon may be reemitted or scattered with a new direction, and it again steadily marches down in frequency. If we conceive of the distributions of lines in frequency as stochastic, the whole process looks like a random walk or diffusion. The expansion opacity is defined by identifying the mean free path in this random walk as $1/(\kappa_{\rm exp}\rho)$.

6.9.1 The Karp et al. model

The Karp *et al.* (1977) formulation is slightly different in flavor from the preceding paragraph. Karp *et al.* do not adopt a stochastic picture, and they also include the effect of continuous opacity (electron scattering). Their basic result is

$$\kappa_{\rm exp}(\nu) = \sigma_T \left\{ 1 - \sum_{j=J}^{N} \left[1 - \exp(-\tau_j)\right] (\nu_j/\nu)^s \exp\left(-\sum_{i=J}^{j-1} \tau_i\right)\right\}^{-1}. \quad (6.119)$$

The parameter s that appears as an exponent of ν_j/ν in (6.119) is defined as

$$s = \frac{\sigma_T \rho c}{du/d\ell}, \quad (6.120)$$

where σ_T is the Thomson scattering opacity and $du/d\ell = \mathbf{n}\cdot\nabla\mathbf{u}\cdot\mathbf{n}$. In terms of s the Sobolev optical depth (equation (6.93)) is $\tau(\mathbf{n}) = k_L s/\sigma_T\rho\nu_0$. The list of lines $\{\nu_j,\ j = 1, \ldots, N\}$ is arranged in descending order of frequency, and J is the index of the first line in the list with a frequency less than ν.

In order to gain a little more insight into (6.119), we can work out the simplified case introduced in their paper. We suppose that the lines all have the same strength, so their Sobolev optical depth is a constant τ, which we do *not* assume is small. We suppose the lines are equally spaced by Δ in frequency. We will assume $s\Delta/\nu \ll 1$. Equation (6.119) can be expressed as $\kappa_{\text{exp}}(\nu) = \sigma_T/(1-\epsilon_\nu)$, in which ϵ_ν is the summation in the denominator of the right-hand side. In this simplified case the sum becomes a geometrical series, and leads to

$$\epsilon_\nu = \frac{[1-\exp(-\tau)](\nu_J/\nu)^s}{1-\exp(-s\Delta/\nu-\tau)}. \tag{6.121}$$

To first order in the small quantity $s\Delta/\nu$ this result is the same as

$$\epsilon_\nu = \left[1+\frac{s\Delta}{\nu}\frac{\exp(-\tau)}{1-\exp(-\tau)}+\frac{(\nu-\nu_J)s}{\nu}\right]^{-1}. \tag{6.122}$$

This result leads to the following expression for the expansion opacity:

$$\kappa_{\text{exp}} = \sigma_T + \frac{\sigma_T\nu}{s}\frac{\exp(\tau)-1}{\Delta+(\nu-\nu_J)[\exp(\tau)-1]}. \tag{6.123}$$

In the large-τ limit this is the result given by Karp *et al.* The expansion opacity in this limit is not dependent on τ, nor on the spacing of the lines, but solely on the frequency displacement from the next-lower line in the list: $\kappa_{\text{exp}} \approx \sigma_T[1+\nu/(\nu-\nu_J)/s]$. If τ is finite but we set $\nu=\nu_J$ the results for the line contribution to κ_{exp} would be $\sigma_T\nu/(s\Delta)[\exp(\tau)-1]$.

Blinnikov (1996) gives a criticism of Karp *et al.* (1977), based on a Boltzmann equation solution. He finds that the proper mean free path calculation of the photon should average over its history in the *upwind* direction, not in the downwind direction as in Karp *et al.* This is pointed out by Pinto and Eastman (2000). But, as a matter of fact, Blinnikov's result (20) for the expansion opacity, when evaluated for the case of equally-spaced lines of uniform strength, and with $s\Delta/\nu \ll 1$ and $s \gg 1$ as before, leads to the same result for ϵ_ν as in Karp *et al.*'s model, except that $\nu-\nu_J$ is replaced by $\nu_L-\nu$; the index L is that of the closest line in the list with frequency larger than ν, just as J is the index of the closest line with frequency smaller than ν. Blinnikov's model and that of Karp *et al.* are equivalent apart from reversing the order of the line list.

The next step in the use of the monochromatic expansion opacity is to evaluate its Rosseland mean over a frequency band that is large compared with the line spacing; this is the quantity that enters diffusion calculations in the Sobolev regime. We find, following Karp *et al.*, that

$$\frac{\sigma_T}{\kappa_{R,\text{exp}}} = 1-\langle\epsilon_\nu\rangle, \tag{6.124}$$

where the brackets signify the average over the frequency band. Using (6.122) for equally-spaced constant-strength lines leads to

$$\langle\epsilon_\nu\rangle = \frac{\nu}{s\Delta}\log\left[\frac{1-\exp(-\tau)+s\Delta/\nu}{1-\exp(-\tau)+(s\Delta/\nu)\exp(-\tau)}\right]. \tag{6.125}$$

The corresponding value of the Rosseland mean depends on how large τ is:

$$\kappa_{R,\mathrm{exp}} \approx \begin{cases} 2\dfrac{\nu|du/d\ell|}{\rho c\Delta} & \tau \gg 1 \\ \dfrac{k_L}{\rho\Delta} & \dfrac{\sigma_T\rho c\Delta}{\nu|du/d\ell|} \ll \tau \ll 1 \\ \sigma_T & \tau \ll \dfrac{\sigma_T\rho c\Delta}{\nu|du/d\ell|} \end{cases}. \tag{6.126}$$

The first case here is when the lines are optically thick in the Sobolev sense. The Rosseland absorption coefficient, per unit length, becomes twice the probability of encountering a line per unit path length. This factor 2 is related to the choice of equally-spaced lines. We will comment on this factor below, in the discussion of Wehrse, Baschek, and van Waldenfels (2003). The second case applies if the lines are optically thin in the Sobolev sense, yet the smeared-out line opacity is still greater than the continuum absorption coefficient. This only applies if the continuum optical depth of the typical path length between line encounters, $\Delta c/(\nu du/d\ell)$, is small, the usual case (and assured if $s\Delta/\nu \ll 1$). In the third case the smeared-out line opacity is less than the continuum opacity, and the Rosseland mean is unaffected by lines.

6.9.2 Friend and Castor model

The stochastic approach to line transfer with a forest of lines treated in the Sobolev approximation is presented by Friend and Castor (1983). The statistical model of the line distribution is that the lines with various values of the line strength k_L have independent Poisson distributions in frequency, so that the mean number of lines in a frequency interval $[\nu, \nu+\Delta\nu]$ and in a strength interval $[k_L, k_L+\Delta k_L]$ is $\mu(k_L,\nu)\Delta k_L\Delta\nu$. If we now consider a photon of frequency ν traveling in the direction $\mathbf{n}$, the probability that it will encounter a line in this strength range in traveling a distance $d\ell$ is

$$\mu(k_L,\nu)\Delta k_L\frac{\nu d\ell}{c}\mathbf{n}\cdot\nabla\mathbf{u}\cdot\mathbf{n}. \tag{6.127}$$

The probability that there will be an interaction, given that a line is encountered, is $1-\exp[-\tau(\mathbf{n})]$, with the Sobolev optical depth given by (6.93) in terms of k_L and $\mathbf{n}\cdot\nabla\mathbf{u}\cdot\mathbf{n}$. Summing over the line strength distribution gives the total probability of encountering a line and being absorbed or scattered in $d\ell$:

$$\mu(k_L, \nu) dk_L \frac{\nu}{c} \mathbf{n} \cdot \nabla \mathbf{u} \cdot \mathbf{n} \{1 - \exp[-\tau(\mathbf{n})]\} d\ell, \tag{6.128}$$

from which the effective opacity, the same as we mean by the expansion opacity and including now the continuous opacity, is seen to be

$$\kappa_{\text{exp}}(\nu) = \frac{\nu}{\rho c} \mathbf{n} \cdot \nabla \mathbf{u} \cdot \mathbf{n} \int_0^\infty \mu(k_L, \nu) \{1 - \exp[-\tau(\mathbf{n})]\} \, dk_L + \sigma_T, \tag{6.129}$$

where, in the integrand, $\tau(\mathbf{n})$ should be substituted using (6.93). The result (6.129) is expressed in terms of the hypothetical Poisson distribution of lines of different strengths. That turns out to be very useful when the line statistics have been expressed in analytic form, and the further development in Friend and Castor (1983) makes use of this to evaluate the overlapping line effect on radiatively-driven stellar winds (cf. Section 12.5). But we can also estimate the line densities by using the actual line list: the density is approximated by summing the lines within a bandwidth $\Delta\nu$ and dividing by the bandwidth. This gives equation (9) of Friend and Castor (1983), which takes this form in the present notation:

$$\kappa_{\text{exp}}(\nu) = \frac{\nu}{\rho c \Delta\nu} \mathbf{n} \cdot \nabla \mathbf{u} \cdot s\mathbf{n} \sum_{\Delta\nu} \{1 - \exp[-\tau(\mathbf{n})]\} + \sigma_T. \tag{6.130}$$

6.9.3 *Eastman and Pinto model; Wehrse, Baschek, and von Waldenfels model*

In an ambitious study of non-LTE spectrum modeling of supernovae, Eastman and Pinto (1993) employ expansion opacity to treat, in an approximate way, the effect of the forest of weak lines for which a detailed transfer solution is not feasible. They independently derive the expansion opacity, formula (4.2) in their paper, which is identical with (6.130). The more general result, in which the Sobolev approximation has not been made, is their equation (23), in which the left-hand side should read κ_{R}^{-1}. This equation agrees with Blinnikov's equation (14).

Wehrse *et al.* (2003) discusses methodically the calculation of the expansion opacity and the general issue of the diffusion approximation for a flow with velocity gradients. They consider both a deterministic distribution of lines, and a stochastic distribution of lines, as well as both infinitely narrow lines (Sobolev limit) and lines with finite widths. Their results for the monochromatic opacity with infinitely narrow lines reproduce Blinnikov's expressions, and with finite line widths the Wehrse *et al.* expressions are essentially the same as Blinnikov's; both imply a local harmonic mean opacity

$$\kappa_{\text{eff}}(\nu)^{-1} = \frac{s}{\sigma_T}\left\langle \int_0^\infty \exp\left[-\frac{s}{\sigma_T}\int_{\xi-\eta}^{\xi} \kappa(\zeta)\, d\zeta\right] d\eta \right\rangle, \tag{6.131}$$

in which $\xi = -\log \nu$, the variable η is defined in terms of the path length ℓ by $\eta = \sigma_T \rho \ell / s$, and the angle brackets signify an average over a moderate bandwidth centered at ν. In the inner integral $\kappa(\zeta)$ is the monochromatic opacity at $\nu = \exp(-\zeta)$ calculated without the fluid velocity but including all the other line broadening mechanisms.

For the stochastic, infinitely narrow line case, in which the lines form a Poisson point process as described above, Wehrse *et al.* give a result for the effective opacity that is equivalent to (6.129) or (6.130), and thus also in agreement with the Eastman and Pinto (1993) formula and Pinto and Eastman (2000), equation (9). Wehrse *et al.* give results using the stochastic model in the particular case that the line strengths follow a power-law model $\mu(k_L, \nu) \propto k_l^{-\alpha}$, just as in the Castor, Abbott, and Klein (1975) stellar wind model (cf. Section 12.5), and as used in Friend and Castor (1983). The exponent α in Wehrse *et al.* is $1 - \alpha$ as used by Castor *et al.* Wehrse *et al.*, use high and low cutoffs in k_L in evaluating the effective opacity; this is unnecessary if $0 < \alpha < 1$.

Some clarification is needed on the differences between the monochromatic Karp *et al.* opacity and the stochastic model opacity. The former is a strong function of frequency in the spaces between lines. The latter, which is a constant in a given medium-scale frequency band,[2] actually corresponds to the *expectation value* of the mean free path with respect to realizations of the Poisson process that produces the line spectrum. Friend and Castor (1983) argue that the expectation of the monochromatic intensity can be calculated from the opacity as defined by (6.129). The calculation by Wehrse *et al.* of the diffusion flux for infinitely narrow Poisson lines gives a result consistent with (6.129), in confirmation of this argument. Thus the local Rosseland mean expansion opacity, in the infinitely narrow stochastic line case, is the same quantity given by (6.129) or (6.130), Eastman and Pinto (1993) and Pinto and Eastman (2000), equation (9).

We now return to the factor 2 difference between the Rosseland average of the expansion opacity for equally-spaced strong lines, and the Poisson average opacity for strong lines with the same mean density in frequency. In fact, this is the difference between the Poisson statistics and equally-spaced lines. A property of the Poisson process is that if a frequency is selected at random, and the frequency displacement is found to the next higher frequency in a realization of the Poisson

[2] As in (6.130); a band large enough to include many lines, but small in comparison with the frequency itself or the interval over which the line statistics change substantially.

distribution, then the mean value of that interval is just equal to the reciprocal of the line density. But if this same frequency is compared with the list of equally-spaced lines, then the mean displacement is one half of the line spacing in the list. This is the factor of 2. These results are verified by a simple Monte Carlo calculation, which also shows that (6.119) does lead to $\langle 1 - \epsilon_\nu \rangle = s\Delta/\nu$ for strong lines with a mean spacing in frequency of Δ, twice the result for equally spaced lines.

In summary, the stochastic model is a useful tool for simulations of spectra when there is a dense forest of lines and the Sobolev approximation is valid, such as in novae and supernovae, stellar winds, and high-velocity laser-plasma experiments.

6.9.4 Expansion opacity example: the iron spectrum

The ideas of expansion opacity are best illustrated using an idealized but realistic example. We have chosen the spectrum of the ions Fe II, Fe III, and Fe IV as calculated in LTE at a temperature $T = 3\,\mathrm{eV}$ and electron density $N_e = 10^{16}\,\mathrm{cm}^{-3}$. Iron is assumed to have the number abundance 4×10^{-5} relative to hydrogen. For simplicity the three ions are assumed to have equal abundance. The data of Kurucz and Bell (1995) are used for the line frequencies, energy levels, oscillator strengths, and the radiative and Stark damping constants.[3] Each of the lines has a Voigt profile based on the thermal Doppler width and the damping constant. The total number of iron lines treated is about 37 000, and of these 836 lie within the range of Figure 6.2(a), which shows a portion of the synthesized opacity spectrum between 213 and 222 nm. In order to produce the expansion opacities a larger region, from 182 to 222 nm, had to be synthesized first, to provide the data for (6.124), (6.130) and (6.131). We calculate the harmonic mean opacity over the band illustrated in Figure 6.2(a), accounting for the velocity gradient, in three ways: (i) using the Karp *et al.*/Blinnikov formula (6.119) with (6.124); (ii) using the Friend–Castor/Eastman–Pinto formula (6.130); and (iii) using the Blinnikov/Wehrse *et al.* formula (6.131) for lines with finite widths. These results for k_{eff} in this band are shown as functions of s in Figure 6.2(b).

It is apparent that the calculations of k_{eff} that omit the intrinsic line widths, i.e., in the Sobolev approximation, considerably underestimate the effective opacity unless the velocity gradient is quite large, i.e., s is less than some amount. In this case that value of s is about 10^3. This is the s for which about half the lines in the frequency band in question have Sobolev optical depths less than unity. The two Sobolev calculations shown in Figure 6.2(b) agree well with each other. Since one is a deterministic formula that makes no assumption about the line statistics,

[3] We are grateful to R. L. Kurucz and B. Bell and the Smithsonian Astrophysical Observatory for making these data available.

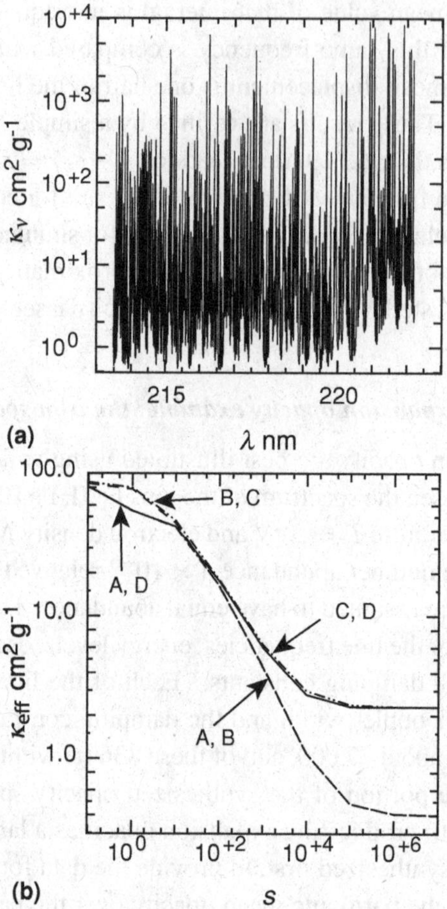

Fig. 6.2 (a) Illustration of the opacity in a synthetic spectrum of Fe II, Fe III, and Fe IV vs wavelength. (b) Mean expansion opacity over the band in (a) using: A (6.119) (dash-dotted line); B (6.130) (dashed line); C (6.131) (solid line); and D (6.130) with the replacement of σ_T with the static Rosseland mean (dash-dot-dot line). The four curves overlie each other in pairs, but differently on the left- and right-hand sides, as indicated in the figure by the labels A, B, C, and D.

and the other assumes that the lines have a Poisson distribution, the agreement indicates that the stochastic approach gives a quite usable answer.

The large-s and small-s limits of k_{eff} have natural interpretations. The large-s limit is the harmonic mean or Rosseland mean for this band of the $du/d\ell = 0$ spectral opacity, or at least it should be if the approximation were accurate. The small-s limit is the arithmetic or Planck mean for this band. As $s \to 0$, the region over which the average is taken becomes larger and larger for (6.124) and (6.131), which accounts for differences that are seen in the different models at small s, owing also to the fact that the line density varies somewhat with frequency. The values of the harmonic and arithmetic means are 2.23 and 86 $\mathrm{cm^2\,g^{-1}}$ for this band.

An *ad hoc* modification of the Sobolev calculations that makes them substantially more usable is to replace the added quantity σ_T in (6.130) with the actual harmonic mean opacity calculated without the velocity gradient. The Sobolev result for the line opacity alone tends to zero as the velocity gradient becomes small, while we know that with the intrinsic line widths taken into account the harmonic mean can be significantly larger than σ_T owing to line blanketing, i.e., bandwidth constriction. The fourth curve plotted in Figure 6.2(b) shows the result using (6.130) with σ_T replaced by $\kappa_{\mathrm{eff}}(du/d\ell = 0)$. The agreement with (6.131) is good over the entire range.

7

Hydrodynamics with radiation: waves and stability

The goal of this chapter is to explore the effect on the hydrodynamic equations of the terms representing exchange of energy and momentum between matter and radiation. The present discussion will be rather general; most of the specific examples will be taken up after discussing velocity effects on transport and numerical methods. Background information on the properties of thermal equilibrium radiation is presented by, for example, Cox and Giuli (1968).

7.1 Imprisoned equilibrium radiation

We note one suspicious thing about the overall energy equation (4.47), which is that the flux term includes an enthalpy flux for matter, but there is no enthalpy flux for the radiation. Since our claim is that there is no intrinsic difference between material particles and photons, why is there this apparent difference that would persist even when the opacity is so great that the flux vanishes? The explanation was found in Chapter 6. In brief, it is that the flux depends on the reference frame. The same is true for the energy density and the radiation pressure, but for these variables the corrections are small. We have seen that

$$\mathbf{F} = \mathbf{F}_0 + \mathbf{u}E_0 + \mathbf{u} \cdot \mathsf{P}_0 \tag{7.1}$$

to first order in $\mathbf{u}/c$, where E_0, $\mathbf{F}_0$, and P_0 are evaluated in the comoving frame. In the limit of vanishing mean free path it is $\mathbf{F}_0$ that goes to zero, not $\mathbf{F}$. What $\mathbf{F}$ tends to in that limit is the convective radiation enthalpy flux. If we make use of this in (4.47), we see that the symmetry between matter and radiation is restored.

This is the stellar interiors model of radiation. The intensity equals the Planck function at the local temperature, which implies that the energy density and radiation pressure are given by

$$E = aT^4 \tag{7.2}$$

and

$$P = \frac{1}{3}aT^4. \tag{7.3}$$

The radiation enthalpy whose flux is added to the material enthalpy flux in the total energy equation is $(4/3)aT^4$. The energy and enthalpy here are per unit volume; the specific internal energy and enthalpy contributions are obtained by dividing by ρ. The specific radiation entropy s_r comes from the first law of thermodynamics:

$$T\,ds_r = d\left(\frac{aT^4}{\rho}\right) + \frac{aT^4}{3}d\left(\frac{1}{\rho}\right) = 4a\frac{T^3dT}{\rho} - \frac{4}{3}aT^4\frac{d\rho}{\rho^2}, \tag{7.4}$$

so

$$s_r = \frac{4}{3}\frac{aT^3}{\rho}. \tag{7.5}$$

All these radiation contributions can just be added to their material counterparts. The radiation makes a difference to the equation of state. The γ for radiation is $4/3$, less than that for the non-relativistic monatomic ideal gas, which is $5/3$. Thus radiation softens the equation of state. In massive stars Eddington showed years ago that a large part of the weight of the stellar material is supported by radiation pressure, and the effective spring constant of the star for radial oscillations, which is proportional to $\gamma - 4/3$, becomes smaller and smaller as the star's mass increases. Above some mass the stellar oscillations are so easily excited by the modulation of nuclear energy generation in the interior that the star becomes unstable to pulsations and is either disrupted or becomes an unusual kind of object instead of a quiescent main sequence star. Radiation pressure is also a several percent effect in the pulsating cepheid variables. The reduction of $\gamma - 4/3$ due to radiation appreciably lengthens their pulsation periods, which are very well studied and one of the bases of the extragalactic distance scale.

7.2 Nonadiabatic waves

We will consider some examples of waves that couple to a radiation field that is neither optically thick nor optically thin. The first example will be the cooling mode that was discussed earlier using a Newton's cooling model. For now we will omit the fluid motion part of that problem and just consider the zero sound speed limit. This discussion draws on Mihalas and Mihalas (1984), Sections 97, 101 and Castor (1972). What we are interested in is the thermal response, i.e., what is the decay rate of temperature fluctuations of different wavelengths. Earlier on we found the decay rate $1/\tau$ for the cooling mode, where $1/\tau$ was the coefficient in the cooling law (equation (2.52)). We write the energy coupling term $-g^0$ as

$$-g^0 = k_P(cE - acT^4), \tag{7.6}$$

where the emissivity has been expressed in terms of the thermal equilibrium energy density, and the frequency integration has been performed using an appropriate average value (Planck mean) of the absorptivity, k_P. If E is held fixed while T is perturbed, then the perturbation of $-g^0$ is $-k_P 4acT^3\Delta T$. Dividing the coefficient of ΔT, namely $k_P 4acT^3$, by ρC_v, where C_v is the specific heat at constant volume, gives the damping constant $1/\tau$. Thus we identify τ with $\rho C_v/(k_P 4acT^3)$.

In the general case in which E responds to the temperature fluctuations the linearized coupling term is

$$-\delta g^0 = k_P(c\delta E - 4acT^3\delta T). \tag{7.7}$$

We use the frequency integrated form of the combined moment equation (4.42), and take for the unperturbed state an infinite homogeneous medium in which $E = aT^4$. The right-hand side of the integrated moment equation is just g^0, thus

$$\frac{\partial E}{\partial t} - \nabla \cdot \left(\frac{c}{3k_R}\nabla E\right) = g^0. \tag{7.8}$$

The frequency integral of the flux divergence term is approximated here using the Rosseland mean (q.v. Section 6.7) k_R of k_ν. We drop the time derivative term, which is to say we neglect the photon time of flight. When the equation is linearized and when spatial dependence $\exp(i\mathbf{k}\cdot\mathbf{r})$ is assumed it gives

$$\frac{k^2 c}{3k_R}\delta E = -k_P(c\delta E - 4acT^3\delta T). \tag{7.9}$$

The response of the radiation field to the temperature fluctuations is therefore given by

$$\delta E = \frac{4aT^3\delta T}{1 + k^2/(3k_R k_P)}. \tag{7.10}$$

This says that if the wavelength is very long, the energy density tracks the temperature perfectly, but that if the wavelength is short the energy density hardly varies. The roll-over occurs where the wavenumber is comparable to the geometric mean of the two different mean absorption coefficients, in other words, where the wavelength is about one mean free path. The value of $-\delta g^0$ that results when δE is substituted into (7.9) is

$$-\delta g^0 = -k_P 4acT^3\delta T \frac{k^2/(3k_R k_P)}{1 + k^2/(3k_R k_P)}. \tag{7.11}$$

Thus the cooling time is modified to

$$\tau \to \tau\left(1+\frac{3k_R k_P}{k^2}\right) = \frac{\rho C_v}{4acT^3}\left(\frac{1}{k_P}+\frac{3k_R}{k^2}\right) \tag{7.12}$$

by the response of the radiation field to the temperature fluctuations.

It is interesting to consider the radiative cooling time in different parts of a star in response to a fixed wavelength. In the tenuous upper atmosphere the Planck mean absorption coefficient will become very small, and the cooling time will be long; the coupling between the matter and radiation is simply weak. In the deep interior the Rosseland mean absorption coefficient becomes very large and the cooling time again becomes long, but this time because the leakage of radiation through the opaque material is sluggish. The minimum cooling time is attained if the wavenumber is comparable with the geometric mean, as just mentioned. The minimum value of it is approximately

$$\tau_{\min} = \frac{2\rho C_v \lambda}{4acT^3}, \tag{7.13}$$

where λ is the spatial wavelength. This is comparable to the time it would take the radiation flux σT^4 to radiate the internal energy content of a wavelength-thick slab of material since $\sigma = ac/4$.

7.3 Atmospheric oscillations with radiation pressure

An extension of the wave propagation ideas of the previous section is the problem of the oscillations of a hydrostatic isothermal atmosphere, for which the equilibrium condition is an exponential stratification of the density, $\rho \propto \exp(-x/H)$. This problem is treated, with respect to the vertical oscillations, by Lamb (1945). It becomes more interesting when the horizontal motions are included, and more interesting still when the effect of the radiation pressure is included as a momentum coupling term in the optically thin approximation. This idea for atmospheric instability in luminous stars was advanced by Hearn (1972, 1973) under the name "radiation-driven sound waves." It provides an opportunity to look critically at what is meant by instability of an unbounded system, such as a stellar atmosphere.

Consider first the case without the radiation pressure term. The discussion of sound waves in Section 2.7 needs to be modified to account for the stratification. In this discussion the notation will also be modified: the variable ρ' will denote the perturbation of ρ divided by ρ, i.e., the logarithmic perturbation; likewise for p'.

The continuity equation (2.2) when linearized now becomes

$$\frac{\partial \rho'}{\partial t} + \nabla \cdot \mathbf{u} + \mathbf{u} \cdot \nabla \ln(\rho) = 0$$

or

$$\frac{\partial \rho'}{\partial t} + \nabla \cdot \mathbf{u} + \frac{\gamma}{a^2} \mathbf{g} \cdot \mathbf{u} = 0. \tag{7.14}$$

Here $\mathbf{g}$ is the downward-directed gravity vector, a is the adiabatic speed of sound, and γ is the ratio of specific heats, so $a/\sqrt{\gamma}$ is the isothermal speed of sound, and $a^2/(\gamma g)$ is H, the scale height of the static atmosphere. The perturbed momentum equation (2.3) becomes

$$\frac{\partial \mathbf{u}}{\partial t} - \rho' \mathbf{g} + p' \mathbf{g} + \frac{a^2}{\gamma} \nabla p' = 0, \tag{7.15}$$

and the linearization of the internal energy equation, including Newton's cooling as in Section 2.7 with a time constant τ, is

$$\frac{\partial p'}{\partial t} + \frac{\gamma}{a^2} \mathbf{g} \cdot \mathbf{u} - \gamma \frac{\partial \rho'}{\partial t} - \frac{\gamma^2}{a^2} \mathbf{g} \cdot \mathbf{u} = -\frac{p' - \rho'}{\tau}. \tag{7.16}$$

The system of equations (7.14)–(7.16) for the unknowns ρ', $\mathbf{u}$, and p' has constant coefficients, and therefore it is relatively simple to eliminate two of the unknowns to obtain a PDE for the remaining one, say ρ'. This is

$$\left\{ \left[\partial_t \left(\partial_t + \frac{1}{\gamma\tau} \right) + \frac{(\gamma - 1)a^2}{\gamma^2 H^2} \right] (\partial_x^2 + \partial_y^2) + \partial_t \left(\partial_t + \frac{1}{\gamma\tau} \right) \partial_z^2 - \frac{1}{H} \partial_t \left(\partial_t + \frac{1}{\gamma\tau} \right) \partial_z - \frac{1}{a^2} \partial_t^3 \left(\partial_t + \frac{1}{\tau} \right) \right\} \rho' = 0. \tag{7.17}$$

The approach to solving this equation is to first take the Laplace transform with respect to time, which gives an equation of this form:

$$(\nabla^{\mathrm{T}} \mathbf{A} \nabla + 2\mathbf{b}^{\mathrm{T}} \nabla + c)\tilde{\rho}' = S, \tag{7.18}$$

in which S is a certain cubic polynomial in the transform variable s that contains the initial conditions for ρ', $\mathbf{u}$, and p' – cubic because (7.17) is fourth order in time

– and the coefficients A, **b**, and c are defined by

$$\mathsf{A} = \begin{pmatrix} s\left(s+\dfrac{1}{\gamma\tau}\right)+\dfrac{(\gamma-1)a^2}{\gamma^2H^2} & 0 & 0 \\ 0 & s\left(s+\dfrac{1}{\gamma\tau}\right)+\dfrac{(\gamma-1)a^2}{\gamma^2H^2} & 0 \\ 0 & 0 & s\left(s+\dfrac{1}{\gamma\tau}\right) \end{pmatrix}, \tag{7.19}$$

$$\mathbf{b} = \begin{pmatrix} 0 \\ 0 \\ -\dfrac{1}{2H}s\left(s+\dfrac{1}{\gamma\tau}\right) \end{pmatrix}, \tag{7.20}$$

and

$$c = -\frac{s^3(s+1/\tau)}{a^2}. \tag{7.21}$$

The variable $\tilde{\rho}'$ is the transform of ρ', the operator $\boldsymbol{\nabla}$ represents the column vector $(\partial_x, \partial_y, \partial_z)^{\mathrm{T}}$, and ^{T} denotes the transpose. The same equation would be found for $\mathbf{u}$ and p'; only the initial condition function S would differ.

Some general conditions that can be imposed on the coefficients in (7.18) are: (1) A is symmetric; (2) $\mathsf{A} \sim s^n\mathsf{B}$ for $s \to \infty$ for some n, where B is positive definite; (3) $c = O(s^2||\mathsf{A}||)$ for $s \to \infty$; (4) $||\mathbf{b}||^2/(||\mathsf{A}||c) = o(s)$ for $s \to \infty$. These conditions are sufficient to ensure that the original system of PDEs is hyperbolic, and are met in this example.

If a solution for $\tilde{\rho}'$ is in hand, then the solution of (7.17) for ρ' is given by the Laplace inversion formula

$$\rho'(x, y, z, t) = \frac{1}{2\pi i}\int_{d-i\infty}^{d+i\infty} e^{st}\,\tilde{\rho}'\,ds, \tag{7.22}$$

where d is a positive real constant sufficiently large to ensure that the contour passes to the right of all the singularities of the integrand in the complex s plane. For large, real, positive s, A is positive definite and can be factored:

$$\mathsf{A} = \mathsf{M}^{\mathrm{T}}\mathsf{M}, \tag{7.23}$$

where M is a nonsingular real matrix. The coordinate vector $\mathbf{r} = (x, y, z)^{\mathrm{T}}$ can be transformed using M to a new set of coordinates ξ defined by $\mathbf{r} = \mathsf{M}^{\mathrm{T}}\xi$. In this way (7.18) may be turned into Helmholtz's equation in ξ-space, for which the Green's function is known. The steps of the transformation will be omitted; the result for the solution $\tilde{\rho}'$ is found to be

$$\tilde{\rho}' = \iiint_V G(\mathbf{r} - \mathbf{r}')S(\mathbf{r}')\,dV, \tag{7.24}$$

where the integration volume includes that part of 3-D space where the initial conditions are nonvanishing, and the Green's function is

$$G(\mathbf{r}) = -\frac{1}{4\pi}\frac{\exp\left\{-\mathbf{b}^{\mathrm{T}}\mathsf{A}^{-1}\mathbf{r} - \left[(\mathbf{b}^{\mathrm{T}}\mathsf{A}^{-1}\mathbf{b} - c)\mathbf{r}^{\mathrm{T}}\mathsf{A}^{-1}\mathbf{r}\right]^{1/2}\right\}}{\left[\det(\mathsf{A})\mathbf{r}^{\mathrm{T}}\mathsf{A}^{-1}\mathbf{r}\right]^{1/2}}. \tag{7.25}$$

We now argue as follows about the behavior of the solution ρ' obtained from applying (7.22) and (7.25). The Green's function given by (7.25) is analytically continued from large, real, positive s in the negative real direction. At some value of $\Re(s)$ singularities will be encountered. The constant d in the Laplace inversion can be set to any number larger than the largest $\Re(s)$ of any of the singularities. If d turns out to be negative, then ρ' tends to zero for $t \to \infty$ if the initial conditions vanish outside a compact volume. If d is positive, then ρ' will become exponentially large for $t \to \infty$. If the singularities of G lie on the real axis a more careful analysis is required to decide whether ρ' grows or not.

This notion of stability is called *absolute stability.* A dynamical problem is absolutely stable if the response to an initial disturbance at a point in space eventually dies away at any other *fixed* point in space. It may still be true that no matter how large t may be, there is a point in space where the response at this time is large, and that as t increases the maximum response becomes larger and larger. This is consistent with absolute stability if the location of the maximum response moves further and further from the site of the initial disturbance. *Convective instability* is the term applied to this situation.[1] A dynamical problem is convectively stable or unstable depending on whether the maximum response over all space to an initial disturbance decays or grows in time. Convective stability guarantees absolute stability, but the reverse is not true. A third kind of stability, called global stability, pertains to dynamical systems in a bounded domain, with specific boundary conditions. Imposing the boundary conditions turns (7.18) into an eigenvalue problem. The signs of the real parts of all the discrete eigenvalues determine stability and instability in this case.

In other words, *the possible singularities of G determine whether the problem is absolutely stable or unstable.* Convective stability can be diagnosed by

[1] The term "convective instability" should not be confused with the instability associated with thermal convection. In fact, the latter could perhaps be called an absolute instability of the gravity wave modes.

finding solutions of the dispersion relation for all real propagation vectors, $\tilde{\rho}' \sim \exp(i\mathbf{k}\cdot\mathbf{r})$. It may be possible to relate global stability to absolute stability.

By inspection of (7.25) we can derive three conditions when G would be singular:

$$\text{(a)} \qquad \det(\mathsf{A}) = 0; \tag{7.26}$$

$$\text{(b)} \qquad \mathbf{b}^{\mathrm{T}}\mathsf{A}^{-1}\mathbf{b} - c = 0; \tag{7.27}$$

$$\text{(c)} \qquad \mathbf{r}^{\mathrm{T}}\mathsf{A}^{-1}\mathbf{r} = 0. \tag{7.28}$$

We can establish absolute stability or instability by examining the real parts of all the values of s that obey any of (7.26)–(7.28).

We return to our specific problem, with the definitions (7.19)–(7.21). Condition (a) above gives

$$s\left(s + \frac{1}{\gamma\tau}\right)\left[s\left(s + \frac{1}{\gamma\tau}\right) + \frac{(\gamma - 1)a^2}{\gamma^2 H^2}\right]^2 = 0, \tag{7.29}$$

condition (b) gives

$$s^3 + \frac{1}{\tau}s + \frac{a^2}{4H^2}s + \frac{a^2}{4\gamma H^2 \tau} = 0 \tag{7.30}$$

and condition (c)

$$\frac{x^2 + y^2}{s[s + 1/(\gamma\tau)] + (\gamma - 1)a^2/(\gamma^2 H^2)} + \frac{z^2}{s[s + 1/(\gamma\tau)]} = 0. \tag{7.31}$$

Condition (c) determines roots s that depend on the vector $\mathbf{r} = (x, y, z)$. This is interpreted as follows. If one of these roots has a positive real part, then $G(\mathbf{r})$ will grow exponentially for that $\mathbf{r}$. Absolute stability requires G to decay at *all* $\mathbf{r}$, and therefore one of the conditions is that *all* the roots of condition (c) should have a nonpositive real part, whatever the values of x, y, and z.

The roots of (7.29) are $s = 0$, $s = -1/(\gamma\tau)$ and the roots of the quadratic $s[s + 1/(\gamma\tau)] + (\gamma - 1)a^2/(\gamma^2 H^2) = 0$. Since there are no sign changes in the coefficients of the latter, those roots have negative real parts. Thus all the roots for condition (a) indicate stability except for the marginally stable trivial root $s = 0$.

The cubic equation (7.30) also has only positive coefficients, and therefore its roots have negative real parts, and condition (b) also indicates absolute stability. It is interesting to examine the roots of the cubic in the realistic limit that τ is small compared with H/a. (In the atmosphere of a typical hot star the dynamic time H/a is of order 10^3 s while the cooling time is of order 10 s.) A helpful tool for doing this is to sketch the overall power of τ that each term in the equation represents versus the exponent n in a hypothetical relation $s = \tau^n$, as in Figure 7.1. A possible

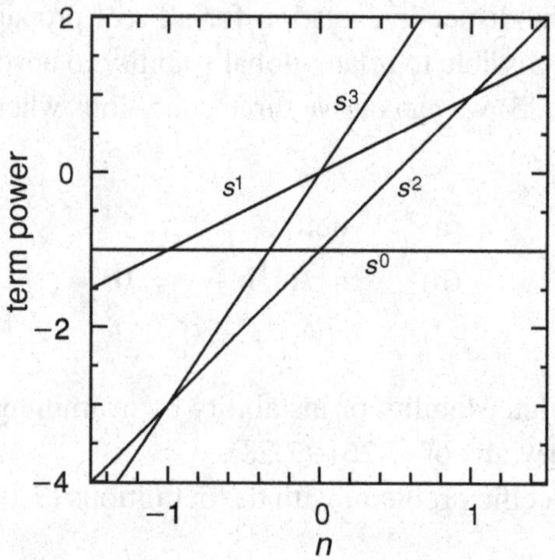

Fig. 7.1 Overall power of τ of each term in (7.30) if $s \propto \tau^n$ vs n.

value of n must be where two or more terms in the cubic have the same overall power of τ, since otherwise the terms cannot balance each other. Furthermore, since τ is a *small* parameter, the overall power these terms have must be smaller than the powers of the remaining terms, otherwise the terms that match will not be the dominant ones. Figure 7.1 shows us that the crossing points at $n = 0$, where the quadratic term balances the constant term, and at $n = -1$, where the cubic term balances the quadratic term, must describe the roots. The first crossing point gives two roots and the second one gives a single root, thus accounting for the three roots. When the dominant part of a given root has been determined in this way, the remaining terms can be evaluated to provide a first order correction to the dominant one, if desired. The roots are thus found to be

$$s \approx \pm \frac{ia}{2\sqrt{\gamma}H} - \frac{(\gamma - 1)a^2\tau}{8\gamma H^2}, \qquad s \approx -\frac{1}{\tau}. \tag{7.32}$$

These roots can be physically identified with endpoints of branches of the wave spectrum. The complex conjugate pair are the endpoints of the acoustic branches, and the real root is an endpoint of the thermal mode branch. Since $\tau \ll a/H$, the damping of the thermal mode is very large, while the damping of the acoustic mode is slight.

Condition (c) in (7.31) can be reexpressed as this quadratic equation

$$s\left(s + \frac{1}{\gamma\tau}\right) + \frac{(\gamma - 1)a^2\cos^2\theta}{\gamma^2 H^2} = 0, \tag{7.33}$$

where θ is the angle between $\mathbf{r}$ and the vertical. The roots all have a negative real part, except that $s = 0$ is a root if $\theta = \pi/2$. As θ ranges from 0 to

$\pi/2$ these roots fill in the space between the roots derived from condition (a). They are physically associated with the gravity mode waves, but, like the acoustic mode, they are mixed with the thermal mode. The roots are complex only if $\tau > H/(2a\sqrt{\gamma-1})|\sec\theta|$. If the cooling is too efficient the buoyancy force that provides the "spring" for gravity waves is suppressed and the waves do not oscillate. In the adiabatic limit, $\tau \gg a/H$, the $\theta = 0$ endpoints of the gravity branches are at $s = \pm i\sqrt{\gamma-1}a/(\gamma H)$, of which the imaginary part is the Brunt–Väisälä frequency. These are similar to, but just slightly smaller in magnitude than, the endpoints $\pm ia/(2H)$ of the acoustic branches in the adiabatic limit.

In summary, the oscillations of the exponential atmosphere are absolutely stable. The atmosphere is certainly not convectively stable, since the pulse produced by the initial disturbance increases in amplitude as $\exp[z/(2H)]$ as it rises through the atmosphere at the speed a.

We turn now to the case that the body force on the matter due to absorption or scattering of radiation is included in the material momentum balance. We saw earlier that this force should be the negative of $\mathbf{g}$ given by (4.44). When expressed per unit mass of material it is

$$\mathbf{g}_{\mathrm{R}} = \frac{1}{\rho c}\int d\nu \int_{4\pi} d\Omega\, \mathbf{n}(k_\nu I_\nu - j_\nu). \tag{7.34}$$

If the emissivity j_ν is isotropic that term in $\mathbf{g}_{\mathrm{R}}$ vanishes, and if k_ν is isotropic it can be taken out of the integral over angle, which becomes the total monochromatic flux, $\mathbf{F}_\nu$. For a stellar atmosphere with slab symmetry, the flux is a vector in the $+z$ direction. For the present purpose we assume that the frequency-dependent absorption coefficient can be replaced by the flux-weighted mean, $\kappa_{\mathrm{F}}\rho$. Thus we will use

$$\mathbf{g}_{\mathrm{R}} = \frac{\kappa_{\mathrm{F}} F}{c}\mathbf{e}_z, \tag{7.35}$$

where κ_{F} is the flux-mean opacity and F is the total radiative flux, which we will assume to be constant. The ratio of $g_{\mathrm{R}} = |\mathbf{g}_{\mathrm{R}}|$ to the normal gravity g will be denoted by a new variable Γ:[2]

$$\Gamma = \frac{\kappa_{\mathrm{F}} F}{gc}, \tag{7.36}$$

and therefore the momentum equation can be written

$$\frac{\partial \mathbf{u}}{\partial t} + \mathbf{u}\cdot\nabla\mathbf{u} + \frac{1}{\rho}\nabla p = (1-\Gamma)\mathbf{g}. \tag{7.37}$$

[2] The Eddington luminosity for a star is that value for which the radiation force balances gravity; it is $L_{\mathrm{Edd}} \equiv 4\pi G\mathcal{M}c/\kappa_{\mathrm{F}}$, and so $\Gamma = L/L_{\mathrm{Edd}}$. Normal stars approach but do not exceed $L = L_{\mathrm{Edd}}$.

The flux-mean opacity consists of a constant Thomson scattering part, denoted by σ_e, and an absorption part due to processes such as bound–free and free–free absorption, as well as line absorption. The absorption term varies with temperature and density, so it will be approximated by $\kappa_1 \rho^n T^{-q}$, where n and q are constant exponents, and therefore Γ can be written

$$\Gamma = \frac{F}{gc}\left(\sigma_e + \kappa_1 \rho^n T^{-q}\right). \tag{7.38}$$

The variation of the opacity with height, due to its ρ dependence, should be included, but that will be neglected here. However, the effect of density perturbations on the opacity *will* be included. The linearized form of the momentum coupling term $(1 - \Gamma)\mathbf{g}$ becomes

$$-\Gamma'\mathbf{g} = -\frac{\kappa_1 F}{gc}\rho^n T^{-q}[(n+q)\rho' - qp']\mathbf{g} = -\Gamma_{\mathrm{e}}[(n+q)\rho' - gp']\mathbf{g}_{\mathrm{eff}}, \tag{7.39}$$

where Γ_{e} is defined by

$$\Gamma_{\mathrm{e}} = \frac{1}{1-\Gamma}\frac{\kappa_1 F}{gc}\rho^n T^{-q}, \tag{7.40}$$

and $\mathbf{g}_{\mathrm{eff}} = (1 - \Gamma)\mathbf{g}$, in terms of Γ in the unperturbed atmosphere. Γ_{e} will be treated as a constant in the problem, in a Boussinesq-like approximation. The perturbed momentum equation, replacing (7.15), is now

$$\frac{\partial \mathbf{u}}{\partial t} - \rho'\mathbf{g}_{\mathrm{eff}} + p'\mathbf{g}_{\mathrm{eff}} + \frac{a^2}{\gamma}\nabla p' = -\Gamma_{\mathrm{e}}[(n+q)\rho' - qp']\mathbf{g}_{\mathrm{eff}}. \tag{7.41}$$

The occurrences of $\mathbf{g}$ in (7.14) and (7.16) will also be replaced by $\mathbf{g}_{\mathrm{eff}}$, since the scale height of the static atmosphere is now $H = a^2/(\gamma g_{\mathrm{eff}})$.

With these changes the equation satisfied by $\tilde{\rho}'$ remains of the form (7.18), but now A, $\mathbf{b}$, and c are given by

$$\mathsf{A} = \begin{pmatrix} s\left(s + \frac{1}{\gamma\tau}\right) + \frac{(\gamma-1)a^2}{\gamma^2 H^2}[1 - (n+q)\Gamma_{\mathrm{e}}] & 0 & 0 \\ 0 & s\left(s + \frac{1}{\gamma\tau}\right) + \frac{(\gamma-1)a^2}{\gamma^2 H^2}[1 - (n+q)\Gamma_{\mathrm{e}}] & 0 \\ 0 & 0 & s\left(s + \frac{1}{\gamma\tau}\right) \end{pmatrix}, \tag{7.42}$$

$$\mathbf{b} = \begin{pmatrix} 0 \\ 0 \\ -\frac{1}{2H}s\left\{s\left[1+\frac{\Gamma_e}{\gamma}[n-(\gamma-1)q]\right]+\frac{1+n\Gamma_e}{\gamma\tau}\right\} \end{pmatrix}, \tag{7.43}$$

and

$$c = -\frac{s}{a^2}\left(s+\frac{1}{\tau}\right)\left(s^2-\frac{n\Gamma_e a^2}{\gamma H^2}\right). \tag{7.44}$$

Condition (a) now leads to

$$s\left(s+\frac{1}{\gamma\tau}\right)\left\{s\left(s+\frac{1}{\gamma\tau}\right)+\frac{(\gamma-1)a^2}{\gamma^2 H^2}[1-(n+q)\Gamma_e]\right\}^2 = 0. \tag{7.45}$$

Condition (b) becomes

$$\frac{s}{4H^2[s+1/(\gamma\tau)]}\left(s\left\{1+\frac{\Gamma_e}{\gamma}[n-(\gamma-1)q]\right\}+\frac{1+n\Gamma_e}{\gamma\tau}\right)^2 + \frac{s}{a^2}\left(s+\frac{1}{\tau}\right)\left(s^2-\frac{n\Gamma_e a^2}{\gamma H^2}\right) = 0. \tag{7.46}$$

Condition (c) becomes

$$s\left(s+\frac{1}{\gamma\tau}\right)+\frac{(\gamma-1)a^2\cos^2\theta}{\gamma^2 H^2}[1-(n+q)\Gamma_e] = 0, \tag{7.47}$$

where θ, as before, is the angle between $\mathbf{r}$ and the z axis.

The change in conditions (a) and (c) is that the Brunt–Väisälä frequency is modified, and becomes

$$\omega_{BV}^2 = \frac{(\gamma-1)a^2}{\gamma^2 H^2}[1-(n+q)\Gamma_e]. \tag{7.48}$$

This has the important implication that if Γ_e exceeds the critical value $\Gamma_e = 1/(n+q)$, then ω_{BV} becomes imaginary. In this case one of the endpoints of a gravity mode branch will produce absolute instability. It is entirely possible for this condition to be met in the atmospheres of the most luminous stars, in which Γ approaches unity.

Condition (b) is more complicated with the addition of the radiation force term. Equation (7.46) can be rearranged as a quartic equation

$$s^4 + a_1 s^3 + a_2 s^2 + a_3 s + a_4 = 0 \tag{7.49}$$

after discarding the factor s, with the coefficients

$$a_1 = \frac{\gamma+1}{\gamma\tau},$$
$$a_2 = \frac{a^2}{H^2}\left\{\frac{1}{\gamma}\left(\frac{H}{a\tau}\right)^2 + \frac{1}{4} - \frac{n+(\gamma-1)q}{2\gamma}\Gamma_e + \frac{[n-(\gamma-1)q]^2}{4\gamma^2}\Gamma_e^2\right\},$$
$$a_3 = \frac{a^2}{H^2\tau^2}\frac{1}{2\gamma^2}\left\{\gamma - [(\gamma+1)n + (\gamma-1)q]\Gamma_e + n[n-(\gamma-1)q]\Gamma_e^2\right\}, \tag{7.50}$$
$$a_4 = \frac{a^2}{4\gamma^2 H^2\tau^2}(1-n\Gamma_e)^2.$$

The Hurwitz–Routh criterion applied to this quartic equation leads to these conditions for stability:[3]

$$a_1 \geq 0, \tag{7.51}$$
$$a_1a_2 - a_3 \geq 0, \tag{7.52}$$
$$a_1a_2a_3 - a_1^2a_4 - a_3^2 \geq 0, \tag{7.53}$$
$$a_4(a_1a_2a_3 - a_1^2a_4 - a_3^2 \geq 0. \tag{7.54}$$

The inequalities (7.52) and (7.54) can be replaced by

$$a_3 \geq 0, \tag{7.55}$$
$$a_4 \geq 0 \tag{7.56}$$

without changing the results. Inspecting the coefficients shows that conditions (7.51) and (7.56) are always satisfied. Condition (7.55) may be violated for some Γ_e, depending on n and q. If $n < (\gamma-1)q$, the statement that the absorption opacity decreases in an adiabatic compression, the condition is certainly violated for sufficiently large Γ_e. That leaves condition (7.53) to discuss. It can be rearranged in this way:

$$2(\gamma+1)E\left(\frac{H}{a\tau}\right)^2 \geq -FG, \tag{7.57}$$

with three new coefficients that are polynomials in Γ_e:

$$E = (1+n\Gamma_e)[1-(n+2q)\Gamma_e], \tag{7.58}$$
$$F = \gamma - [(\gamma+1)n + (\gamma-1)q]\Gamma_e + n[n-(\gamma-1)q]\Gamma_e^2, \tag{7.59}$$
$$G = \gamma\left[1 + \frac{n-(\gamma-1)q}{\gamma}\Gamma_e\right]\left[1 - \frac{n+(\gamma+1)q}{\gamma}\Gamma_e\right]. \tag{7.60}$$

We note that $a_3 \propto F$ and that condition (7.55) is the same as $F \geq 0$.

[3] The roots of an algebraic equation are the same as the eigenvalues of a certain matrix derived from the coefficients, and the real parts of the eigenvalues will all lie in the left half-plane provided the principal minors of the matrix are nonnegative. See Ralston (1965).

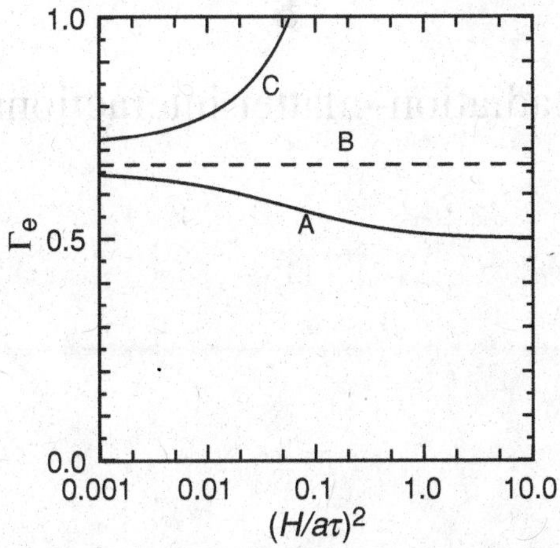

Fig. 7.2 Stability domain in radiation force vs adiabacy parameter. Acoustic modes are unstable between curves A and C. Gravity modes are unstable above dashed line B. Curves A and C and line B are described in the text.

The summary of the absolute stability for this problem is that there is stability in the isothermal limit $H/(a\tau) \to \infty$ for $\Gamma_e < 1/(n+2q)$. There is stability in the adiabatic limit $H/(a\tau) \to 0$ for Γ_e less than the smaller of the smallest positive zeroes of F and G. These limits turn out to be tighter than the limit $\Gamma_e < 1/(n+q)$ derived from the gravity modes. Depending on whether $n < (\gamma - 1)q$ or not, either F has one positive zero and G has two or none, or the reverse. The product FG has either one or three positive zeroes. With specific values of n, q, and γ it is a simple matter to map the stability domain in Γ_e as a function of $H/(a\tau)$. Figure 7.2 shows the domain for the choices $\gamma = 5/3$, $n = 1$, and $q = 1/2$. All pairs $(H/(a\tau), \Gamma_e)$ that produce equality in (7.57) define the locus in the $\Gamma_e, H/(a\tau)$ diagram that separates regions of stability and instability. For this case the locus has two disconnected branches, curves A and C in the diagram. Stability according to (7.57) occurs below curve A and above curve C. But the gravity wave criterion indicates instability above the line B, $\Gamma_e = 1/(n+q)$. Thus the region that is stable for both acoustic and gravity modes is the one below curve A, which goes smoothly between the isothermal limit $\Gamma_e = 1/(n+2q) = 1/2$ and the adiabatic limit, which is $\Gamma_e \approx 0.6492$. The implication is that the atmospheres of the most luminous stars may be unstable on this account, as suggested by Hearn (1972, 1973).

8

Radiation–matter interactions

The interaction of radiation and matter is a problem, or rather a set of problems, in quantum physics. This is because the spatial scale of the atoms and the typical momenta of the atomic electrons exactly obey the uncertainty limit, $\Delta x \Delta p \gtrsim \hbar$. If this were not true, then the atoms could shrink and find a state of lower energy. Therefore all the atomic properties needed for radiative transfer come from the application of quantum mechanics to the atoms. The quantum mechanical theory presented below is based mostly on Messiah (1962), volume 2. The angular algebra leading to the final results in terms of reduced matrix elements draws on Sobel'man (1979) and Condon and Shortley (1951).

8.1 QED for dummies

The one-paragraph sketch of the quantum mechanics of radiation processes is this: the non-relativistic Hamiltonian of the system comprising an atom and a radiation field is

$$H = -\sum_i \frac{Ze^2}{r_i} + \sum_{i<j} \frac{e^2}{r_{ij}} + \sum_i \frac{1}{2m}\left[\mathbf{p}_i - \frac{e}{c}\mathbf{A}(\mathbf{r}_i)\right]^2 + \int dV \frac{1}{8\pi}\left(\mathcal{E}^2 + \mathcal{H}^2\right), \tag{8.1}$$

where the index i labels the individual atomic electrons, $\mathbf{p}_i$ is the momentum of the ith electron, and $\mathbf{r}_i$ is its position. The vector function $\mathbf{A}(\mathbf{r})$ is the vector potential for the electromagnetic field (in the radiation gauge), and $\mathcal{E}$ and $\mathcal{H}$ are the fields. The electromagnetic field is quantized by expressing $\mathbf{A}$ in terms of the creation and annihilation operators for quanta of the plane-wave basis states,

$$\mathbf{A}(\mathbf{r}) = \sum_{\mathbf{k}\varpi} \left(\frac{2\pi\hbar c^2}{\omega V}\right)^{1/2} \mathbf{e}_{\mathbf{k}\varpi} \left[a^\dagger_{\mathbf{k}\varpi} \exp(-i\mathbf{k}\cdot\mathbf{r}) + a_{\mathbf{k}\varpi} \exp(i\mathbf{k}\cdot\mathbf{r})\right], \quad (8.2)$$

where $a^\dagger$ is the creation operator and a is annihilation operator. The wavevectors $\mathbf{k}$ belong to a discrete set corresponding to the modes in a large but finite volume V with periodic boundary conditions. The vectors $\mathbf{e}_{\mathbf{k}\varpi}$, for $\varpi = 1, 2$, are the two electric field polarization directions perpendicular to $\mathbf{k}$, and $\omega = kc$ is the angular frequency associated with $\mathbf{k}$. The interaction terms between radiation and matter are those involving $\mathbf{A}$ that result when the squared terms in (8.1) are expanded, in other words

$$H_{\text{int}} = -\frac{e}{mc}\sum_i \mathbf{p}_i \cdot \mathbf{A}(\mathbf{r}_i) + \frac{e^2}{2mc^2}\sum_i \mathbf{A}(\mathbf{r}_i)^2. \quad (8.3)$$

The transition rates, and therefore cross sections, for various radiative processes are found by treating H_{int} using perturbation theory, which entails finding the matrix element of H_{int} between an initial state $|a\mathbf{n}\rangle$ and a final state $|b\mathbf{n}'\rangle$. Here a and b are labels of atomic states, and the initial state of the radiation field is specified by a vector $\mathbf{n}$ giving the number of photons present in each basis state, and similarly for the final radiation state. The first term in H_{int} is linear in $\mathbf{A}$, therefore this term can allow exactly one photon to be created or alternatively, to be destroyed. The second term is quadratic, therefore this term allows processes that create two photons, destroy two photons, or destroy one and create another. In other words, absorption and emission come from the $\mathbf{p}\cdot\mathbf{A}$ term, and scattering comes from the $\mathbf{A}^2$ term (and also from $\mathbf{p}\cdot\mathbf{A}$ in second order perturbation theory, as it turns out).

The part of the radiation field belonging to one particular basis state, i.e., to one wave vector $\mathbf{k}$ and polarization $\mathbf{e}_{\mathbf{k}\varpi}$, is exactly like a harmonic oscillator. The states are labeled by the occupation number $n_{\mathbf{k}\varpi}$, which takes on the values $0, 1, \ldots$. The occupation number is the eigenvalue of the operator $n_{\mathbf{k}\varpi} = a^\dagger_{\mathbf{k}\varpi} a_{\mathbf{k}\varpi}$. The energy of the state is $(n_{\mathbf{k}\varpi} + 1/2)\hbar\omega$, where ω is the angular frequency for this $\mathbf{k}$. When the matrix elements of the interaction are calculated, we end up with something proportional to $\langle n+1|a^\dagger|n\rangle$ for a process that produces emission, $\langle n-1|a|n\rangle$ for a process that represents absorption, and a product of matrix elements $\langle n_2+1|a_2^\dagger|n_2\rangle\langle n_1-1|a_1|n_1\rangle$ for a scattering process in which a type-1 photon goes away and a type-2 photon is created. All these matrix elements come from the standard harmonic oscillator theory, where $a^\dagger$ and a are the raising and lowering operators. The matrix elements are simple:

$$\langle n+1|a^\dagger|n\rangle = \sqrt{n+1}, \tag{8.4}$$

$$\langle n-1|a|n\rangle = \langle n|a^\dagger|n-1\rangle = \sqrt{n}. \tag{8.5}$$

When we recall that transition rates in first order perturbation theory are proportional to the matrix element of the interaction squared, we see that emission rates are proportional to 1 plus the number of photons initially present; absorption rates are proportional to the number of photons initially present. What is the implication for scattering? We square the product of the two matrix elements and find something proportional to $n_1(n_2+1)$, i.e., proportional to the number of photons in the out-scattering state, and 1 plus the number of photons in the in-scattering state. The reverse scattering process, in which a type-2 photon goes away and a type-1 photon is created has a rate proportional to $n_2(n_1+1)$. You might think that the $n_1 n_2$ part of these rates just cancels out, and this is almost true. There is a correction because the frequencies of the photons are not exactly equal, as we will discuss later in connection with the Kompaneets equation.

8.2 Emission and absorption; Einstein coefficients

We consider the emission of a photon by an atom in an excited state b as it makes a transition to a lower state a. The transition probability per unit time is given by Fermi's Golden Rule as

$$w_{ba} = \frac{2\pi}{\hbar}|\langle a, n+1|H_{\text{int}}|b, n\rangle|^2 \rho(E), \tag{8.6}$$

where $\rho(E)$ is the number of photon basis states per unit energy near $E = \hbar\omega = E_a - E_b$, the amount of energy lost by the atom; the volume V has been assumed so large that the photon states form a pseudo-continuum. The occupation number n of the states has been assumed to be smooth around this energy. The steps we take next to reduce (8.6) are to: substitute the $a^\dagger$ term from the expansion of $\mathbf{p}\cdot\mathbf{A}$ for H_{int}; impose the dipole approximation which is obtained on setting $\exp(i\mathbf{k}\cdot\mathbf{r}) \approx 1$ and is valid when $kr \ll 1$, i.e., when the radiation wavelength is much larger than the atom; substitute

$$\rho(E) = \frac{Vk^2\,dk\,d\Omega}{(2\pi)^3\hbar c\,dk} = \frac{Vk^2}{8\pi^3\hbar c}d\Omega; \tag{8.7}$$

sum the results over the two possible polarizations of the emitted photon; and integrate the answer over all directions of emission. One tricky bit is this step:

$$\int_{4\pi}\sum_{\varpi}\langle a|\mathbf{p}|b\rangle^* \cdot \mathbf{e}_{\mathbf{k}\varpi}\mathbf{e}_{\mathbf{k}\varpi}\cdot\langle a|\mathbf{p}|b\rangle = \frac{8\pi}{3}|\langle a|\mathbf{p}|b\rangle|^2, \tag{8.8}$$

which follows because the sum of the two tensors $\mathbf{e}_{\mathbf{k}1}\mathbf{e}_{\mathbf{k}1}$ and $\mathbf{e}_{\mathbf{k}2}\mathbf{e}_{\mathbf{k}2}$ is $\mathsf{I} - \hat{\mathbf{k}}\hat{\mathbf{k}}$, where $\hat{\mathbf{k}}$ is the unit vector along $\mathbf{k}$. The angle average of $\hat{\mathbf{k}}\hat{\mathbf{k}}$ is $\mathsf{I}/3$, which leads directly to the result. Carrying out the steps mentioned leads to this:

$$w_{ba} = \frac{4}{3}\frac{e^2\omega}{\hbar c^3 m^2}|\langle a|\mathbf{p}|b\rangle|^2(n+1). \tag{8.9}$$

A transformation of the matrix element in (8.9) is possible using the commutator relation

$$\mathbf{p} = \frac{i}{\hbar}m[H_{\text{atom}}, \mathbf{r}], \tag{8.10}$$

which turns (8.9) into

$$w_{ba} = \frac{4}{3}\frac{e^2\omega^3}{\hbar c^3}|\langle a|\mathbf{r}|b\rangle|^2(n+1). \tag{8.11}$$

Finally we need to allow for degeneracy of the atomic states corresponding to the possible orientations of a state with nonzero total angular momentum. If the a state can have g_a different angular momentum projections m_a, and likewise for the b state, we get the transition rate for nonoriented atoms by averaging over m_b and summing on m_a:

$$w_{ba} = \frac{4}{3}\frac{\omega^3}{g_b\hbar c^3}\mathsf{S}_{ab}(n+1), \tag{8.12}$$

where S_{ab} is the electric dipole *line strength* defined by

$$\mathsf{S}_{ab} = \sum_{m_a, m_b} |\langle a\, m_a|e\mathbf{r}|b\, m_b\rangle|^2. \tag{8.13}$$

The occupation number n is the same as the number of photons per mode introduced earlier (see (6.16)). Thus the transition rate for emission is

$$w_{ba} = A_{ba} + B_{ba}I_\nu, \tag{8.14}$$

where A_{ba} and B_{ba} are two of the *Einstein coefficients*, the ones for spontaneous and stimulated emission, and are defined by

$$A_{ba} = \frac{32\pi^3}{3}\frac{\nu^3}{g_b\hbar c^3}\mathsf{S}_{ab}, \tag{8.15}$$

$$B_{ba} = \frac{A_{ba}}{4\pi\hbar\nu^3/c^2} = \frac{8\pi^2}{3g_b\hbar^2 c}\mathsf{S}_{ab}. \tag{8.16}$$

The absorption process can now be handled immediately. Nothing changes in the quantum mechanics except the factor involving n in the matrix element, which changes from $\sqrt{n+1}$ to $\sqrt{n}$ since the number in the initial state, n, is the larger of the initial and final occupations, and that the averaging is over m_a and summation is taken over m_b. The transition rate turns out to be

$$w_{ab} = B_{ab} I_\nu, \tag{8.17}$$

with

$$B_{ab} = \frac{g_b}{g_a} B_{ba} = \frac{8\pi^2}{3 g_a \hbar^2 c} \mathrm{S}_{ab}, \tag{8.18}$$

which defines the Einstein coefficient for absorption.

In a parcel of material in which state a is populated by N_a atoms per unit volume and state b by N_b atoms per unit volume, we can inquire what the net rate of transitions is between the two states for this parcel. We see that the rate is

$$N_a B_{ab} I_\nu - N_b (A_{ba} + B_{ba} I_\nu) \tag{8.19}$$

and that it would vanish if the intensity had the value

$$I_\nu = S_\nu = \frac{N_b A_{ba}}{N_a B_{ab} - N_b B_{ba}}. \tag{8.20}$$

This is therefore the atomic expression for the source function, since the definition of that is the intensity for which emission and absorption are in balance. Substituting the relations between the Einstein coefficients leads to an expression independent of the atomic constants,

$$S_\nu = \frac{2h\nu^3/c^2}{g_b N_a / g_a N_b - 1}. \tag{8.21}$$

We see that the source function will be the Planck function if and only if the Boltzmann relation is obeyed by the atomic state populations:

$$\frac{N_b}{N_a} = \frac{g_b}{g_a} \exp(-h\nu/kT). \tag{8.22}$$

The circumstance that the atomic state populations are determined by a balance of the rates of a variety of different processes, and that therefore the Boltzmann relation is not necessarily obeyed, leads to a source function that is not the same as the Planck function. When the equation of radiative transfer is solved using this source function, the intensity that is found does not agree with the Planck function either. This intensity may govern the rates of some of the atomic processes that, in turn, lead to the non-Boltzmann values of the state populations. Since the intensities are non-Planckian, the non-Boltzmann character of the populations is confirmed. The self-consistent calculation of atomic populations and the radiation field, neither of which agrees with thermodynamic equilibrium, is the subject of the theory of non-LTE, i.e., of systems not in local thermodynamic equilibrium.

The remaining quantities to discuss are the *oscillator strength* and the *absorption cross section*. The oscillator strength is defined by factoring out of B_{ab} all the dimensional quantities to leave a number of order unity; this is

$$f_{ab} = \frac{m\hbar c\nu}{2\pi e^2} B_{ab} = \frac{4\pi m\nu}{3g_a\hbar e^2} \mathrm{S}_{ab} = \frac{g_b}{g_a} \frac{mc^3}{8\pi^2 e^2 \nu^2} A_{ba}. \tag{8.23}$$

The emission oscillator strength is defined by a similar relation that is obtained from (8.23) by interchanging a and b and noting that the frequency changes sign; it is a negative quantity and obeys

$$f_{ba} = -\frac{g_a}{g_b} f_{ab}. \tag{8.24}$$

The usefulness of the oscillator strength stems from the *oscillator strength sum rule*, which says that the algebraic sum of all oscillator strengths, absorption and emission, beginning from a given state, equals the number of electrons in the atom. If the sum is restricted to certain subsets of the totality of states, sometimes there can be a sum rule relating the restricted sum to, for example, the number of outer-shell electrons. Absorption oscillator strengths, and sometimes dipole line strengths in atomic units of $(ea_0)^2$, are the quantities most often found in the tables.

The absorption transition rate given in (8.17) is based on the angle-average of the intensity. The part deriving from a particular solid angle $d\Omega$ would be $d\Omega/(4\pi)$ times the total. But $I_\nu d\Omega/(h\nu)$ is the photon particle flux per unit frequency in this solid angle, and therefore $h\nu B_{ab}/(4\pi)$ is related to the absorption cross section. But the transition rate is the total over frequency, while $I_\nu d\Omega/(h\nu)$ is the photon flux per unit frequency. Thus $h\nu B_{ab}/(4\pi)$ must be the frequency integral of the cross section:

$$\int d\nu\, \sigma_\nu = \frac{h\nu}{4\pi} B_{ab} = \frac{\pi e^2}{mc} f_{ab}. \tag{8.25}$$

The actual cross section σ_ν is a sharply peaked function of frequency around $\nu = (E_b - E_a)/h$, but as we see there must be an inverse relation between the peak value of the cross section and its width in frequency, since its integral is fixed by the atomic constants. The width of the feature is actually produced by a variety of processes; some of these are the thermal motion of the atoms, the Stark effect produced by nearby positive ions, the perturbation by plasma electrons, the Zeemann effect of disordered magnetic fields, a small-scale random turbulent velocity in the gas, and on and on. These are all subsumed in a normalized distribution function $\phi(\nu)$,

$$\int d\nu\, \phi(\nu) = 1. \tag{8.26}$$

Thus we have the actual cross section

$$\sigma_\nu = \frac{h\nu}{4\pi} B_{ab}\phi(\nu) = \frac{\pi e^2}{mc} f_{ab}\phi(\nu). \tag{8.27}$$

The absorptivity and emissivity follow from the Einstein coefficients and the line profile function. We note immediately that the absorptivity is the *difference* between absorption and stimulated emission, and that the emissivity comes from the spontaneous emission alone. The results are

$$\begin{aligned} k_\nu &= \frac{h\nu}{4\pi}\phi(\nu)(N_a B_{ab} - N_b B_{ba}) \\ &= N_a \sigma_\nu \left(1 - \frac{N_b g_a}{N_a g_b}\right) \end{aligned} \tag{8.28}$$

and

$$\begin{aligned} j_\nu &= \frac{h\nu}{4\pi}\phi(\nu) N_b A_{ba} \\ &= \frac{N_b g_a}{g_b} \sigma_\nu \frac{2h\nu^3}{c^2}. \end{aligned} \tag{8.29}$$

In LTE, when the Boltzmann equation is obeyed, the relations become

$$k_\nu = N_a \sigma_\nu \left[1 - \exp(-h\nu/kT)\right] \tag{8.30}$$

and

$$j_\nu = N_a \sigma_\nu \frac{2h\nu^3}{c^2} \exp(-h\nu/kT) = k_\nu B_\nu. \tag{8.31}$$

8.3 Scattering

Our treatment of scattering follows Messiah (1961, 1962). Compton scattering and the Klein–Nishina cross section are discussed by Cox and Giuli (1968). Rayleigh scattering is considered by Sobel'man (1979). The simplest kind of scattering to consider is scattering by free electrons in a plasma, and in the nonrelativistic approximation this is Thomson scattering. No atom is involved in the process, and the initial and final states both include a free electron and a photon. The changes of energy and momentum of the photon are just balanced by the changes for the electron. In this case the $\mathbf{p} \cdot \mathbf{A}$ term in the interaction does not contribute anything since the matrix element vanishes for a free particle. We treat the $\mathbf{A}^2$ term in first order perturbation theory, and select from the sums a destruction term for $\mathbf{k}_1\varpi_1$ and a creation term for $\mathbf{k}_2\varpi_2$. Let the initial electron momentum be $\hbar\mathbf{K}_1$ and the final momentum $\hbar\mathbf{K}_2$. The tricky bit is deciding on the appropriate density of states expression for Fermi's Golden Rule, and handling the matrix element, since

$$\int d^3\mathbf{r}\,\exp[i(\mathbf{k}_1+\mathbf{K}_1-\mathbf{k}_2-\mathbf{K}_2)\cdot\mathbf{r}] = (2\pi)^3\delta(\mathbf{k}_1+\mathbf{K}_1-\mathbf{k}_2-\mathbf{K}_2) \quad (8.32)$$

when taken over all space, where $\delta(\mathbf{k})$ is the 3-D Dirac delta function. This delta function ensures total momentum conservation for the scattering event. The strategy is something like this. If the electron states are normalized to one electron in the volume V, then each of the initial and final wavefunctions contains a factor $1/\sqrt{V}$ besides the free-wave exponential. We integrate over the factor $d^3\mathbf{K}_2/(2\pi)^3$ in the density of final states that accounts for the outgoing electron. That will use up one of the two factors of the integral (8.32). If momentum conservation is obeyed precisely, then the integral has the value V. Thus the second factor of the integral cancels with the factor $1/V$ that comes from the normalization of the electron states. Then we have to deal with the normalization of the photon states. We put a factor $Vk_2^2/(8\pi^3\hbar c)d\Omega_2$, the same as for spontaneous emission, in the density of final states to account for the outgoing photon. We in effect adjust the outgoing photon frequency to satisfy conservation of energy, having already adjusted the outgoing electron momentum to satisfy conservation of momentum. The transition rate has one surviving factor $1/V$ that comes from the expression for $\mathbf{A}$. One additional detail to notice is that the expansion of $\mathbf{A}^2$ contains the $a_{\mathbf{k}_1\varpi_1}a^\dagger_{\mathbf{k}_2\varpi_2}$ cross term twice, which becomes a factor 4 when the matrix element is squared.

After carrying out the substitutions as outlined, the following transition rate is found:

$$\left(\frac{e^2}{mc^2}\right)^2 (\mathbf{e}_{\mathbf{k}_1\varpi_1}\cdot\mathbf{e}_{\mathbf{k}_2\varpi_2})^2 \frac{k_2^2c^3}{\omega_1\omega_2 V} n_{\mathbf{k}_1\varpi_1}(n_{\mathbf{k}_2\varpi_2}+1)d\Omega_2. \quad (8.33)$$

As discussed above, the transition rate is proportional to the number of photons per mode in the initial state, and one plus the number of photons per mode in the final state. This will be used later in the detailed discussion of Compton scattering see (12.42). Since there is exactly one electron in the volume V, this transition rate should be the scattering cross section times the total incoming photon flux. The latter is $n_{\mathbf{k}_1\varpi_1}c/V$ since the photon states are normalized in the volume V. Thus we find a differential cross section

$$\frac{d\sigma}{d\Omega_2} = \left(\frac{e^2}{mc^2}\right)^2 (\mathbf{e}_{\mathbf{k}_1\varpi_1}\cdot\mathbf{e}_{\mathbf{k}_2\varpi_2})^2 \frac{\omega_2}{\omega_1}(n_{\mathbf{k}_2\varpi_2}+1), \quad (8.34)$$

in which the stimulated scattering factor $n_{\mathbf{k}_2\varpi_2} + 1$ has been retained for the moment.

The factor $(\mathbf{e}_{\mathbf{k}_1\varpi_1} \cdot \mathbf{e}_{\mathbf{k}_2\varpi_2})^2$ in the differential cross section is interesting since it describes how polarized light is scattered by electrons, and also how polarized light is created by scattering unpolarized light. But for now we will average over the initial polarizations and sum on the final polarizations to get the differential cross section for unpolarized scattering. As above, we can replace $\mathbf{e}_{\mathbf{k}_1\varpi_1}\mathbf{e}_{\mathbf{k}_1\varpi_1}$ after summing over ϖ_1 with $\mathsf{I} - \hat{\mathbf{k}}_1\hat{\mathbf{k}}_1$, do the same with $\mathbf{e}_{\mathbf{k}_2\varpi_2}\mathbf{e}_{\mathbf{k}_2\varpi_2}$, then form the fully contracted product of those two tensors. This operation yields $1 + (\hat{\mathbf{k}}_1 \cdot \hat{\mathbf{k}}_2)^2$, and therefore the unpolarized differential cross section is

$$\frac{d\sigma}{d\Omega_2} = \left(\frac{e^2}{mc^2}\right)^2 \frac{1 + (\hat{\mathbf{k}}_1 \cdot \hat{\mathbf{k}}_2)^2}{2} \frac{\omega_2}{\omega_1}(n_{\mathbf{k}_2\varpi_2} + 1). \tag{8.35}$$

If Θ is the angle between the incoming and outgoing photon directions, then $\hat{\mathbf{k}}_1 \cdot \hat{\mathbf{k}}_2$ is the same as $\cos\Theta$, and the factor in the differential cross section is $(1 + \cos^2\Theta)/2$. This is the *phase function* for Thomson scattering. Its average over scattering angle is 2/3, which leads to the total Thomson scattering cross section

$$\sigma = \frac{8\pi}{3}\left(\frac{e^2}{mc^2}\right)^2 \frac{\omega_2}{\omega_1}(n_{\mathbf{k}_2\varpi_2} + 1). \tag{8.36}$$

The difference between ω_2 and ω_1 is indeed very small, which is what we will show next. The standard Compton effect considers scattering of a photon from an electron at rest, in other words $\mathbf{K}_1 = 0$. The (relativistic) conservations laws for energy and momentum then require

$$\mathbf{K}_2 = \mathbf{k}_1 - \mathbf{k}_2, \tag{8.37}$$

$$E_2 - mc^2 = \hbar c(k_1 - k_2) \tag{8.38}$$

and the relativistic relation between E_2 and K_2 is

$$E_2^2 - (\hbar \mathbf{K}_2 c)^2 = (mc^2)^2. \tag{8.39}$$

Using the first two equations to eliminate $\mathbf{K}_2$ and E_2 from the third leads to the equation for the Compton effect

$$\frac{1}{k_2} - \frac{1}{k_1} = \frac{\hbar}{mc}(1 - \hat{\mathbf{k}}_1 \cdot \hat{\mathbf{k}}_2). \tag{8.40}$$

This tells us that the fractional change (reduction) in ω is approximately $\hbar\omega/(mc^2)$ times $(1 - \cos\Theta)$. Of course mc^2 is the electron rest energy, 511 keV. So except for hard x-rays and gamma-rays the fractional frequency shift is negligible.

There is a second contribution to the frequency shift when the initial electron is not at rest, but has a random velocity as in a Maxwellian distribution. The Doppler shifts to the rest frame of the incoming electron and back to the laboratory frame again after scattering lead to a random up or down shift of the photon frequency by an amount of order $\omega(1 - \cos\Theta)\sqrt{2kT/(mc^2)}$. Although $\sqrt{2kT/(mc^2)}$ is usually small, it is not nearly as small as $\hbar\omega/(mc^2)$, so in practical cases these up and down shifts are larger than the Compton shift. Even so, they can generally be neglected for scattering by electrons whose temperature is in the eV range. At the temperature of the solar corona, however, around 150 eV, the Doppler shift is about 2.4%, and is responsible for washing out the photospheric absorption lines in the scattered spectrum observed from the corona in eclipse or with a coronagraph.

For ordinary purposes, then, we can drop the factor ω_2/ω_1 in the cross section, which therefore also means that stimulated in-scattering will precisely cancel stimulated out-scattering, and the factor $1 + n$ can be discarded as well. This gives the common Thomson cross section

$$\sigma_T = \frac{8\pi}{3}\left(\frac{e^2}{mc^2}\right)^2. \tag{8.41}$$

For completeness we will note that when the photon energy is no longer small compared with mc^2, in the domain of the Compton effect in other words, the cross section is less than the Thomson value and is given by the Klein–Nishina formula

$$\sigma_{KN} = \frac{3}{4}\sigma_T\left\{\frac{1+u}{u^3}\left[\frac{2u(1+u)}{1+2u} - \log(1+2u)\right] + \frac{\log(1+2u)}{2u} - \frac{1+3u}{(1+2u)^2}\right\}, \tag{8.42}$$

where $u = h\nu/mc^2$, provided the scattering electron is initially at rest; the cross section is shown in Figure 8.1. For moving electrons the cross section in any

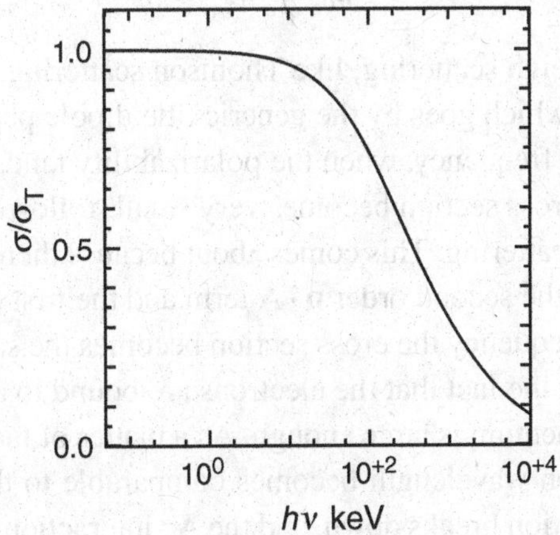

Fig. 8.1 Relativistic (Klein–Nishina) Compton cross section in terms of the Thomson cross section vs photon frequency in keV.

particular direction must be found by transforming the photon frequency into the rest frame of the electron using (6.9), evaluating the Klein–Nishina cross section, then transforming back to the fixed frame with (6.23). Some of the implications of an exact treatment of Compton scattering are considered in Section 12.3.

The other kind of scattering we want to consider is scattering by the electrons bound in atoms. This is called Rayleigh or Raman scattering depending on whether the initial and final atomic states are the same or different. Because the bound-state wave functions occupy a finite volume, unlike the free electron states, the subtleties of the density of states and delta functions are not required. What is required is to find the effective interaction matrix element by combining the $\mathbf{A}^2$ term in first order with the $\mathbf{p} \cdot \mathbf{A}$ term in second order. We will spare the details and just quote the result. This is simple in the Rayleigh case where states a and b are the same and if the total angular momentum is $J = 0$ in the initial state a. The differential cross section for specific polarization states in that case is

$$\frac{d\sigma}{d\Omega} = \left(\frac{e^2}{mc^2}\right)^2 \left| \sum_c \frac{\omega^2 f_{ac}}{\omega_{ac}^2 - \omega^2} \right|^2 (\mathbf{e}_{\mathbf{k}_1\varpi_1} \cdot \mathbf{e}_{\mathbf{k}_2\varpi_2})^2, \tag{8.43}$$

where the frequencies of the scattered photon and the initial photon must be the same in this case, $\omega_1 = \omega_2 = \omega$. The summation is taken over all the members c of a complete set of atomic states, and ω_{ac} is the excitation energy from state a to the intermediate state c. The cross section can also be written as

$$\frac{d\sigma}{d\Omega} = k^4 [\alpha(\omega)]^2 (\mathbf{e}_{\mathbf{k}_1\varpi_1} \cdot \mathbf{e}_{\mathbf{k}_2\varpi_2})^2, \tag{8.44}$$

where $\alpha(\omega)$ is the AC polarizability of the atom in the state a:

$$\alpha(\omega) = \frac{e^2}{m} \sum_c \frac{f_{ac}}{\omega_{ac}^2 - \omega^2}. \tag{8.45}$$

We observe that Rayleigh scattering, like Thomson scattering, has the phase function $(1 + \cos^2 \Theta)/2$, which goes by the generic title dipole phase function.

You see that at low frequency, when the polarizability tends to the constant DC value, the scattering cross section becomes very small, following then the λ^{-4} law for optical Rayleigh scattering. This comes about because there is almost complete cancellation between the second order $\mathbf{p} \cdot \mathbf{A}$ term and the first order $\mathbf{A}^2$ term in the interaction. At high frequency the cross section becomes the same as the Compton cross section, because the fact that the electrons are bound to an atom is irrelevant when the photon momentum is large enough. As a matter of fact, (8.43) is inappropriate when the photon wavelength becomes comparable to the size of the atom. The dipole approximation breaks down, and the $\mathbf{A}^2$ interaction term should then be multiplied by a structure factor that depends on the momentum transfer, $\mathbf{k}_2 - \mathbf{k}_1$.

As the wavelength decreases, the momentum transfer increases and the structure factor becomes very small, which makes this part of the interaction negligible. At the same time a Raman scattering term becomes important that corresponds to leaving the atom in a final state with an ejected electron that has the momentum $\hbar(\mathbf{k}_1 - \mathbf{k}_2)$. If the atom has N electrons, then at photon energies large compared with all the atomic binding energies but small compared with $\hbar c/r$, all N electrons add coherently in the polarizability, which makes the scattering cross section N^2 times as large as that for a single electron. But when the photon energy is large compared with $\hbar c/r$ the Compton scattering contributions of the electrons add incoherently, so the cross section is N times that for a single electron.

8.4 Photoionization and recombination

When the final state of the photoabsorption process considered earlier belongs to a continuum, as it does when a photoelectron is ejected, the calculation needs to be modified since the density of states factor can be based on the electron continuum instead of the continuum of absorbed photon energies. This discussion is partly based on Sobel'man (1979). Since there is a reciprocal relation between the free-wave normalization and the density of states, we will make a specific normalization choice for the free waves $\psi(\mathbf{r})$, which is

$$\psi(\mathbf{r}) = \psi_{\mathbf{K}}(\mathbf{r}) \sim \exp(i\mathbf{K}\cdot\mathbf{r}) \quad \text{for } r \to \infty, \tag{8.46}$$

which corresponds to a unit density of free electrons. The corresponding density of final states per unit energy is

$$\rho = \frac{mK\, d\Omega_{\mathbf{K}}}{(2\pi)^3\hbar^2}, \tag{8.47}$$

and the incoming photon flux is $n_{\mathbf{k}\varpi}c/V$. With these replacements the transition rate for photoabsorption becomes

$$\begin{aligned} w_{ab\mathbf{K}} &= \frac{2\pi}{\hbar}\left(\frac{e}{mc}\right)^2 \frac{2\pi\hbar c^2}{\omega V} |\langle b\mathbf{K}| \exp(i\mathbf{k}\cdot\mathbf{r})\mathbf{p}\cdot\mathbf{e}_{\mathbf{k}\varpi}|a\rangle|^2 n_{\mathbf{k}\varpi} \frac{mK\, d\Omega_{\mathbf{K}}}{(2\pi)^3\hbar^2} \\ &= \frac{1}{2\pi}\frac{e^2 K}{\hbar^2 m V\omega} |\langle b\mathbf{K}| \exp(i\mathbf{k}\cdot\mathbf{r})\mathbf{p}\cdot\mathbf{e}_{\mathbf{k}\varpi}|a\rangle|^2 n_{\mathbf{k}\varpi}, \end{aligned} \tag{8.48}$$

which gives the differential cross section upon being divided by the flux times $d\Omega_{\mathbf{K}}$:

$$\frac{d\sigma_{PI}}{d\Omega_{\mathbf{K}}} = \frac{1}{2\pi}\frac{e^2 K}{\hbar^2 mc\omega} |\langle b\mathbf{K}| \exp(i\mathbf{k}\cdot\mathbf{r})\mathbf{p}\cdot\mathbf{e}_{\mathbf{k}\varpi}|a\rangle|^2. \tag{8.49}$$

In the dipole approximation we can drop the exponential in the matrix element and then use the commutator relation (8.10) to obtain

$$\frac{d\sigma_{PI}}{d\Omega_{\mathbf{K}}} = \frac{1}{2\pi}\frac{me^2Kk}{\hbar^2}|\langle b\mathbf{K}|\mathbf{r}\cdot\mathbf{e}_{\mathbf{k}\varpi}|a\rangle|^2. \tag{8.50}$$

In order to get the total cross section corresponding to this differential cross section it is most convenient to expand in spherical waves for the final state:

$$\psi_{\mathbf{K}}(\mathbf{r}) = \frac{4\pi}{K}\sum_{lm} i^l (Y_l^m(\theta_{\mathbf{K}}, \phi_{\mathbf{K}}))^* \psi_{Klm}(\mathbf{r}) \tag{8.51}$$

with

$$\psi_{Klm}(\mathbf{r}) = Y_l^m(\theta, \phi) R_{Kl}(r). \tag{8.52}$$

Here θ and ϕ are the angular coordinates of the ejected electron, and $\theta_{\mathbf{K}}$ and $\phi_{\mathbf{K}}$ are the coordinates for the direction of $\mathbf{K}$. The radial function has the asymptotic form

$$R_{Kl}(r) \sim \frac{\sin(Kr - l\pi/2 + \delta)}{r}. \tag{8.53}$$

The result of inserting the expansion for each factor $\psi_{\mathbf{K}}(\mathbf{r})$ and integrating over $d\Omega_{\mathbf{K}}$, where the orthogonality of the $Y_l^m(\theta_{\mathbf{K}}, \phi_{\mathbf{K}})$ can be used, is

$$\sigma_{PI} = 8\pi\frac{me^2k}{\hbar^2K}\sum_{lm}|\langle bKlm|\mathbf{r}\cdot\mathbf{e}_{\mathbf{k}\varpi}|a\rangle|^2. \tag{8.54}$$

The average over the polarization modes introduces a factor $(\mathsf{I} - \hat{\mathbf{k}}\hat{\mathbf{k}})/2$. Averaging over the direction of the incoming photon, or, which is equivalent, averaging over the m_a values for the initial state and summing over the m_b values for the residual ion state, and furthermore summing over the two possible free-electron spin states, allows us to replace this tensor with $\mathsf{I}/3$. That gives finally

$$\sigma_{PI} = \frac{8\pi}{3}\frac{me^2k}{g_a\hbar^2K}\sum_{lmm_sm_am_b}|\langle bm_bKlmm_s|\mathbf{r}|am_a\rangle|^2. \tag{8.55}$$

The cross section in (8.55) is for the process of photoionization. For a simple atom such as hydrogen the values of K and the matrix elements are of order unity when expressed in atomic units ($\hbar = m = e = 1$), but $k = \omega/c$, and while ω is of order unity in atomic units, $c = 1/\alpha \approx 137$ in those units. Thus the cross section is of order $\alpha\pi a_0^2$, where a_0 is the Bohr radius, the atomic unit of length. This makes the cross section 10^{-18} cm^2 or less. This is still much larger than the Thomson cross section, which is $8\alpha^4/3 \approx 10^{-8}$ times πa_0^2, or 6.65×10^{-25} cm^2.

The inverse process, radiative capture or radiative recombination, is described by exactly the same matrix element, but the density of final states should be that for the photon state, not the free electron, and the flux by which the rate is divided to obtain the cross section is the free-electron flux, not the photon flux. Making these adjustments leads to this differential cross section for the radiative capture process:

$$\frac{d\sigma_{RC}}{d\Omega_{\mathbf{k}}} = \frac{1}{2\pi}\frac{me^2k^3}{\hbar^2 K}|\langle b\mathbf{K}|\mathbf{r}\cdot\mathbf{e}_{\mathbf{k}\varpi}|a\rangle|^2. \tag{8.56}$$

Putting in the spherical state expansion in place of the plane wave state, summing over the two polarizations and forming the average over incoming electron direction and the angle integral of the outgoing photon direction, as well as summing over the possible values of m_a and averaging over the possible values of m_b and of the free-electron spin m_s, as before, gives the result for the total cross section for radiative capture:

$$\sigma_{RC} = \frac{8\pi}{3}\frac{me^2k^3}{g_b\hbar^2K^3}\sum_{lmm_sm_am_b}|\langle bm_bKlmm_s|\mathbf{r}|am_a\rangle|^2. \tag{8.57}$$

We have not carried along the factor $1 + n_{\mathbf{k}\varpi}$ that enters from the matrix element of $a^\dagger_{\mathbf{k}\varpi}$, but it should be included as well to give the total recombination including stimulated recombination. The stimulation factor can as before be expressed as $1 + I_\nu/(2h\nu^3/c^2)$, and is included as a subtraction in the absorptivity for photoionization. The cross section σ_{RC} is for spontaneous recombination.

Evidently the reciprocity relation between photoionization and radiative capture or radiative recombination is

$$\frac{\sigma_{RC}}{\sigma_{PI}} = \frac{k^2}{K^2}\frac{g_a}{g_b} = \frac{h^2\nu^2}{m^2v^2c^2}\frac{g_a}{g_b}. \tag{8.58}$$

This reciprocity relation is called the *Milne relation* and it is the equivalent for photoionization/recombination of the relation between the Einstein A and B coefficients. The statistical weight ratio here is simply that between the initial atom state and the residual ion state; the electron spin factor 2 has been canceled by a factor 2 for the number of photon polarization states. The recombination cross section is of order $\alpha^2 \approx 10^{-4}$ times the photoionization cross section.

If the intensity I_ν equals the source function S_ν for photoionization then the rate of photoionizations by photons in a range $[\nu, \nu + d\nu]$ must balance the rate of radiative recombinations due to free electrons in a velocity range $[v, v + dv]$, where the frequencies and velocities are in correspondence according to $h\nu = E_b - E_a + mv^2/2$. The relation of $d\nu$ and dv will therefore be $h\,d\nu = mv\,dv$. The photon flux in $d\nu$ with $I_\nu = S_\nu$ is $4\pi S_\nu/d_\nu/(h\nu)$. The free-electron flux in dv

comes from the Maxwellian velocity distribution, and is

$$\sqrt{\frac{2}{\pi}}\left(\frac{m}{kT}\right)^{3/2} N_e v^3 dv\, e^{-mv^2/2kT}, \tag{8.59}$$

in which N_e is the total density of free electrons. The recombination rate has to include the stimulation factor corresponding to S_ν. Therefore the balance equation is

$$N_a \sigma_{PI} \frac{4\pi}{h\nu} S_\nu\, d\nu = N_b \sigma_{RC} \sqrt{\frac{2}{\pi}}\left(\frac{m}{kT}\right)^{3/2} N_e v^3 dv\, e^{-mv^2/2kT} (1 + c^2 S_\nu/(2h\nu^3)). \tag{8.60}$$

The solution of this equation for S_ν is

$$S_\nu = \frac{2h\nu^3/c^2}{\mathcal{F} - 1}, \tag{8.61}$$

where $\mathcal{F}$ is defined by the expression

$$\begin{aligned} \mathcal{F} &= \frac{N_a \sigma_{PI} (8\pi \nu^2/c^2) d\nu}{N_b \sigma_{RC} \sqrt{2/\pi}(m/kT)^{3/2} N_e v^3 dv \exp(-mv^2/2kT)} \\ &= \frac{N_a}{N_e N_b} \frac{2g_b}{g_a} \left(\frac{2\pi mkT}{h^2}\right)^{3/2} e^{mv^2/2kT}. \end{aligned} \tag{8.62}$$

This expression for S_ν will become the Planck function, as it must in thermal equilibrium, if $\mathcal{F} = \exp(h\nu/kT)$, and therefore if

$$\frac{N_e N_b}{N_a} = \frac{2g_b}{g_a} \left(\frac{2\pi mkT}{h^2}\right)^{3/2} e^{-(E_b - E_a)/kT}, \tag{8.63}$$

which is Saha's equation for ionization balance in thermal equilibrium.

For the ideal nondegenerate and nonrelativistic electron gas, the chemical potential, the Gibbs free energy per particle, in this case per free electron, is given by

$$\mu_e = kT \log\left[\frac{N_e}{2}\left(\frac{h^2}{2\pi mkT}\right)^{3/2}\right]. \tag{8.64}$$

The factor 2 dividing N_e in (8.64) arises from the spin multiplicity of the electron. In terms of the chemical potential Saha's equation becomes

$$\frac{N_b}{N_a} = \frac{g_b}{g_a} e^{-(E_b - E_a + \mu_e)/kT}. \tag{8.65}$$

In this form the equation remains valid even if the electrons are degenerate, provided they can still be treated as an ideal Fermi gas. The nonrelativistic formula connecting μ_e to the temperature and electron density becomes

$$\frac{2}{\sqrt{\pi}} F_{1/2}\left(\frac{\mu_e}{kT}\right) = \frac{N_e}{2}\left(\frac{h^2}{2\pi mkT}\right)^{3/2}, \tag{8.66}$$

where the function $F_{1/2}(x)$ is the Fermi function defined by

$$F_{1/2}(x) = \int_0^\infty \frac{t^{1/2}\,dt}{e^{t-x}+1}. \tag{8.67}$$

For a hydrogen-like system the initial state for photoionization is a hydrogenic wavefunction with quantum numbers n and l, and the final state is a Coulomb wave. The exact calculation of the photoionization cross section in this case gives

$$\sigma_{PI}(\nu) = \frac{64}{3\sqrt{3}}\alpha\pi a_0^2 \frac{n g_{II}}{Z^2}\left(\frac{\nu_0}{\nu}\right)^3, \tag{8.68}$$

(see Allen (1973)) where Z is the charge of the residual ion ($Z = 1$ for hydrogen), α is the fine-structure constant, 1/137.036, g_{II} is the bound–free Gaunt factor that depends on n, l, and ν. The frequency ν_0 is the edge value for level n, Z^2/n^2 times the Rydberg frequency. The constant $64/(3\sqrt{3})\alpha\pi a_0^2$ is 7.90707×10^{-18} cm^2. The Gaunt factor is not *very* far from 1, especially when averaged over l at a given n. Replacing it with 1 is an approximation (Kramer's law) that is qualitatively correct. Accurate tables and subroutines for it exist today. The hydrogenic results can be extended in an approximate way to nonhydrogenic ions by using in place of Z in the hydrogenic formulae the expression $Z - N + 1$ in which N is the number of bound electrons.

The total rate coefficient for radiative recombination in a plasma in which the electrons have a Maxwellian distribution at a temperature T can be calculated, using the Milne relation, as an integral over the photoionization cross section times the Planck distribution corrected for stimulated emission. That is, the rate of spontaneous recombinations per unit volume is given by

$$N_e N_b \alpha_{RR}(T) = N_a \int_{\nu_0}^\infty d\nu\, \sigma_{PI}(\nu) \frac{8\pi\nu^2}{c^2} \exp(-h\nu/kT), \tag{8.69}$$

where it is understood that the Saha equation is to be used to express N_a in terms of $N_e N_b$. Here $\alpha_{RR}(T)$ is the recombination coefficient. When the Saha equation is substituted the expression for it is

$$N_e \alpha_{RR}(T) = \int_{\nu_0}^\infty d\nu\, \sigma_{PI}(\nu) \frac{8\pi\nu^2}{c^2} \exp\left(\frac{\mu_e - h\nu}{kT}\right). \tag{8.70}$$

In the nondegenerate case this becomes

$$\alpha_{RR}(T) = \frac{g_a}{2g_b}\left(\frac{h^2}{2\pi mkT}\right)^{3/2}\int_{\nu_0}^{\infty} d\nu\, \sigma_{PI}(\nu)\frac{8\pi\nu^2}{c^2}\exp(-h\nu/kT). \quad (8.71)$$

The recombination coefficient to a hydrogen-like ion, for the principal quantum number n with all the l values summed, is given by

$$\alpha_n(T) = \frac{2.06505\times 10^{-11}}{\sqrt{T}}\frac{\bar{g}_{II}}{n}[X_n e^{X_n} E_1(X_n)] \quad \text{cm}^3\,\text{s}^{-1}, \quad (8.72)$$

where $X_n = Z^2 Ryd/(n^2 kT)$ is the binding energy of level n in units of kT. The expression in the brackets in (8.72) is about unity for levels more tightly bound than kT and tends to 0 as $1/n^2$ as n increases above that point. This means that at low temperature there are sizeable contributions to the total recombination from many values of n. The total recombination rate varies with T close to the power $T^{-1/2}$.

8.5 Free–free absorption – bremsstrahlung

Free–free absorption, also called inverse bremsstrahlung, is the next step past bound–free absorption, which is another name for photoionization. Here not only does the final state of the process have a free electron, but so does the initial state. So the process has a formula like this:

$$e + X + h\nu \rightarrow X' + e'. \quad (8.73)$$

Most often the case we are interested in is when the final state of the atom X is the same as the initial state, and so the energy of the photon goes into the energy of the outgoing electron. (The case that X changes state too is an example of a two-electron transition, i.e., two or more electrons have to change from one orbital to another, and since the radiative interaction $\mathbf{p}\cdot\mathbf{A}$ is a one-particle operator, such transitions do not occur without configuration mixing, which makes them much less likely.)

The quantum mechanics of this process is only altered a little from that for photoionization. We expect a cross section for absorbing the photon that is proportional to the density of electrons with the initial momentum $\hbar\mathbf{K}_1$. We get the cross section for unit density by taking the initial free-electron wave-function to be $\exp(i\mathbf{K}_1\cdot\mathbf{r})$. The density of final states in energy and the photon flux are unchanged. The total cross section for all angles of the photon and electrons is found as before by introducing expansions (8.51) in spherical waves for both the incoming and outgoing electrons. The result is

$$\sigma(K_1\omega; K_2) = \frac{16\pi^2}{3}\frac{me^2k}{g_a\hbar^2 K_1^2 K_2} \times \sum_{m_a m_a' l m m_s l' m' m_s'} |\langle a m_a' K_2 l' m' m_s' | \mathbf{r} | a m_a K_1 l m m_s\rangle|^2. \quad (8.74)$$

For the bremsstrahlung process the colliding particle is an electron, and therefore we are seeking a cross section for electron collisions, but since there are two continuously distributed particles in the final state, we can assign a range $d\omega$ to the photon frequency and find the differential cross section for that range. What we find is

$$\frac{d\sigma(K_2; K_1\omega)}{d\omega} = \frac{16}{3}\frac{m^2e^2k^3}{g_a\hbar^3 K_1 K_2^3} \times \sum_{m_a m_a' l m m_s l' m' m_s'} |\langle a m_a' K_2 l' m' m_s' | \mathbf{r} | a m_a K_1 l m m_s\rangle|^2. \quad (8.75)$$

The reciprocity relation between $\sigma(K_1\omega; K_2)$ and $d\sigma(K_2; K_1\omega)/d\omega$ is that $\sigma(K_1\omega; K_2)$ times the electron density times the fraction of the electrons that have kinetic energy lying between $E_1 = \hbar^2 K_1^2/2m$ and $E_1 + dE$ times the photon flux lying in the range ω to $\omega + d\omega$ should be equal, in thermodynamic equilibrium, to $d\sigma(K_2; K_1\omega)/d\omega$ times $d\omega$ times the electron density times the fraction of the electrons that have kinetic energy lying between $E_2 = \hbar^2 K_2^2/2m$ and $E_2 + dE$ times the velocity $\hbar K_2/m$.

There is a problem with the matrix element that appears in (8.74) and (8.75), and that is that the integral of r times the product of two Coulomb waves does not converge in $r \to \infty$, but oscillates with increasing amplitude. So rather than using the commutator relation to express the matrix element of $\mathbf{p}$ in terms of the matrix element of $\mathbf{r}$, we should go the other way, and express the matrix element of $\mathbf{p}$ in terms of the matrix element of the central force that acts on the electron. If we now let V be the spherically symmetric potential energy for the initial and final free electrons, forgetting about its previous meaning as the volume of a box, the commutator relation we need is

$$[H, \mathbf{p}] = -\frac{\hbar}{i}\frac{dV}{dr}\hat{\mathbf{r}}, \quad (8.76)$$

so that an alternative form of (8.74) is

$$\sigma(K_1\omega; K_2) = \frac{16\pi^2}{3}\frac{e^2}{g_a m\hbar^2 c\omega^3 K_1^2 K_2} \times \sum_{m_a m_a' l m m_s l' m' m_s'} \left| \left\langle a m_a' K_2 l' m' m_s' \left| \frac{dV}{dr}\hat{\mathbf{r}} \right| a m_a K_1 l m m_s \right\rangle \right|^2, \quad (8.77)$$

and a similar form exists for $d\sigma(K_2; K_1\omega)/d\omega$. At large r, dV/dr varies as r^{-2}, which ensures convergence of the radial integral.

If the target ion, or the residual ion in the final state, is a bare nucleus, then we have hydrogen-like initial and final states, which for free electrons are Coulomb waves. Like the bound–bound and bound–free transitions, the free–free transitions can be expressed in terms of a Gaunt factor. We let Z be the charge of the target ion, represent the frequency in terms of ν instead of ω, and set $\hbar K_1/m = v_1$. We find the following result for the absorption cross section:

$$\sigma(v_1\omega; v_2) = \frac{4\pi}{3\sqrt{3}}\frac{Z^2 e^6 g_{III}}{hcm^2 v_1 \nu^3}. \quad (8.78)$$

We must keep in mind that this is the cross section for photoabsorption given that there is one electron per unit volume with the velocity v_1. The quantity g_{III} is the free–free gaunt factor, and is generally around unity. It is given by a formula very similar to the one for the bound–free gaunt factor g_{II}, and also a bound–bound Gaunt factor g_I we have not discussed. When the free–free absorption for a thermal plasma is the quantity we want, then we have to average (8.78) over the Maxwellian distribution for v_1. The average of $1/v_1$ becomes $\sqrt{2m/\pi kT}$. As a result the absorptivity at frequency ν by the free–free process, but not yet corrected for stimulated emission, for ions that have a charge Z and an ionic density N_i, is given by

$$\begin{aligned} k_{\nu,ff} &= \frac{4\sqrt{2\pi}}{3\sqrt{3}}\frac{Z^2 e^6 \bar{g}_{III}}{hc\sqrt{m^3 k}\nu^3}\frac{N_e N_i}{\sqrt{T}} \\ &= 3.69234 \times 10^8 \frac{Z^2 \bar{g}_{III}}{\sqrt{T}\nu^3} N_e N_i, \end{aligned} \quad (8.79)$$

where in the numerical form the densities are in cm^{-3}, the temperature is in K, the frequency is in Hz and the result is in cm^{-1}.

The radio-frequency case, i.e., when ν is small, is an exception to the general rule that the Gaunt factor is close to unity. An expression that covers this case is the following, from Allen (1973):

$$g_{III} \approx \max\left[1, \frac{\sqrt{3}}{\pi}\log\left(\frac{4\pi v/\nu}{\max(Ze^2/mv^2, \hbar/mv)}\right)\right], \tag{8.80}$$

which becomes logarithmically large when $\nu \to 0$. Here v is the final electron velocity. In the classical picture this logarithm, as in other Coulomb integrals, comes from an integral over the impact parameter b for electron–ion trajectories that to begin with looks like

$$\int_0^\infty \frac{db}{b},$$

which obviously diverges unless cutoffs $b_{\rm min}$ and $b_{\rm max}$ are applied. Quantum mechanics supplies one value for $b_{\rm min}$, which is $\hbar/mv$, since the quantization of angular momentum is equivalent to quantizing b in units of $\hbar/mv$. An alternative lower limit comes from the geometry of the hyperbolic classical orbits: when b is larger than Ze^2/mv^2 the deflection of the colliding electron is by an angle less than $\pi/2$ and therefore the general direction of the acceleration is consistent through the entire collision. But when b is smaller than Ze^2/mv^2 the orbit tends to double back on itself. The result is that the radiation of the second half of the collision partly cancels that from the first half. This cancellation factor allows the integral to converge. The correct $b_{\rm min}$ is approximately equal to whichever is larger of $\hbar/mv$ and Ze^2/mv^2. The quantum limit is effective at very high velocity and the $\pi/2$ deflection limit applies at somewhat lower velocity. The upper cutoff is less obvious. The case of an isolated ion, far from any other ion, is treated with the quantum mechanical method as sketched above, and the upper cutoff comes from the requirement that the collision produce some radiation at the frequency ν; if b is too large, then the time-dependent fields produced by the acceleration of the electron do not have an appreciable component at frequency ν. The collision time at impact parameter b is of order b/v and therefore the Fourier spectrum of the radiated energy cuts off above $\nu = \omega/2\pi = v/2\pi b$. These results for $b_{\rm min}$ and $b_{\rm max}$ lead to (8.80) when it is further observed that the classical arguments involving b fail altogether if they lead to a Gaunt factor less than unity, which is the typical non-classical value. It should be noted that the exact quantum mechanical calculations of the free–free Gaunt factor, based on Coulomb wave functions, implicitly include the cutoffs described in this paragraph; these accurate results are roughly approximated by (8.80).

In a denser plasma it can happen that the cutoff $v/2\pi\nu$ is larger than the Debye length, r_D. Since the target ion's field is considered fully shielded at $r > r_D$, the integral should be cut off at that value. This will happen for $\omega < v/r_D$. If v is taken to be $\sqrt{kT/m}$ then the frequency below which the Debye limit is in effect is $\sqrt{kT/mr_D^2} \approx \omega_{pe}$, the electron plasma frequency. When the frequency is the same order as the electron plasma frequency the effects of plasma dispersion must also be considered, and we have to deal with a nonunit refractive index. We will touch on this topic some more below.

8.6 Opacity calculations

For the astrophysical plasma the total absorptivity is obtained by adding up the processes outlined in the last four sections. The bound–bound (Section 8.2) bound–free (Section 8.4) and free–free (Section 8.5) cross sections are evaluated for each of the possible states of all the ion species in the plasma, multiplied by the ion density appropriate for that state, then summed. This sum is corrected for stimulated emission, in local thermodynamic equilibrium, by multiplying it by $1 - \exp(-h\nu/kT)$. Thomson scattering contributes an amount $N_e\sigma_T$ to the absorptivity. Rayleigh scattering, like bound–bound absorption, is the sum of ion state densities multiplied by the Rayleigh scattering cross section. The total of Thomson and Rayleigh scattering is added to the absorptivity as-is, since the stimulated scattering correction is negligible. The total absorptivity (dimension inverse length) is converted to opacity (dimension length^2/mass) by dividing by the mass density ρ. Thus the opacity, denoted by κ_ν, is given by an expression like

$$\kappa_\nu\rho = \left[\sum_{ijk} N_{ij}\frac{\pi e^2}{mc} f_{jk}\phi_{jk}(\nu) + \sum_{ij} N_{ij}\sigma_{ij}^{PI}(\nu) + \sum_{ij} N_{ij}N_e\bar{\sigma}_{ij}(v;v'\nu)\right]\left(1 - e^{-h\nu/kT}\right) + N_e\sigma_T + \sum_{ij} N_{ij}\sigma_{ij}^{RS}(\nu), \quad (8.81)$$

in which the index i denotes a particular charge state for a particular element and the index j denotes a particular electronic excitation state of that ion. The quantity $\bar{\sigma}_{ij}(v;v'\nu)$ is the thermally-averaged free–free absorption cross section for a unit density of colliding electrons, and $\sigma_{ij}^{RS}(\nu)$ is the Rayleigh scattering cross section for ions i in excitation state j. The Rayleigh term is most often approximated using the ground-state contribution alone. Opacity calculations in general, and the construction of mean opacities, are discussed by Cox and Giuli (1968).

As indicated earlier, diffusion theory leads to a frequency-integrated flux that is proportional to the gradient of the temperature, *viz.*,

$$\mathbf{F} = -\frac{16\sigma T^3}{3\kappa_R\rho}\nabla T, \quad (8.82)$$

where here we have replaced the absorptivity in terms of the opacity. To repeat how we have arrived at this formula, we have the diffusion formula for the monochromatic flux,

$$\mathbf{F}_\nu = -\frac{4\pi}{3\kappa_\nu\rho}\nabla B_\nu(T) = -\frac{4\pi}{3\kappa_\nu\rho}\frac{dB_\nu}{dT}\nabla T, \tag{8.83}$$

which we integrate over ν. The integral of the derivative of the Planck function is just the derivative of the integral, so

$$\int_0^\infty d\nu \,\frac{dB_\nu(T)}{dT} = \frac{d}{dT}\int_0^\infty d\nu\, B_\nu(T) = \frac{d}{dT}\left(\frac{\sigma T^4}{\pi}\right) = \frac{4\sigma T^3}{\pi}. \tag{8.84}$$

The Rosseland mean opacity κ_R is defined by

$$\frac{1}{\kappa_R} = \frac{\displaystyle\int_0^\infty d\nu \,\frac{1}{\kappa_\nu}\frac{dB_\nu}{dT}}{\displaystyle\int_0^\infty d\nu\,\frac{dB_\nu}{dT}}, \tag{8.85}$$

and combining the last three results gives the desired expression. We also repeat for emphasis that the Rosseland mean is computed from the sum of absorption and scattering in which the absorption part is corrected for stimulated emission.

We also may sometimes want a mean opacity with which to express the total emission rate by the material, as in (4.43). If there is a mixture of absorption and scattering, so that $k_\nu = k_\nu^a + k_\nu^s$, and if the source function for the absorption part is B_ν and that for the scattering part is J_ν, as discussed in the previous chapter, then the expression for the energy coupling rate g^0 reduces to this:

$$\begin{aligned} g^0 &= \int d\nu \int d\Omega\, \left[k_\nu^a B_\nu + k_\nu^s J_\nu - (k_\nu^a + k_\nu^s) I_\nu\right] \\ &= 4\pi \int d\nu\, \left[k_\nu^a B_\nu + k_\nu^s J_\nu - (k_\nu^a + k_\nu^s) J_\nu\right] \\ &= 4\pi \int d\nu\, k_\nu^a (B_\nu - J_\nu). \end{aligned} \tag{8.86}$$

We have assumed that the absorptivity is isotropic, so this reduction is appropriate for the comoving frame. What we see is that the scattering process disappears from the net energy exchange rate. To be more precise, the energy exchange due to scattering comes from the terms, neglected here, that allow the frequency of a photon to change in the scattering process. We focus now on the B_ν term in the integral; this is from the emission rate. If we define the Planck mean opacity by

$$\kappa_P = \frac{\displaystyle\int_0^\infty d\nu\,\kappa_\nu B_\nu}{\displaystyle\int_0^\infty d\nu\, B_\nu}, \tag{8.87}$$

then the energy emission rate is

$$4\pi\kappa_P\rho B = 4\sigma\kappa_P\rho T^4. \tag{8.88}$$

Unfortunately, the spectral distribution J_ν is not generally known unless a detailed multifrequency calculation is done, so the integral of the absorption term cannot be evaluated exactly. What might be done in a gray approximation is to assume that the spectrum of J_ν has the same shape as that of B_ν, and hope for the best. Thus the same mean, the Planck mean, might be taken for both terms. If this is done, the result for the energy coupling (in the comoving frame, in general) is

$$g^0 = 4\pi\kappa_P\rho(B - J), \tag{8.89}$$

where B is $\sigma T^4/\pi$ and J stands for the frequency integral $\int d\nu\, J_\nu$. This is the relation that was used in Section 7.2.

We note that the Planck mean is computed by omitting the scattering from the opacity and using the stimulated emission correction on the absorption part. It is possible (just for amusement, perhaps) to cancel the factor $1 - \exp(-h\nu/kT)$ in the corrected opacity with the similar factor in the denominator of the Planck function, and thereby calculate the numerator integral using the *uncorrected* absorption opacity and the Wien approximation $(2h\nu^3/c^2)\exp(-h\nu/kT)$ to the Planck function. This is to say that the emissivity does not care whether the photons are bosons or not.

Some general remarks on the two types of mean opacity are in order. When the spectral distribution of κ_ν is very complex, with a great many strong lines, photoionization edges, and so on, the dynamic range of κ_ν may be very large. In the center of a strong line we are not surprised to find opacity values of $10^{10}\,\mathrm{cm^2\,g^{-1}}$, while in the windows between lines, and below absorption edges, the opacity may become as low as the scattering part, which is around $0.2\,\mathrm{cm^2\,g^{-1}}$, giving a range of ten orders of magnitude! In this circumstance the Planck mean, since it is a linear mean, may be orders of magnitude larger than the Rosseland mean, which is like a harmonic, i.e., reciprocal, mean. In very tenuous plasmas the scattering can be much larger than the absorption opacity, with the result that the Rosseland mean, which includes scattering, can be much larger than the Planck mean, which does not. These gross variations should be kept in mind and appropriately applied to the problem at hand.

Another remark concerns the relative importance of different elements in a mixed plasma such as the astrophysical mixture, which has large amounts of H and He, and progressively smaller amounts of heavier elements C, N, O, Ne, Na, Mg, Al, Si, and on to Fe and Ni. The atomic fraction of Fe is only about 4×10^{-5} compared with H. Nonetheless, Fe is the dominant opacity contributor under some plasma conditions. The reason is that at high temperature all the lighter elements, and abundant H and He in particular, are fully stripped: the mean number of bound electrons is quite small. This eliminates bound–bound and bound–free absorption for these elements. What is left is free–free absorption, which is relatively weak except at low frequency. The heavier elements still have several bound electrons under the same conditions, and therefore their opacity contribution is considerable. Even if Fe, say, were fully stripped its opacity contribution would be large in comparison to its abundance because of the Z^2 factor in (8.79); accounting for this factor we see that Fe free–free contributes 3% as much as H free–free.

The procedure for computing opacities runs like this. A theory of atomic structure is first used to determine the energy levels and ionization potentials of all the relevant ions of all the elements thought to make significant contributions to the total opacity, and, if necessary, the wave functions that go with each energy level. Then the oscillator strengths of all the possible transitions are computed from these wave functions, and the photoionization cross sections as well. If the atomic structure is being treated in a simplified way, then perhaps simple formulae or just the hydrogenic values may be used, thereby avoiding having to calculate large numbers of radial integrals by numerical quadrature. The photoionization cross sections and perhaps the free–free cross sections are found by other quadratures or other formulae.

The final step before the opacity can be computed is to determine the broadening of the spectral lines. The Doppler broadening is easy: a Gaussian distribution with a standard deviation $\Delta\nu = \nu\sqrt{kT/Mc^2}$, where M is the mass of the atom. Electron collisions, by themselves, generally make a Lorentzian profile $(\Gamma/4\pi^2)/[(\nu - \nu_0)^2 + (\Gamma/4\pi)^2]$, where Γ is the rate of electron collisions with the radiating atom. Ions in the plasma have a different effect: they do not move rapidly enough to interrupt the process of radiation, which is what leads to the Lorentzian profile. Instead, the radiation occurs while there is a reasonably constant configuration of perturbing ions. For each configuration there is a certain value of the electric field at the location of the radiating atom, which shifts the energy levels through the Stark effect. The probability distribution of the electric field (plasma microfield), folded with the relations that determine the level shifts in terms of the field, gives the line broadening contribution due to ions. (In more sophisticated calculations the ions are not treated as completely stationary, nor are the electrons treated with a simple Lorentzian profile.) The Doppler broadening, electron impact broadening, and ion broadening are the major mechanisms, but other mechanisms must be included in special cases.

The calculation of opacities for astrophysical use has evolved greatly over the years, as we might have expected. The first calculations applied the hydrogenic formulae to all elements, and in fact omitted the line contribution entirely. Such were the opacity calculations by Keller and Meyerott (1955), for example. (Although some elements were represented using Hartree-type wave functions.) In the next generation, the Cox and Stewart opacities of the 1960s (Cox, Stewart, and Eilers, 1965, Cox and Stewart, 1965), the hydrogenic formulae were still used, but with scaling factors to account for the shielding of the nucleus by other electrons in a multielectron ion. The line absorption was included for the first time. At around the same time the Thomas–Fermi method was developed as an alternative to the hydrogenic model. In this approach single-electron wave functions were actually computed numerically using for the central potential the solution of the Thomas–Fermi model atom. For technical reasons this approach gave opacities similar in accuracy to the scaled hydrogenic method. The advances in opacity theory since the 1970s have included using the Hartree–Fock self-consistent field method to get much more accurate values of the energy levels and the wave functions. (The Fock part of the name means that antisymmetry of the wave functions was explicitly built in, by using both a direct and an exchange part of the electron–electron potential function.) It should be noted that even the very best of these atomic structure calculations is far less accurate than the experimental determinations of atomic energy levels. One might ask: why not just use the experimental data? The problem is that the experimental data are not complete enough to supply all the levels that are needed, and the energy level information does not help much with determining the oscillator strengths and cross sections. A saving fact is that the mean opacity is really not very sensitive to the precise values of the transition frequencies. If the calculated frequencies are statistically correct then the opacity will be about right, when the average is taken, even though some of the lines have been shifted around compared with the true frequencies. There have been advances beyond Hartree–Fock as well. Some opacity calculations (e.g., the Opacity Project (Seaton, 1995)) have used the method of configuration interaction, in which the multielectron Hamiltonian is computed for a large basis set of somewhat arbitrarily selected states, then diagonalized to determine energy levels and eigenfunctions. This method can be very accurate provided the basis set is large enough.

So far the methods described have treated the atoms as isolated entities, so the wave functions are calculated as extending to $r = \infty$. This is a reasonable approximation when the mean distance between ions is much larger than the radius of the Bohr orbits of interest, which is about $n^2 a_0 / Z^*$ in terms of the principal quantum number n and the ion charge Z^*. It is not a good approximation when the mean ion spacing is small, nor is it a good model for highly excited states. Calculating the atomic states and their occupation probabilities for dense plasmas is a significant challenge. The OPAL opacity code uses a less elaborate atomic structure method than the configuration-interaction method adopted for the Opacity Project – OPAL uses the parametric-potential method, a fitting technique – but applies quite

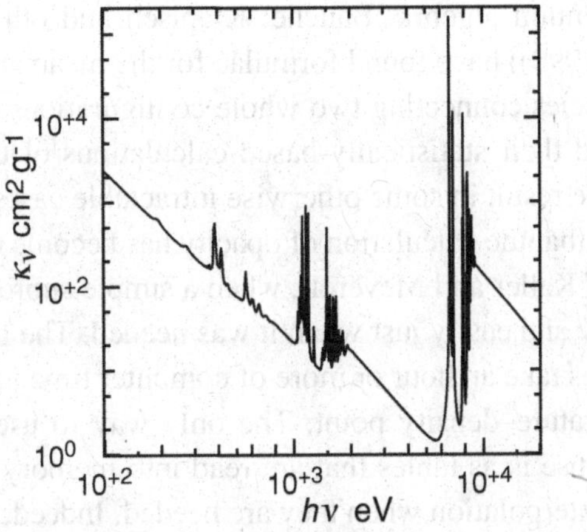

Fig. 8.2 The monochromatic opacity of iron calculated with OPAL is shown for the conditions $T = 1\,\text{keV}$, $\rho = 1\,\text{g cm}^{-3}$.

a sophisticated treatment of the electron–electron, electron–ion and ion–ion correlations that arise because the charges are close enough to interact strongly. OPAL calculates all the individual lines from the various configurations in detail, including the relativistic fine structure, and uses a careful treatment of ion microfield broadening and electron impact broadening. Thus even though the individual line frequencies are not given to high accuracy by the parameterized-potential method, the OPAL opacities have been shown to be free from major systematic errors. A sample OPAL opacity calculation for iron is shown in Figure 8.2 and illustrates the universal properties of a general ν^{-3} dependence of the bound–free and free–free opacity between edges and lines, the high value attained by the free–free opacity at low ν, the jumps at bound–free edges that are ever-larger as ν increases until the last one (the K edge) is reached, and the extremely large values of opacity attained within the lines, which tend to cluster on the low-ν side of the edges. This example is neither unusually simple nor unusually complex in terms of line structure.

The OPAL and Opacity Project opacity databases are nicely reviewed in Seaton (1995), Berrington (1995), and Rogers and Iglesias (1995). The opacity project is also described by Seaton, Yu, Mihalas, and Pradhan (1994), and the latest OPAL results by Iglesias and Rogers (1996).

An approach such as that used in OPAL and in the Opacity Project, in which every line is computed individually to form a spectrum, fails badly when there are atomic shells, perhaps several of them, that contain several electrons with larger angular momenta, such as $3d^6 4s 4p^3$, for example. Some success has been achieved in these complicated atoms by using the ideas of unresolved transition arrays (UTAs) and super transition arrays (STAs). Using powerful modern methods for angular momentum algebra, Bauche, Klapisch and others (see Bauche and Bauche-Arnoult (1990)) have found formulae for the mean and standard deviation of the line frequencies connecting two whole configurations or even small sets of configurations, and their statistically-based calculations of the spectrum give an acceptably accurate result in some otherwise intractable cases.

It must be clear that the calculation of opacity has become much more elaborate than in the days of Keller and Meyerott, when a simple subroutine could compute the opacity quickly and easily just when it was needed. The present-day elaborate opacity calculations take an hour or more of computer time just to find the opacity at a single temperature–density point. The only way to use such opacities that makes practical sense is as tables that are read into memory and that are used to supply values by interpolation when they are needed. Indeed, the use of larger and larger tables of opacity has been common since the production of the Los Alamos Opacity Library in the 1970s (Huebner, Merts, Magee, and Argo, 1977). In some of today's advanced stellar atmosphere modeling the frequency-dependent opacity is needed as well as, or instead of, the Rosseland mean, and these tables are now saved when the opacity codes are run. The size of these tables is very large, but, as we will see shortly, all the other aspects of the radiation hydrodynamic calculations that use multifrequency opacities are large as well.

9

Spectral line transport

9.1 Non-LTE

The title non-LTE refers to a method for analyzing the interaction between a gas and the radiation field that accounts for the modification of the excitation and ionization state of the atoms in the gas by the influence of the radiation field. If the radiation field is weak or the density is large, then the occupation probabilities of the atomic states are governed by the dominant collisional processes. In a wide variety of cases the rates of electron–electron and ion–ion collisions, and usually electron–ion collisions as well, are so great that the relaxation of all these species to a single kinetic temperature is essentially complete. So when collisional processes dominate the atomic excitation rates, the result is occupation probabilities that agree with thermodynamic equilibrium, no matter what the radiation intensities may be. This is LTE – local thermodynamic equilibrium. So non-LTE is the other situation. In non-LTE the relaxation of the velocity distributions of the electrons and atoms to a single kinetic temperature is still supposed to be true. But what is *not* true in non-LTE is that collisional processes are generally dominant over the competing radiative processes for populating and depopulating the atomic states.

Non-LTE is a large subject. Good references for non-LTE line transport are Mihalas's *Stellar Atmospheres* (1978), Athay (1972b) and Ivanov (1973). Approximations to the electron-impact excitation rate are discussed by Sobel'man (1979) and by Sampson and Zhang (1992, 1996).

9.1.1 Kinetic equations

As a system for doing calculations, non-LTE consists of solving for the radiation field I_ν and the set of atomic occupations N_{ij}, also often called the "populations." Here i is an index for elements, and j is an index for an energy level belonging to that element. The set of equations that determines these are the transport equation, which fixes the intensity, and the atomic kinetic equations. The kinetic equations have this generic form:

$$\rho \frac{D(N_{ij}/\rho)}{Dt} = \sum_k \left[N_{ik}(P_{kj} + C_{kj}) - N_{ij}(P_{jk} + C_{jk}) \right]. \tag{9.1}$$

The Ps and Cs are rate coefficients. The Ps in particular are radiative rates, such as spontaneous and stimulated decay, photoabsorption, and photoionization, and the Cs are collisional rates, most commonly the ones for inelastic electron collisions. The elastic rates are ignored, since they do not change the level populations (although they count in electron impact line broadening). The electron rates can be expressed as N_e times a collisional rate coefficient, and the latter is a quantity like $\langle v\sigma_{jk} \rangle$, where the brackets indicate an average over the electron Maxwellian distribution of velocities. The radiative rates are combinations of spontaneous downward rates, stimulated downward rates that are written as integrals of appropriate cross sections over the photon flux spectrum, and upward radiative rates that are likewise integrals of photon fluxes times cross sections.

Every rate in (9.1) appears twice, once as a positive term in the equation for the receiving level, and a second time as a negative term in the equation for the originating level. As a result, if all the equations for the levels of all the ions of one element are added up, everything in the sum cancels out. The time derivative on the left turns into ρ times the convective time derivative of the atomic fraction of that element, which vanishes unless there is diffusion (of atoms!) in the fluid.

Of course, time scales vary enormously, but it helps to think about the magnitude of things. A typical flow time scale in the sun or some other star might be around 10^3 s, or perhaps in some cataclysm as short as 1 s. The lifetime for spontaneous decay in a typical strong spectral line in the optical region is something like 10^{-8} s. Radiative recombination lifetimes in stellar atmospheres are on the order of $1/(N_e \alpha_{RR})$ which comes out 10^{-3} s if the electron density is 10^{16} cm^{-3}, a fairly typical value for photospheres. Electron collisional rate coefficients are something like 10^{-8} cm^3 s^{-1}, which makes the collision rate around 10^8 s^{-1}. Some of the photoionization rates, those that involve threshold frequencies in the far-UV, will be vastly smaller than the radiative recombination rates. Thus we see that there is a very large dynamic range of the rate values, from a rate that is perhaps comparable with the flow rate, to processes such as spontaneous bound–bound rates and electron collisions that may be 10^8 times faster. In other environments the comparisons will differ, but the large dynamic range is a common property. We notice that in this case the rates are (perhaps all?) larger than the flow rate, and that no great error would be committed by dropping the time derivative term. This is often, but not always, a reasonable approximation, called statistical equilibrium.

Let us think for a minute what is entailed in solving a non-LTE problem. We are trying to solve simultaneously (4.23) and (9.1). In both of these the characteristic time for the right-hand side terms may be very short – the light-travel time for the transport equation and the time scale of the fastest rates for the kinetics. A straightforward explicit time differencing of either equation would be subject to a stability condition that restricts the time step to a value smaller than these characteristic times. This is usually prohibitively small. Thus some kind of implicit time differencing must be used, which amounts to solving the steady-state equations for the advanced-time variables, with the time derivative term as a small perturbation. This situation is described by saying that these differential equations (in time) are *stiff*. As the numbers just quoted indicate, the kinetic equations are very stiff indeed, since there may be a factor 10^8 separating the fastest relaxation time from the time scale on which we would like to integrate. (Those who are familiar with the association between stiffness and the magnitude of the eigenvalues of the Jacobian matrix of the derivatives with respect to the unknowns may want to know that the largest eigenvalue of the rate matrix is comparable with the fastest rate.) Thus we want to consider the worst case of stiffness, which is the steady-state limit, and reality is often close to it.

9.1.2 Two-level atom

In order to develop some more of the non-LTE ideas, let us consider the prototype non-LTE problem, which is the two-level atom in a steady state. That is, we restrict ourselves to a single ion, and ignore all but two energy levels, the ground state and an excited state connected to the ground state by a strong line. We use the labels 1 and 2 for the levels, and account for the radiative processes of radiative excitation, spontaneous and stimulated emission, collisional excitation, and collisional deexcitation (superelastic collisions). There is just one rate equation, since the second one is guaranteed to be the same as the first. This is

$$N_1\left(B_{12}\bar{J}_{12} + N_e C_{12}\right) = N_2\left(A_{21} + B_{21}\bar{J}_{12} + N_e C_{21}\right). \tag{9.2}$$

The radiative rates have been expressed in terms of the Einstein coefficients. Because the intensity may, indeed will, have some variation within the line bandwidth, the absorption and stimulation emission rates are proportional to a cross-section-weighted angle-averaged intensity:

$$\bar{J}_{12} = \int d\nu\, \phi_{12}(\nu) J_\nu. \tag{9.3}$$

The expressions for the absorptivity and emissivity are as follows (see (8.28), (8.29)):

$$k_\nu = \frac{h\nu}{4\pi}(N_1 B_{12} - N_2 B_{21})\,\phi_{12}(\nu), \tag{9.4}$$

$$j_\nu = \frac{h\nu}{4\pi} N_2 A_{21} \phi_{12}(\nu). \tag{9.5}$$

Notice that we *cannot* assume that the stimulated correction factor is $1 - \exp(-h\nu/kT)$. If we divide the emissivity by the absorptivity we get the line source function,

$$S_\nu = \frac{N_2 A_{21}}{N_1 B_{12} - N_2 B_{21}} = \frac{2h\nu^3/c^2}{N_1 g_2/N_2 g_1 - 1}, \tag{9.6}$$

as we described earlier in (8.21). One thing we notice here is that the shape function $\phi_{12}(\nu)$ has canceled out, and the source function is very nearly independent of frequency. If we inquire a little more deeply into this result, it is a consequence of the assumption that the quantum state of the excited atom has lost any *coherence* it may have acquired in a photoexcitation, and has the same random phase that it would have if it were collisionally excited, for example. Reality is more complicated. If the probability of a radiative decay vastly exceeds the probability of any collisional process, then the coherence is *not* lost, which means that the frequency profile of the emitted radiation will contain traces of the spectral distribution of the radiation that produced the excitation. This situation is called *partial redistribution* and will be discussed in the next section. We note that scattering of a photon in the wing of a resonance line takes two forms, which correspond to the two possibilities for coherence. If the coherence is retained, then the emitted photon has essentially the same frequency as the absorbed photon, and this process is called *Rayleigh scattering*. When the coherence is lost, the emitted photon frequency is near line center, and the process is called *resonance fluorescence*. When the incoming photon's frequency is quite far in the wing, the Rayleigh scattering process predominates. A general term for scattering in which the coherence is lost is *complete redistribution*, and this is what we shall assume for the present.

The rate equation can be solved for N_1/N_2 which is then substituted into the formula for S_ν. Before doing so we note that the collisional rate coefficients must themselves be consistent with LTE. The upward and downward electron-collision rate coefficients are determined by the velocity distribution of the electrons, which we assume to be Maxwellian at the temperature T. The upward and downward collision rates would then balance if the atomic populations obeyed the Boltzmann excitation formula at the same temperature. This leads to

$$C_{12} = \frac{g_2}{g_1} e^{-h\nu_0/kT} C_{21} \tag{9.7}$$

in view of the Boltzmann excitation formula, since $h\nu_0$ is the excitation energy. Performing the substitutions leads to this expression when the radiative rates are also included:

$$S_\nu = (1-\epsilon)\bar{J}_{12} + \epsilon B_\nu, \tag{9.8}$$

with

$$\frac{\epsilon}{1-\epsilon} = \frac{N_e C_{21}}{A_{21}} \left[1 - \exp(-h\nu_0/kT)\right]. \tag{9.9}$$

The formula (9.8) strongly resembles the scattering source function considered in Milne's second problem (see Section 5.5), and $1-\epsilon$ is analogous to the scattering albedo. We can regard the line scattering event in this way. When a photoexcitation occurs, the excited atom can decay in two ways. Either it decays radiatively, producing another photon, or it decays by a superelastic collision, in which case we say that the excitation has been quenched. In the first case we can say that the photon has been reborn, and begins its next flight, while in the second case we say that the collision has killed the photon. The branching ratio for reemission is the albedo. Apart from the factor $1 - \exp(-h\nu_0/kT)$ this is simply given by the ratio of the A value to the sum of the A value and the collisional deexcitation rate. To repeat: this is not scattering in the sense of Thomson or even Rayleigh scattering, since in those cases the amplitude for emission combines coherently with the amplitude for absorption. Here we assume that the excited atom lingers a while, so that its phase is scrambled before reemission. However, in the far wings of the line, the uncertainty principle for time vs energy requires that the emission occur promptly, which makes the process coherent. Thus line scattering as we describe it here goes over into Rayleigh scattering in the far wings of the line. Complete redistribution is inappropriate in that case.

There is a useful approximate relation between C_{21} and A_{21} due to Van Regemorter (1962) which is found when the process of electron impact excitation is treated in a large-r dipole approximation. It is

$$\frac{C_{21}}{A_{21}} \approx \frac{0.2\lambda^3 g}{\sqrt{T/10^4\,\mathrm{K}}}, \tag{9.10}$$

in which g is formally the same as the free–free Gaunt factor $\bar{g}_{III}(\nu_0)$, but is better regarded as a fitting constant. Comparing (9.10) with accurate electron excitation calculations suggests that $g \approx 0.2$ is a good choice for collisions with positive ions, except that transitions between states of the same principal quantum number

n require a larger value, more like unity. Using the Van Regemorter formula gives this simple result for ϵ:

$$\frac{\epsilon}{1-\epsilon} \approx \frac{0.2 N_e \lambda^3 g \left[1 - \exp(-h\nu_0/kT)\right]}{\sqrt{T/10^4\,\mathrm{K}}}. \qquad (9.11)$$

The wavelength λ that appears here is the wavelength of the spectral line being considered. Just so we can appreciate the order of magnitudes involved, we suppose that T is 10^4 K and the line is in the visible part of the spectrum, so $\lambda \approx 5000$ Å. The result for $\epsilon/(1-\epsilon)$ is $N_e/4 \times 10^{13}\ \mathrm{cm}^{-3}$. The place in the solar atmosphere where the electron density is 4×10^{13} is near the photosphere; the electron density there is much smaller than the value 10^{16} mentioned earlier since the degree of ionization is low. In the atmosphere of a hot star this electron density comes at approximately $\tau = 10^{-2}$. The resonance lines we are considering are formed higher in the atmosphere than either of these estimates, and so we generally expect ϵ to be small.

We will describe the solution of the simplest problem of resonance line transfer, namely finding the source function and the emergent spectrum when the temperature and ϵ are uniform in a half-space. We need to introduce an optical depth scale that describes the line as a whole in a meaningful way, and the first step toward this is to introduce a frequency scale measured in units of the Doppler width from the center of the line. Here the Doppler width is $\sqrt{2}$ times the standard deviation of the Gaussian profile, $\Delta\nu_D = \nu\sqrt{2kT/Mc^2}$. Our scaled frequency is

$$x = \frac{\nu - \nu_0}{\Delta\nu_D}. \qquad (9.12)$$

We then use a cross section profile function that is normalized on the x scale:

$$\int_{-\infty}^{\infty} dx\, \phi(x) = 1. \qquad (9.13)$$

Next we define the optical depth scale τ, without a frequency suffix, so that $\tau_\nu = \tau\phi(x)$. This means that τ is calculated from

$$d\tau = \frac{h\nu_0}{4\pi\,\Delta\nu_D}(N_1 B_{12} - N_2 B_{21})\,dz. \qquad (9.14)$$

For simplicity we assume that $\phi(x)$ is independent of depth, which is reasonable for a constant temperature atmosphere.

Now we can obtain J_ν from (5.25) applied frequency-by-frequency,

$$J_\nu(\tau) = \frac{1}{2}\int_0^\infty \phi(x)\,d\tau'\, E_1(\phi(x)|\tau' - \tau|)\,S(\tau'). \qquad (9.15)$$

Next we multiply by $\phi(x)$ and integrate over x to obtain $\bar{J}$:

$$\bar{J}(\tau) = \int_0^\infty d\tau\, K_1(\tau' - \tau) S(\tau'), \tag{9.16}$$

where $K_1(\tau)$ is a new kernel function defined by

$$K_1(\tau) = \frac{1}{2} \int_{-\infty}^{\infty} dx\, \phi^2(x) E_1(\phi(x)|\tau|). \tag{9.17}$$

The K_1 function is normalized to unity, as can easily be shown by integrating over τ under the integral over x, then using the normalization of ϕ. But this function is very different in character from those kernels, like $E_1(\tau)$, that fall off exponentially at large τ. In fact, in the important Doppler broadening case K_1 falls off only as $1/\tau^2$ for large τ. That means that *the mean free path for line scattering is infinite*. This is a major difference between line scattering and Thomson scattering, say. If a photon does 10^4 Thomson scatterings, then its net displacement is about $\Delta\tau = 100$, since each flight represents a rms step of 1 in τ, and the flights add in the square. Line transport does not work that way. On each flight in a line transfer problem the photon chooses a frequency by sampling the distribution $\phi(x)$, then it goes a distance of order $1/\phi(x)$ in τ. What happens after many flights is actually this: *the net displacement is contributed almost entirely by the one flight for which the sampled frequency was farthest from line center*. Line photons do not diffuse when there is complete frequency redistribution.

Consider those 10^4 line scatterings again and assume the line profile is Doppler, $\phi(x) = \exp(-x^2)/\sqrt{\pi}$. The largest frequency x of that many samples will be about the value for which the integral of the tail of ϕ from x to ∞ is around 10^{-4}. But the integral is approximately $\phi(x)/2x$. Thus the value of ϕ for this x is roughly the same as the tail probability, namely 10^{-4}, so the mean free path *at this frequency* is 10^4 in τ units, and that is how far the photon goes on that one flight. So for line scattering with a Doppler line profile, the net displacement after a large number N of scatterings is of order N, not $\sqrt{N}$ as it would be in the usual random walk.

The integral with respect to τ of the K_1 function defines the K_2 function:

$$K_2(\tau) = 2\int_\tau^\infty d\tau'\, K_1(\tau') = \int_{-\infty}^{\infty} dx\, \phi(x) E_2(\phi(x)\tau) \qquad \tau > 0. \tag{9.18}$$

If, in (9.16), $S(\tau')$ is evaluated at $\tau' = \tau$ and taken out of the integral, the result is

$$\bar{J}(\tau) \approx [1 - p_{\rm esc}(\tau)] S(\tau), \tag{9.19}$$

in which the two-sided escape probability is now given by

$$p_{\rm esc}(\tau) = \frac{1}{2}K_2(\tau). \tag{9.20}$$

This is analogous to the monochromatic equation (5.24) discussed earlier, except that now we are considering the escape of line photons from the medium when the frequency can change randomly at each scattering. The nature of the kernels $K_1(\tau)$ and $K_2(\tau)$ is all-important in setting the qualitative nature of the escape of line radiation. The different shapes of line profile $\phi(x)$ lead to different kernel behavior for large τ, as shown by Avrett and Hummer (1965). For $K_1(\tau)$:

$$K_1(\tau) \sim \begin{cases} \dfrac{1}{4\tau^2\sqrt{\ln(\tau/\sqrt{\pi})}} & \text{Doppler} \\ \dfrac{\sqrt{a}}{6\tau^{3/2}} & \text{Voigt, } a\tau \gg 1. \\ \dfrac{1}{6\tau^{3/2}} & \text{Lorentzian} \end{cases} \tag{9.21}$$

For $K_2(\tau)$:

$$K_2(\tau) \sim \begin{cases} \dfrac{1}{2\tau\sqrt{\ln(\tau/\sqrt{\pi})}} & \text{Doppler} \\ \dfrac{2\sqrt{a}}{3\sqrt{\tau}} & \text{Voigt, } a\tau \gg 1. \\ \dfrac{2}{3\sqrt{\tau}} & \text{Lorentzian} \end{cases} \tag{9.22}$$

The Milne equation to be solved for the source function for the constant-temperature half-space problem is the following:

$$S(\tau) = \epsilon B_\nu + (1-\epsilon)\int_0^\infty d\tau' K_1(\tau'-\tau)S(\tau'). \tag{9.23}$$

We will skip the analytical solution methods, which can just be copied over from the single-frequency or gray case, and go to the results. The value of S at the surface $\tau = 0$ is $\sqrt{\epsilon}B_\nu$, an exact result. Then S rises from that value with increasing τ more or less as $\tau^{1/2}$ and eventually reaches B_ν when τ is approximately the thermalization depth. The thermalization depth can be found from the root of a transcendental equation similar to (5.42). It also comes from a physical argument. When ϵ is a small number, the photon will survive about $N = 1/\epsilon$ flights before being destroyed. The thermalization depth is about equal to the net distance the photon will go in that many flights, which as we have seen is about $\Delta\tau = N$. Therefore the thermalization depth is approximately $1/\epsilon$ on the τ scale for the

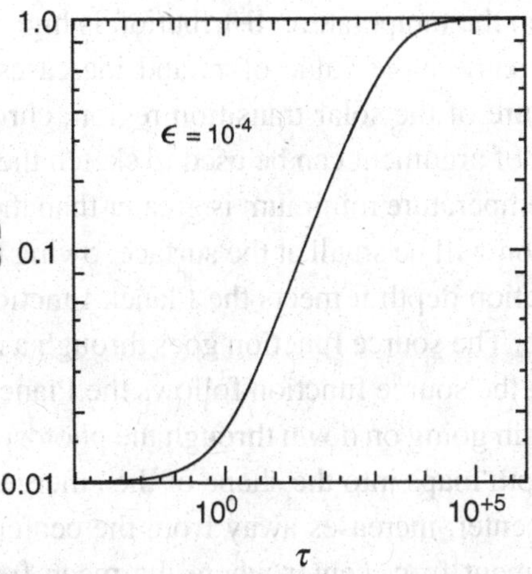

Fig. 9.1 Source function S of a line in a semi-infinite isothermal atmosphere formed with complete redistribution over a Doppler profile, with $\epsilon = 10^{-4}$, in units of Planck function B vs optical depth scale of (9.14).

case of Doppler broadening. This does vary with the shape of the line profile, so for example, for a Lorentzian profile the thermalization depth is roughly $1/\epsilon^2$. A more precise definition of the thermalization depth Λ that is consistent with this argument is given by the statement

$$K_2\left(\frac{1}{2}\Lambda\right) = \epsilon. \tag{9.24}$$

When scattering is quite dominant, so ϵ is very small, the relations for the thermalization depth with the various profiles are

$$\Lambda \sim \begin{cases} \dfrac{1}{\epsilon\sqrt{-\ln(\sqrt{\pi}\epsilon)}} & \text{Doppler} \\ \dfrac{8}{9}\dfrac{a}{\epsilon^2} & \text{Voigt, } a \gg \epsilon. \\ \dfrac{8}{9}\dfrac{1}{\epsilon^2} & \text{Lorentzian} \end{cases} \tag{9.25}$$

The source function for the constant-temperature half-space with Doppler broadening and for $\epsilon = 10^{-4}$ is illustrated in Figure 9.1.

The shape of the line in the spectrum of the emergent flux follows from an application of the ordinary Eddington–Barbier relation. Thus at line center the flux is about π times the surface source function, or $\pi\sqrt{\epsilon}B_\nu$, then away from line center it rises in proportion to $1/\phi(x) \propto \exp(x^2)$ until it reaches the level of πB_ν at $x \approx \sqrt{-\log(\epsilon)/2}$. This gives a trough-shaped profile with quite a black center and steep edges. When the temperature distribution is high at $\tau = 0$, goes through a broad minimum at some large value of τ, and increases again at yet larger τ, which is a rough picture of the solar transition region, chromosphere, and photosphere, the same kind of argument can be used to sketch the emergent spectrum. If the value of τ at the temperature minimum is greater than the thermalization depth, then the source function will be small at the surface, owing to $\sqrt{\epsilon}$, then rise inward until at the thermalization depth it meets the Planck function, which has been declining over this range. The source function goes through a maximum at this point. Going further inward, the source function follows the Planck function down to the minimum, then up again going on down through the photosphere. This shape of the source function vs depth maps into the shape of the emergent flux vs frequency: it is very small at line center, increases away from the center, goes through a maximum at the displacement from center where the mean free path corresponds to the thermalization depth, then drops again farther from line center, goes through a very broad minimum, and finally rises toward the continuum when the mean free path reaches all the way to the photosphere. This is the picture of the Ca II H and K lines and the Mg II h and k lines in the solar spectrum.

9.1.3 Ivanov's approximation

Owing to the nondiffusive character of line transport with complete frequency redistribution, a good approximation for the spatial variation of the source function in a finite or semi-infinite medium with slab symmetry is not simple to obtain. Sobolev's Russian school, and notably V. V. Ivanov, have been successful at doing this. In an analysis that is too long to reproduce here, Ivanov (1973) obtains this approximate solution of the finite slab problem with a uniform Planck function B_ν:

$$S \approx \frac{\epsilon B_\nu}{\sqrt{[\epsilon + (1-\epsilon)K_2(\tau)][\epsilon + (1-\epsilon)K_2(\tau_0 - \tau)]}}, \tag{9.26}$$

where τ is the optical depth measured from one face of the slab and τ_0 is the optical thickness of the slab. $K_2(\tau)$ is the kernel given in (9.18). It has the limits $K_2(\tau) \to 1$ for $\tau \to 0$, and $K_2(\tau) \to 0$ for $\tau \to \infty$. Equation (9.26) has all the right limits: $S \sim \epsilon B_\nu$ when $\tau_0 \to 0$, $S \sim B_\nu$ for τ, $\tau_0 - \tau \to \infty$, and $S(0) = \sqrt{\epsilon} B_\nu$ for $\tau = 0$ and $\tau_0 \to \infty$. In Section 8.11 of Ivanov (1973) the accuracy of (9.26) is computed, and it is always within a factor 2 and often within 20%.

9.1.4 Multilevel atom

The non-LTE problems of practical importance involve more than two levels, indeed they may involve a great many levels. An important stepping-stone to solving such multilevel problems is to treat the spectral lines one at a time, in each case accounting for the response of the populations of the upper and lower levels of this one line to changes in its own radiation field, but omitting the changes that may occur in the other levels. This is the *equivalent two-level atom* method (ETLA) that has been developed by Avrett (1965), Avrett and Kalkofen (1968), Avrett and Loeser (1987), and others. Let the line in question be formed in the transition between lower level ℓ and upper level u. The source function expression (9.6) becomes for this line

$$S_\nu = \frac{2h\nu^3}{c^2} \frac{1}{(b_\ell/b_u)\exp(h\nu/kT) - 1}, \tag{9.27}$$

in which b_ℓ and b_u are the *departure coefficients* of the level populations from LTE:

$$b_n \equiv \frac{N_n}{N_n^*}, \tag{9.28}$$

in which the quantity N_n^* is the population of level n calculated on the basis of thermodynamic equilibrium at the local electron temperature. One technical point to note about (9.27) is that ν need *not* be treated as constant across the line profile. In fact, a more careful analysis of the absorptivity and emissivity profiles shows that on the assumption of complete frequency redistribution in the line scattering, the frequency dependences of absorption, spontaneous emission, and stimulated emission are related in just the way needed for (9.27). This is discussed by Castor, Dykema, and Klein (1992). That reference will be used for the following discussion.

The ratio b_ℓ/b_u can be found by solving the kinetic equations for the two populations N_ℓ and N_u, assuming the remaining populations are given. The full kinetic equations for the two levels ℓ and u, in the steady-state case with $D(N_n/\rho)/Dt = 0$, become

$$\begin{aligned} (P_{\ell u} + C_{\ell u} + a_1)N_\ell - (P_{u\ell} + C_{u\ell})N_u &= a_2, \\ -(P_{\ell u} + C_{\ell u})N_\ell + (P_{u\ell} + C_{u\ell} + a_3)N_u &= a_4. \end{aligned} \tag{9.29}$$

The terms including N_ℓ and N_u are put on the left-hand sides of these equations, and the remaining terms are put on the right. The rate coefficients connecting levels ℓ and u are broken into their radiative and collisional parts, P and C. The photoexcitation rate is given by

$$P_{\ell u} = \int_0^\infty \frac{4\pi\sigma_\nu}{h\nu} J_\nu \, d\nu, \tag{9.30}$$

and the photo deexcitation rate is

$$P_{u\ell} = \left(\frac{N_\ell}{N_u}\right)^* \int_0^\infty \frac{4\pi\sigma_\nu}{h\nu}\left(\frac{2h\nu^3}{c^2} + J_\nu\right)\exp\left(-\frac{h\nu}{kT_e}\right)d\nu, \tag{9.31}$$

in which T_e is the local electron temperature and $(N_\ell/N_u)^*$ is the LTE population ratio at that temperature. The photoabsorption cross section σ_ν can be put back in terms of the oscillator strength and the line profile function, if desired. The quantities a_1–a_4 are terms representing rates from ℓ or u to other levels (a_1 and a_3) and terms representing population flow into levels ℓ or u from other levels (a_2 and a_4). They are given by

$$\begin{aligned} a_1 &= \sum_{i\neq\ell,u} R_{\ell i}, \qquad a_2 = \sum_{i\neq\ell,u} R_{i\ell}N_i, \\ a_3 &= \sum_{i\neq\ell,u} R_{ui}, \qquad a_4 = \sum_{i\neq\ell,u} R_{iu}N_i; \end{aligned} \tag{9.32}$$

The rate coefficients R represent the sum of radiative and collisional coefficients.

The two equations (9.29) are readily solved for N_ℓ and N_u, and the ratio of these gives the source function:

$$S_\nu = \gamma_\nu P_{\ell u} + \epsilon_\nu, \tag{9.33}$$

in which the coefficients γ_ν and ϵ_ν depend on the rates:

$$\gamma_\nu = \frac{2h\nu^3/c^2}{(c_2 + c_4)\exp(h\nu/kT_e) - c_1 - c_3}, \tag{9.34}$$

$$\epsilon_\nu = \frac{2h\nu^3}{c^2}\frac{c_3}{(c_2 + c_4)\exp(h\nu/kT_e) - c_1 - c_3}, \tag{9.35}$$

and, furthermore, the coefficients c_1–c_4 are given by

$$\begin{aligned} c_1 &= P_{\ell u}, \qquad c_2 = P_{u\ell}\left(\frac{N_u}{N_\ell}\right)^*, \\ c_3 &= C_{\ell u} + \frac{a_1a_4}{a_2 + a_4}, \qquad c_4 = \left(C_{u\ell} + \frac{a_2a_3}{a_2 + a_4}\right)\left(\frac{N_u}{N_\ell}\right)^*. \end{aligned} \tag{9.36}$$

These expressions do not require that the line be narrow, so that $\nu \approx$ constant could be assumed; thus they work equally well for photoionization continua. A complexity is that (9.33) is nonlinear in the radiation intensity J_ν in the line because of the appearance of $P_{\ell u}$ and $P_{u\ell}$ in the denominators of (9.34) and (9.35). For narrow spectral lines the exponential factor in the integral defining $P_{u\ell}$ can be taken out of the integral and evaluated at line center, so that when γ_ν and ϵ_ν are calculated at line center the stimulated emission part of $c_2 \exp(h\nu/kT_e)$ precisely cancels c_1, and this nonlinearity disappears.

Continuing with the narrow-line approximation, we simplify (9.33) by putting the cross section in terms of the Einstein A coefficient and the profile function, in which we make use of (8.23) and (8.27). We finally get

$$S_\nu = \frac{\bar{J} + \epsilon' B_\nu + \eta B^*}{1 + \epsilon' + \eta}, \tag{9.37}$$

with the definitions

$$\epsilon' = \frac{C_{u\ell}[1 - \exp(-h\nu/kT_e)]}{A_{u\ell}}, \tag{9.38}$$

$$\eta = \frac{a_2 a_3 - (g_\ell/g_u) a_1 a_4}{A_{u\ell}(a_2 + a_4)}, \tag{9.39}$$

$$B^* = \frac{2h\nu^3/c^2}{g_u a_2 a_3/(g_l a_1 a_4) - 1}. \tag{9.40}$$

This form of the source function expression differs from the simple two-level form (9.8) by the presence of the indirect terms η and ηB^* that involve coupling to other levels. The value of η can be interpreted as a quenching rate, involving an upper-level atom going to some other level and then returning to the lower level, corrected in some way for stimulated emission, in terms of the direct spontaneous decay rate. This indirect process effectively destroys a line photon, and thus it contributes to the denominator in the source function expression. The numerator term ηB^* involves the reverse indirect process: a lower-level atom is taken to some other level, from which it returns to the upper level, thus effectively creating a line photon.

When the transfer equation has been solved for this particular line, the results for the radiative rates $P_{\ell u}$ and $P_{u\ell}$ are saved for the next solution of the kinetic equations to yield the level populations. Avrett and Loeser (1987) and others parameterize the rates in terms of a quantity z that is variously called the Net Radiative Bracket (Thomas, 1960), escape factor (Athay, 1972b) or flux divergence coefficient (Canfield and Puetter, 1981). This is defined for narrow lines by

$$z \equiv 1 - \frac{\bar{J}}{S} = \epsilon' \left(\frac{B_\nu}{S} - 1 \right) + \eta \left(\frac{B^*}{S} - 1 \right). \tag{9.41}$$

The second form follows using (9.37). If we compare (9.41) with (9.19) for the escape probability approximation, we see that the Net Radiative Bracket is the accurate quantity that the escape probability approximates. The net radiative rate for transitions from level u to level ℓ is easily expressed, for narrow lines, in terms of the Net Radiative Bracket as follows:

$$N_u P_{u\ell} - N_\ell P_{\ell u} = N_u A_{u\ell} z, \tag{9.42}$$

so that if the rates $P_{\ell u}$ and $P_{u\ell}$ are replaced by $P_{\ell u}^{\rm eff} = 0$ and $P_{u\ell}^{\rm eff} = A_{u\ell} z$, the self-consistent solution for the level populations will be unchanged. This replacement has been found to significantly improve the convergence of the ETLA iteration method. Castor, Dykema, and Klein (1992) provide the relations that can be used to compute the Net Radiative Brackets and the effective rates without making the narrow-line approximation, and in that paper the ETLA formulation was applied to both lines and continua.

Equation (9.37) has been the basis of much of the discussion of line formation in the solar chromosphere by Thomas, Athay, Jefferies, and coworkers (see Thomas and Athay (1961)). They have found, for instance, that the ϵ' term dominates for the H and K resonance lines of Ca II, while the η term is more important for the hydrogen Balmer lines, with the result that the temperature rise in the chromosphere produces emission features in the profiles of H and K, but not in the Balmer profiles. The hydrogen Lyman lines are found to be similar to the Ca II lines in this respect, and in fact are strongly in emission.

The ETLA method is used by Avrett and coworkers, and was also used by Castor *et al.* (1992), but it is by no means the only method for treating multilevel non-LTE problems.

The complete linearization method of Auer and Mihalas (1969) has been used on a large scale to solve non-LTE problems with tens or more of atomic levels. A good review of the linearization method(s) has been given by Auer (1984). The basic structure of the original (Auer and Mihalas, 1969) linearization method is as follows. The Feautrier variable $j(\mu)$ (see Section 5.6) is used to represent the radiation field at a set of angle and frequency points, with NJ angle-frequency pairs in all. The level populations $N_1, N_2, \ldots, N_{NL}$ are another part of the set of unknowns. The total density or pressure and the temperature may be added to the list if the constraints of hydrostatic equilibrium and radiative energy balance are imposed. All these variables are specified on a mesh of space points, $z_1, z_2, \ldots, z_{ND}$. If X_i represents the unknowns at space point i: $X_i = (j_1, \ldots, j_{NJ}, N_1, \ldots, N_{NL}, N_{\rm tot}, T)$, then the full solution vector is $\mathbf{X} = (X_1, X_2, \ldots, X_{ND})$. Thus there are $ND(NJ + NL + 2)$ unknowns. There are also this many equations to determine them. There is a Feautrier equation (5.49) corresponding to each j and a population kinetic equation corresponding to each N. At the spatial boundaries the Feautrier equation is not used; rather, there is a relation between j on the boundary and the j at the next point away from the boundary that is derived from the relation between h and j at the boundary which expresses the constraint of a certain (or zero) incoming intensity, or reflection symmetry. There are also not N_{NL} independent kinetic equations in the steady-state case, since the sum of all N_{NL} equations identically vanishes in this case. One of the equations must be replaced by the condition of number conservation, $\sum N_\ell = N_{\text{tot}}$. If N_{tot} is not one of the unknowns, then it must be given as data.

In the complete linearization method this system of equations is just treated as the large nonlinear system it is, and the method of solution is the Newton–Raphson method. That is, if $\mathbf{F}(\mathbf{X}) = 0$ represents the set of equations and $\mathbf{X}$ is the solution vector just described, then one step of the iteration process is described by

$$\mathbf{X}^{n+1} = \mathbf{X}^n - \left[\frac{\partial \mathbf{F}(\mathbf{X}^n)}{\partial \mathbf{X}}\right]^{-1} \mathbf{F}(\mathbf{X}^n). \tag{9.43}$$

There are the usual problems with Newton–Raphson, and these arise here. A reasonable first guess is required, or else the iterates quickly diverge. And if the correction to $\mathbf{X}^n$ that is calculated by solving the linear system has excessively large values for some components, such that the estimates of these components of $\mathbf{X}^{n+1}$ would be in unphysical ranges, then all of the corrections may need to be scaled down to keep the new unknowns in the physical range. The special problem in the complete linearization method is the sheer size of the system of equations (9.43). For example, if $NJ = 4000$, $NL = 50$ and $ND = 100$, then the system size is $405\,200 \times 405\,200$. This would be called a small problem. Quite obviously one does not simply apply Gaussian elimination to this system. If it were possible to do so within the constraints of memory, the cost would be of order $ND^3(NJ + NL + 2)^3 \approx 7 \times 10^{16}$ operations. Partitioning methods are applied instead. More recently, iterative methods such as preconditioned Newton–Krylov have been used, but for the present discussion we will stick to direct solution methods. Iterative methods will be considered in Section 11.11.

The partitioning employed by Auer and Mihalas makes use of the sparsity of the Jacobian matrix $\partial\mathbf{F}(\mathbf{X}^n)/\partial\mathbf{X}$ as follows. The Feautrier equations couple the intensity at a given space point to its nearest-neighbor space points on each side. The population kinetics equations and the constraint equations are local in space. Thus the large matrix has a block tri-diagonal structure in which the blocks contain all the unknowns at a single space point. That is, the block size is $NJ + NL + 2$ square. The larger system is solved using the standard recursion scheme for tri-diagonal matrices, except applied here to the blocks. Each step of the (forward) recursion requires of order $(NJ + NL + 2)^3$ operations, and there are ND steps, so the total cost is $ND(NJ + NL + 2)^3$ operations, which is ND^2 times fewer than the ignorant elimination.

Rybicki (1971) suggest a different partitioning of the matrix, which, as Auer (1984) points out, makes the method scale in the same way as radiative transfer solutions using numerical quadrature of the Milne equation (5.37). The idea is simple: the values of $j(\mu, \nu)$ at various angles and frequencies enter the kinetic equations only in the combinations that define $\bar{J}$, namely $\int d\mu \int d\nu \, \phi(\nu) j(\mu, \nu)$. But each $j(\mu, \nu)$ can be found by solving a simple tri-diagonal system, *viz.*,

$$\mathbf{j}_k = \mathsf{T}_k^{-1}\mathbf{S}, \tag{9.44}$$

and then the $\bar{J}$ values are sums:

$$\bar{\mathbf{J}} = \sum_k w_k \mathsf{T}_k^{-1}\mathbf{S}. \tag{9.45}$$

In this way all the intensities can be eliminated from the system, which leaves the smaller system to solve, of which the order is $ND(NL+2)$. The operation of inverting a tri-diagonal matrix is very efficient: $O(ND^2)$ operations, so the cost of eliminating the intensities is $O(NJ\,ND^2)$. The system that is left after the elimination is full, since the inverse of a tri-diagonal matrix is a full matrix. Thus the final solution of this system requires $O(ND^3(NL+2)^3)$ operations. Rybicki's partitioning will be more efficient than the block-tri-diagonal scheme when $ND^2 < [(NJ+NL+2)/(NL+2)]^3$. Thus relatively few level populations and a large number of frequency–angle pairs favors Rybicki; comparable numbers of level populations and frequency–angle pairs favors the block tri-diagonal scheme. For the example dimensions above, the block-tri-diagonal scheme requires about 6.6×10^{12} operations, while in Rybicki's scheme the elimination of the intensities takes 4×10^7 operations and the solution of the resulting system takes 1.4×10^{11} operations. In this case the elimination cost is negligible, and the overall cost favors Rybicki by a factor 47.

The foregoing discussion describes solving the linear equations for the intensities themselves, but exactly the same reasoning also applies to the linear system that must be solved for each Newton–Raphson iteration; the sparsity pattern of the Jacobian matrix has the structure needed to apply Rybicki's elimination scheme. Auer (1984) gives quite a complete description of all the finite-difference methods, the different elimination schemes, and the organization of the complete linearization methods.

9.2 Partial redistribution

One important footnote to non-LTE line transfer concerns what is called *partial redistribution* or PRD, which was alluded to above. Spectral line transfer with partial frequency redistribution is treated well in Ivanov (1973) and Mihalas (1978). This is the subject of line scattering when there is some degree of correlation between the frequency of absorption and the frequency of emission in the scattering process. Let us write the absorptivity as

$$k_\nu = k_L \phi(\nu), \tag{9.46}$$

then we write the emissivity, forgetting about the albedo for the moment, as

$$j_\nu = k_L \int_0^\infty d\nu' \, R(\nu, \nu') J_{\nu'}, \tag{9.47}$$

in which $R(\nu, \nu')$ is the *redistribution function*. The expression for the source function in this conservative scattering (albedo = 1) case is

$$S_\nu = \frac{1}{\phi(\nu)} \int_0^\infty d\nu' \, R(\nu, \nu') J_{\nu'}. \tag{9.48}$$

The redistribution function is the joint probability distribution of the two variables, ν and ν', the frequencies of emission and absorption. In order to conserve photons it must obey the relation

$$\int_0^\infty d\nu \, R(\nu, \nu') = \phi(\nu'). \tag{9.49}$$

Thermodynamic consistency also requires that it be symmetric in ν and ν':

$$R(\nu, \nu') = R(\nu', \nu) \tag{9.50}$$

apart from a small correction when $\nu - \nu_0$ and $\nu' - \nu_0$ are not negligible compared with ν_0 and kT/h. If the frequencies are actually uncorrelated, then R factors into a function of ν times a function of ν', which because of the last two constraints has to be $R = \phi(\nu)\phi(\nu')$. If this is put in, then the source function reduces to the frequency-independent value $\bar{J}$ that we used previously; this is then complete redistribution. The opposite approximation is that the frequency after scattering must be precisely the same as before, and in this case

$$R(\nu, \nu') = \phi(\nu)\delta(\nu' - \nu). \tag{9.51}$$

When the particular kinds of line broadening are considered, it is found that Doppler broadening corresponds to a mild, but nonzero correlation between the absorption and emission frequencies. Solutions taking the proper redistribution function into account show small differences from complete redistribution, which as a practical matter are ignored. Similarly if collisional line broadening is considered, either by ion microfields or electron impact, the redistribution function differs either very little or not at all from complete redistribution. The one case in which there are large differences between the accurate redistribution function and complete redistribution is resonance line scattering with a Lorentzian profile due to natural broadening, perhaps combined with Doppler broadening. The term natural broadening refers to the Lorentzian broadening when the decay rate Γ is due to the radiative decay alone, not to any collisional processes. In resonance line scattering the lower level of the line is a ground state, which lives a relatively long time between scattering events, and therefore its energy uncertainty is quite small. This means that in order to conserve energy the difference between the emitted photon energy and the absorbed photon energy will be of this order and therefore also small. Thus apart from Doppler broadening the frequency shift in the scattering is negligible. When this is combined with Doppler broadening the emission frequency can move in either direction relative to the absorption frequency by about one unit of $\Delta\nu_D$.

The definitive paper on the redistribution functions for the cases just discussed is Hummer (1962). He distinguishes four cases, of which three are the ones mentioned in the previous paragraph. These three are R_I, R_{II}, and R_{III} in Hummer's notation. In R_I the line is broadened solely by the Doppler effect, and redistribution is due to the difference of the atom's velocity as projected on the initial and final photon directions. In the frame of the atom the photon is absorbed and emitted precisely at the line center frequency. In R_{II} there is a Lorentzian absorption profile in the frame of the atom, corresponding to natural broadening by the radiative decay of the excited state. It is supposed that the lifetime of the lower state is much longer, and conservation of energy ensures that the emitted and absorbed frequencies are the same in the frame of the atom. This model is appropriate for resonance lines. In the external frame the absorption profile becomes a Voigt function when Doppler broadening is added to the Lorentzian profile. If there is collisional broadening then the absorption profile in the frame of the atom is also Lorentzian, but the collisions cause the excited atom to lose memory of the initial frequency, and there is complete redistribution in the frame of the atom. When the Doppler effect is included the result is the R_{III} redistribution function. Even though this is derived for complete redistribution in the frame of the atom, it is not quite complete redistribution in the external frame. Finally, Hummer considers a type of redistribution in which both the upper and lower states are broadened by radiative decay processes. With the inclusion of Doppler broadening this becomes R_{IV}. However, Hummer adopts a model from Heitler (1954) for the redistribution in the frame of the atom for this case, which turns out to be in error. The correct formula is in fact given by Weisskopf (1933) and by Wooley and Stibbs (1953). This was clarified by Omont, Smith, and Cooper (1972). Few results using R_{IV} are available. The most rigorous development of the theory of the redistribution function, which relates the redistribution to the atomic kinetics and indicates precisely what mixture of R_I, R_{II}, R_{III}, and R_{IV} is needed for each atomic transition, is by Cooper, Ballagh, Burnett, and Hummer (1982).

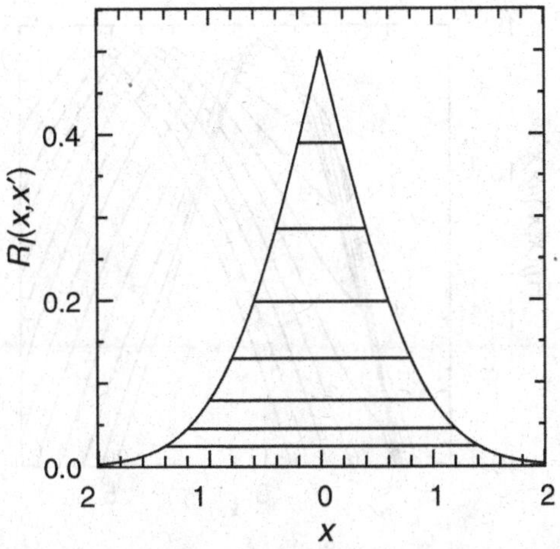

Fig. 9.2 The Doppler redistribution function R_I (x, x') is plotted as a function of the scattered photon frequency displacement from line center, x, for values of the incoming photon frequency displacement from line center, x', from 0 to 1.4 in steps of 0.2; x and x' are expressed in Doppler units.

The nature of R_I is indicated in Figure 9.2. The region with $|x| < |x'|$ for a given x' is rigorously flat; for $|x| > |x'|$ the function falls steeply. The probability of the emission frequency x being near $\pm x'$ is greater with R_I than with complete redistribution. However, x and x' are uncorrelated with this angle-averaged redistribution function. With angle-dependent redistribution x and x' are strongly correlated when $\mathbf{n}$ and $\mathbf{n}'$ are parallel, and strongly anti-correlated when $\mathbf{n}$ and $\mathbf{n}'$ are anti-parallel.

The nature of R_{II} is qualitatively different, as seen in Figure 9.3. The accurate numerical evaluation of R_{II} is somewhat troublesome; a good treatment is given by Adams, Hummer, and Rybicki (1971). While there is little correlation of x and x' in the Doppler core ($|x'| < 2$), in the Lorentzian wings x and x' become strongly correlated. In the far wings the asymptotic form of the marginal distribution is

$$\frac{1}{\phi(x')} R_{II}(x, x') \sim \text{ierfc}\left(\frac{1}{2}|x - x'|\right) \qquad \text{for } x' \gg 1, \tag{9.52}$$

where ierfc(z) is the function defined by

$$\text{ierfc}(z) = \int_z^\infty \text{erfc}(t)\, dt = \frac{1}{\sqrt{\pi}} e^{-z^2} - z\, \text{erfc}(z). \tag{9.53}$$

The relative smallness of $|x - x'|$ in the line wings suggests that a Fokker–Planck treatment may be useful; this is done by Harrington (1973) with some success. A more thorough discussion is given by Frisch (1980). Frisch points out that R_{IV} is like a weighted average of complete distribution with R_{II}, with the weight of

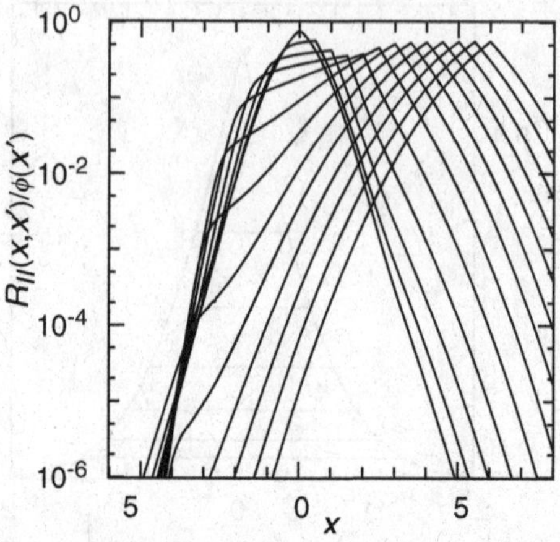

Fig. 9.3 The marginal distribution $R_{II}(x, x')/\phi(x')$ for the Voigt redistribution function for the case of coherence in the frame of the scattering atom is plotted as a function of the scattered photon frequency displacement from line center for values of the incoming photon frequency displacement from the line center, x', from 0 to 4 in steps of 0.5. The Voigt parameter is 0.1.

complete redistribution being proportional to the natural width of the lower level and the weight of R_{II} proportional to the natural width of the upper level. For most subordinate lines (with an excited state as the lower level) these widths are comparable, and thus so are the weights. In this case the complete redistribution part completely dominates the long-range behavior of the scattering problem. In other words, R_{IV}, like R_I and R_{III}, behaves very much the same as complete redistribution. The odd man out is R_{II}, which is the realistic case for resonance-line scattering if the density is low enough that collisional broadening is negligible compared with natural broadening.

In the case of partial redistribution, the two-level-atom source function takes this form:

$$S_\nu \phi(\nu) = (1-\epsilon) \int_0^\infty d\nu' \, R(\nu, \nu') J_{\nu'} + \epsilon B_\nu \phi(\nu). \tag{9.54}$$

For complete redistribution, S_ν is independent of frequency and $R(\nu, \nu') = \phi(\nu)\phi(\nu')$, so a division by $\phi(\nu)$ recovers (9.8).

The problem of resonance line transfer in the situation described above, with quite large values of the optical thickness of the medium, leads to some interesting effects and has been well studied in the laboratory. Line photons in this kind of problem do a kind of double diffusion. The photons diffuse in real space while they simultaneously diffuse in frequency space. Of course the spatial diffusion coefficient is related to $1/\phi(\nu)$, so the diffusion in space is very rapid at frequencies that are far from line center. Since the photons must diffuse in frequency instead of being independently sampled on every scattering, the effect is to keep them more confined toward line center in the partial redistribution case. This means that the typical mean free path is less, and that the photons will not travel as far in a given number of scatterings. Therefore the thermalization depth will be less for a given value of ϵ. In fact it is found for scattering with partial redistribution as described that the thermalization length is roughly just the same as for a Doppler profile, while in complete redistribution, given the Lorentzian wings of the combined profile, the thermalization depth is much larger than that.

9.2.1 Asymptotic resonance line transfer with R_{II}

Harrington's (1973) asymptotic solution of the transfer equation for resonance line scattering with R_{II} is useful for the insight it gives into the PRD transfer process. The action of the scattering kernel is approximated by expanding the function $J_{\nu'}$ inside the integral in a Taylor series about $\nu' = \nu$, then using the known moments of the R_{II} function to obtain this result:

$$\frac{1}{\phi(x)} \int_{-\infty}^{\infty} dx' R_{II}(x, x') J(x')\, dx' \approx J(x) - \frac{1}{x}\frac{dJ(x)}{dx} + \frac{1}{2}\frac{d^2J(x)}{dx^2}, \quad (9.55)$$

so Milne's second equation, in the Eddington approximation, turns into

$$\frac{\partial^2 J}{\partial \tau^2} + \frac{3}{2}\phi^2(x)(1-\epsilon)\left[\frac{d^2J(x)}{dx^2} - \frac{2}{x}\frac{dJ(x)}{dx}\right] = 3\phi^2(x)\epsilon(J - B). \quad (9.56)$$

The optical depth scale τ is the mean over the line profile, given by $d\tau = k_L dz/\Delta\nu_D$. Two more transformations are introduced now. One is to notice that

$$\frac{3}{2}\phi^2(x)\left[\frac{d^2J(x)}{dx^2} - \frac{2}{x}\frac{dJ(x)}{dx}\right] \approx \frac{\partial^2 J}{\partial \sigma^2} \quad (9.57)$$

in terms of a new frequency-related variable σ defined by

$$\sigma = \left(\frac{2}{3}\right)^{1/2} \int_0^x \frac{dx}{\phi(x)}. \quad (9.58)$$

The approximation is accurate in the Lorentzian wings of the Voigt profile, which is where all the transfer will take place in the asymptotic regime we want to consider. The variable σ vanishes at line center, and in the wings $x \to \pm\infty$ it varies as $\sigma \sim \pm\sqrt{2/3}\pi|x|^3/(3a)$. We are also going to make the approximation $\epsilon \ll 1$ so the factor $1-\epsilon$ can be replaced by unity where it multiplies the frequency derivatives. We then have

$$\frac{\partial^2 J}{\partial \tau^2} + \frac{\partial^2 J}{\partial \sigma^2} = 3\phi^2\epsilon(J - B). \quad (9.59)$$

The second clever transformation is the result of observing that ϕ^2, considered as a function of σ, is strongly peaked at $\sigma = 0$, much more so than ϕ is when considered as a function of x. So Harrington's second major approximation is to replace $3\phi^2$ by $\sqrt{6}\delta(\sigma)$. The factor $\sqrt{6}$ preserves the normalization of the profile. Thus finally the transfer equation becomes

$$\frac{\partial^2 J}{\partial \tau^2} + \frac{\partial^2 J}{\partial \sigma^2} = \sqrt{6}\epsilon(J - B)\delta(\sigma). \tag{9.60}$$

The delta function on the right-hand side tells us that line photons are created or destroyed entirely at line center. So except at line center, the intensity obeys a 2-D Laplace's equation in σ–τ space; in other words, there is double diffusion.

The solution $J(\tau, \sigma)$ of (9.60) should be continuous at line center, $\sigma = 0$, and in fact $J(\tau, 0)$ is the value of $\bar{J}$, which sets the population of the excited state. But there is a step discontinuity in $\partial J/\partial\sigma$:

$$\Delta\left(\frac{\partial J}{\partial \sigma}\right) = \sqrt{6}\epsilon[(J(\tau, 0) - B]). \tag{9.61}$$

Let us consider an infinite medium with a source s of line photons confined to the sheet $\tau = 0$, in other words, $\epsilon B = s\delta(\tau)$. We take the Fourier transform of (9.60), using $\tilde{J}(k, \sigma) = \int d\tau \exp(ik\tau)J(\tau, \sigma)$. Then

$$-k^2\tilde{J} + \frac{\partial^2 \tilde{J}}{\partial \sigma^2} = \sqrt{6}\delta(\sigma)(\epsilon\tilde{J}(k, 0) - s). \tag{9.62}$$

The Fourier transform of the jump condition is

$$\Delta\left(\frac{\partial \tilde{J}}{\partial \sigma}\right) = \sqrt{6}(\epsilon\tilde{J}(k, 0) - s). \tag{9.63}$$

The solution must be of the form $\tilde{J} = A\exp(-|k\sigma|)$ with a suitable coefficient A. Putting this into the jump condition leads to

$$A = \frac{\sqrt{6}s}{2|k| + \sqrt{6}\epsilon}. \tag{9.64}$$

The inverse Fourier transform to obtain $J(\tau, \sigma)$ turns out after a little work to give

$$J(\tau, \sigma) = \sqrt{\frac{3}{2}}\frac{s}{\pi}\Re\left\{e^{\sqrt{3/2}(|\sigma|+i|\tau|)\epsilon} E_1[\sqrt{3/2}(|\sigma| + i|\tau|)\epsilon]\right\}. \tag{9.65}$$

The complex exponential integral function is tabulated in Abramowitz and Stegun (1964). We can see that the characteristic scale in τ, and also in σ, is $1/\epsilon$, and thus the thermalization length scales in this way, just as it does for complete redistribution with Doppler broadening. Since the characteristic σ also scales as $1/\epsilon$,

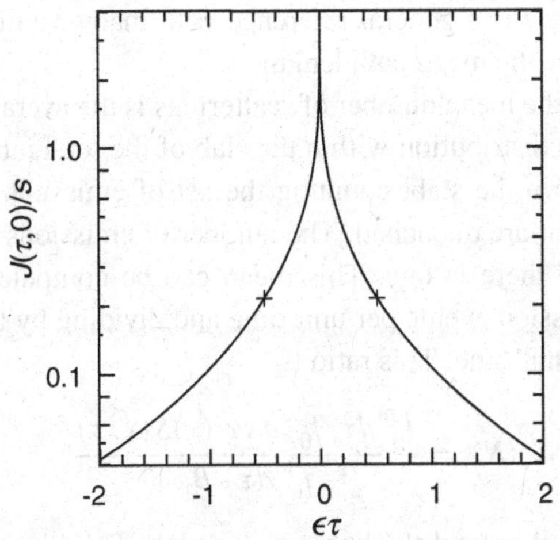

Fig. 9.4 The line-center mean intensity computed in Harrington's asymptotic model of PRD with R_{II} is plotted as a function of $\epsilon\tau$. The crosses on the curve mark the edges of the region in τ that contains half the total population.

the characteristic frequency width in Doppler units is $x \sim (a/\epsilon)^{1/3}$. This asymptotic model is not applicable unless $x \gg 1$, so ϵ must be $\ll a$.

The spatial distribution of the excited state population is proportional to $J(\tau, 0)$, which can be expressed in terms of the auxiliary function for sine and cosine integrals $g(z)$ described by Abramowitz and Stegun (1964), Section 5.2:

$$J(\tau, 0) = \sqrt{\frac{3}{2}} \frac{s}{\pi} g\left(\sqrt{3/2}\epsilon\tau\right). \tag{9.66}$$

This distribution is shown in Figure 9.4. Half the excited atoms are contained in the interval $|\tau| < 0.509/\epsilon$, thus $0.509/\epsilon$ is the thermalization depth for small ϵ.

Harrington's asymptotic theory is useful for aiding understanding, but as a numerical approximation it is not very good. It is based on the smallness of parameters like $(a\tau)^{-1/3}$ or $(\epsilon/a)^{1/3}$. The optical depth τ must be enormous, or ϵ tiny indeed, before the approximation starts to be accurate. For numerical results the calculations of Adams (1972) are much better.

9.3 Mean number of scatterings; mean path length

The mean number of scatterings and the mean path length are ways of describing the transfer of line radiation within a finite medium in an integral sense. There are reasonably accurate estimates available for these quantities, which makes them valuable for rough calculations. The references that develop the principal results for the mean number of scatterings are by Hummer (1964), Adams (1972), and

Ivanov (1973); the last is a general reference with many asymptotic expressions. Ivanov also discusses the mean path length.

The definition of the mean number of scatterings is the average over all photons created in a uniform distribution within the slab of the total number of times they interact with matter in the slab, counting the act of emission, before they either escape the medium or are quenched. The number of emissions is counted, but not a final quenching if there is one. This mean can be computed by counting the total number of emission events per unit time and dividing by the total number of creation events per unit time. This ratio is

$$\langle N \rangle = \frac{\int_0^{\tau_0} d\tau \int_0^{\infty} dx\, \phi(x) S(x,\tau)}{\int_0^{\tau_0} d\tau\, \epsilon B_\nu}, \tag{9.67}$$

in which τ_0 is the full optical thickness of the slab. This simplifies for complete redistribution to

$$\langle N \rangle = \frac{\int_0^{\tau_0} d\tau\, S(\tau)}{\int_0^{\tau_0} d\tau\, \epsilon B_\nu}. \tag{9.68}$$

If the scattering albedo tends to zero, i.e. $\epsilon \to 1$, then S becomes ϵB_ν and $\langle N \rangle = 1$. When ϵ is small compared with the escape probability $K_2(\tau)$ then the quenching probability is negligible and $\langle N \rangle$ tends to a limit that is independent of ϵ but depends on τ_0. Hummer (1964) obtains an upper limit

$$\langle N \rangle \approx \frac{1}{\epsilon + (1-\epsilon) K_2(\tau_0/2)}, \tag{9.69}$$

which he proposes as an estimate. Ivanov's asymptotic methods for large τ_0 sharpen up the estimate to the following, for the respective profile shapes:

$$\langle N \rangle \approx \begin{cases} \frac{1}{2}\tau_0 \sqrt{\ln(\tau_0/\sqrt{\pi})} & \text{Doppler} \\ \dfrac{\Gamma^2(1/4)}{(2\pi)^{3/2}} \sqrt{\dfrac{\tau_0}{a}} & \text{Voigt.} \\ \dfrac{\Gamma^2(1/4)}{(2\pi)^{3/2}} \sqrt{\tau_0} & \text{Lorentz.} \end{cases} \tag{9.70}$$

These estimates are half of Hummer (1964), for the Doppler case, and 21% less than Hummer's for the other two.

The mean number of scatterings in a finite slab in the conservative case $\epsilon = 0$ for PRD with R_{II} is a quantity that is of prime interest. The numerical work by Adams (1972) and asymptotic analysis of Harrington (1973) both agree that the mean number of scatterings with R_{II}, for a photon source that is a thin sheet at the midplane of the slab, is proportional to τ_0 in the asymptotic regime $(a\tau_0)^{1/3} \gg 1$.

Adams suggests a coefficient 0.75, but Harrington's expression

$$\langle N \rangle \sim 0.454658\tau_0 \tag{9.71}$$

fits Adams's numerical results about as well. The proportionality of $\langle N \rangle$ to τ_0 rather than to $\sqrt{\tau_0/a}$ is a significant feature of (R_{II}) PRD; it is much harder to escape the medium with PRD. Harrington's result for a photon source uniformly distributed in the slab is similar, but with a coefficient of 0.332368 rather than 0.454658. This is the result that is comparable to the complete redistribution (CRD) expressions in the previous paragraph.

The mean path length of the scattering photon is defined as the average of the total distance it flies from the place where it is created to the place where it either is quenched or crosses the boundary of the medium. This is equal to the photon's dwelling time within the medium multiplied by the speed of light. The dwelling time can be calculated, in this steady-state problem, by dividing the total photon energy within the medium by the total energy rate of creating photons. This leads to

$$\langle \ell \rangle = \frac{\Delta\nu_D}{k_L} \frac{\int_0^{\tau_0} d\tau \int_0^{\infty} dx\, J(x,\tau)}{\int_0^{\tau_0} d\tau\, \epsilon B_\nu}. \tag{9.72}$$

The integrals for $\langle N \rangle$ and $\langle \ell \rangle$ are very similar; in fact, for a square profile there is the simple relation

$$(1-\epsilon)\langle \ell \rangle = \frac{\Delta\nu_D}{k_L}(\langle N \rangle - 1). \tag{9.73}$$

This says that the average flight path is the mean free path $\Delta\nu_D/k_L$. (Remember that the number of flights is one less than the number of scatterings by the definition we are using.)

Ivanov's (1973) results for the mean path length for $\epsilon = 0$ and $\tau_0 \gg 1$ are the following:

$$\langle \ell \rangle \sim \frac{\Delta\nu_D}{k_L} \times \begin{cases} \tau_0\sqrt{\ln(\tau_0/\sqrt{\pi})} & \text{Doppler} \\ \dfrac{9}{4}\tau_0 & \text{Voigt or Lorentz} \end{cases} \tag{9.74}$$

In PRD, with R_{II}, and with a uniformly distributed photon source in the slab, Harrington's methods lead to a value of the mean path length given by

$$\langle \ell \rangle \sim \frac{\Delta\nu_D}{k_L} 0.692739\, (a\tau_0^4)^{1/3}. \tag{9.75}$$

This exceeds the thickness of the slab by a factor that is $O((a\tau_0)^{1/3})$. The accumulated smaller flights amount to much more than the single longest flight in this case.

If there is both a nonzero quenching probability ϵ and a continuous absorption coefficient k_c, then Ivanov (1973), Section 8.10, points out that the mean escape probability is $1 - \epsilon\langle N\rangle - k_c\langle\ell\rangle$. This is physically obvious once it is recognized that the events of escape, quenching, and continuous absorption all remove the photons, are mutually exclusive, and exhaust all the possible photon fates. The discussion of mean escape probabilities leads directly to the topic of Irons's theorem.

9.3.1 Irons's theorem

Irons's theorem says that the Net Radiative Bracket is equal to the single-flight escape probability in the mean, with the mean defined using a weighting function equal to the emissivity. A good discussion of Irons's theorem, and of escape probability matters in general, is given by Rybicki (1984). The following is a demonstration of this for the plane-parallel slab (optical thickness τ_0) with complete redistribution but no continuous absorption. The transfer equation is

$$\mu\frac{dI_x(\tau,\mu)}{d\tau} = \phi(x)[I_x(\tau,\mu) - S(\tau)]. \tag{9.76}$$

The intensity that emerges at $\tau = 0$ with $\mu > 0$ is given by

$$I_x(0,\mu) = \int_0^{\tau_0}\frac{d\tau}{\mu}S(\tau)\exp\left[-\frac{\tau\phi(x)}{\mu}\right]. \tag{9.77}$$

But an integration of the transfer equation directly, and noting that $I_x(\tau_0,\mu) = 0$ for $\mu > 0$, leads to

$$I_x(0,\mu) = -\int_0^{\tau_0}\frac{d\tau}{\mu}\phi(x)[I_x(\tau,\mu) - S(\tau)]. \tag{9.78}$$

Then we can replace μ with $-\mu$ and develop two expressions for $I_x(\tau_0,-\mu)$:

$$I_x(\tau_0,-\mu) = \int_0^{\tau_0}\frac{d\tau}{\mu}S(\tau)\exp\left[-\frac{(\tau_0-\tau)\phi(x)}{\mu}\right] \tag{9.79}$$

and

$$I_x(\tau_0,-\mu) = \int_0^{\tau_0}\frac{d\tau}{-\mu}\phi(x)[I_x(\tau,-\mu) - S(\tau)]. \tag{9.80}$$

Now we equate the two expressions for $I_x(0,\mu)$, and also the two expressions for $I_x(\tau_0,-\mu)$, and average those equations. This yields

$$\int_0^{\tau_0} \frac{d\tau}{\mu}\phi(x)\frac{1}{2}\left\{\exp\left[-\frac{\tau\phi(x)}{\mu}\right]+\exp\left[-\frac{(\tau_0-\tau)\phi(x)}{\mu}\right]\right\}S(\tau)$$
$$=\int_0^{\tau_0}\frac{d\tau}{\mu}\phi(x)S(\tau)\left[1-\frac{j_x(\tau,\mu)}{S(\tau)}\right]. \qquad (9.81)$$

The quantity $j_x(\tau,\mu)$ is the Feautrier average of the intensity at $\pm\mu$. The next step is to multiply this equation by μ and integrate over all x and over the range [0, 1] of μ. The result is

$$\int_0^{\tau_0} d\tau\, S(\tau)\frac{1}{2}\left[K_2(\tau)+K_2(\tau_0-\tau)\right]=\int_0^{\tau_0} d\tau\, S(\tau)\left[1-\frac{\bar{J}(\tau)}{S(\tau)}\right]. \qquad (9.82)$$

This is a statement of Irons's theorem for this case, since $z = 1-\bar{J}/S$ is the Net Radiative Bracket and the combination $p_{\rm esc}=\frac{1}{2}[K_2(\tau)+K_2(\tau_0-\tau)]$ is the two-sided escape probability for the finite slab, generalizing (9.20). It is fairly clear from this derivation, and from Rybicki's (1984) presentation, that this result can be extended to arbitrary geometries. An immediate corollary that Rybicki draws is the relation between the mean number of scatterings and the escape probability that appears in Irons's theorem. It is

$$\langle N\rangle = \frac{1}{\epsilon+(1-\epsilon)\langle z\rangle}=\frac{1}{\epsilon+(1-\epsilon)\langle p_{\rm esc}\rangle}. \qquad (9.83)$$

The first equality is an identity based on the definition of $z = 1-\bar{J}/S$, and the second equality is Irons's theorem. This relation between the mean number of scatterings and the single-flight escape probability recalls Hummer's relation (9.69). In Hummer's relation the value of $p_{\rm esc}$ at the center of the slab appears instead of the source-function-weighted average over the slab.

9.4 Time-dependent line transport

Our discussion so far has concerned only steady-state line transfer. Now we want to consider the time-dependent effects. Mostly this will be concerned with learning what the important time scales are; doing time-dependent practical problems is a major computational effort not a pedagogical topic.

First we need to consider the two causes of time dependence in spectral line transport: (1) the spontaneous lifetime $t_{\rm sp}=1/A_{u\ell}$, which causes the source function to lag the radiative and collisional rates by a time of order $t_{\rm sp}$; and (2) the time of flight $\lambda_{\rm mfp}/c$ for a mean free path, which causes the mean intensity and the photoexcitation rates to lag the source function in time by about $\lambda_{\rm mfp}/c$. The quantities vary by orders of magnitude in different radiation environments, and therefore so does their ratio. The spontaneous lifetime varies less than the flight time; its order of magnitude is

$$t_{\rm sp} \approx \frac{23}{(h\nu/{\rm eV})^2}\ {\rm ns}. \tag{9.84}$$

This is for a line with unit oscillator strength, and $h\nu/\mathrm{eV}$ is the line energy in electron volts. We see that this time scale is very short in astrophysical terms, when the line energy is indeed around 1 eV, and it is even shorter for high-energy environments when the line is in the keV x-ray range. The flight time of the line photon is harder to estimate in general terms. In many problems of stellar astrophysics the mean free path is of the order of an atmospheric scale height down to a small fraction of a scale height. The flight time for a scale height in the solar photosphere is 0.4 ms. This is four orders of magnitude greater than the spontaneous lifetime. Thus when the time scale is large enough to be of any interest in the astrophysical problem, the flight time will greatly exceed the spontaneous lifetime.

In problems that are on a terrestrial scale, with dimensions from microns (laser targets) up to meters (laboratory scale), the time of flight will range from less than a femtosecond up to a few nanoseconds. In essentially all of these cases the time of flight will be much less than the spontaneous lifetime. Thus in most laboratory-scale problems the dominant time dependence comes from the spontaneous lifetime, while in astrophysical problems the dominant time scale is the flight time.

The time dependence associated with the spontaneous lifetime – in general with the finiteness of all the rates – is incorporated in the two-level-atom model by using the rate for the excited-state population in place of the state–state equation. Thus

$$\frac{dN_u}{dt} = N_\ell(B_{\ell u}\bar{J} + N_{\ell u}) - N_u(A_{u\ell} + B_{u\ell} + C_{u\ell}). \tag{9.85}$$

The important applications of time-dependent transport are to resonance lines, so in that case the time dependence of the lower-level population is neglected. Using (9.85) and the definition of the source function leads to

$$\frac{dS}{dt} = A_{u\ell}\left(1 + \frac{g_\ell + g_u}{g_u}\frac{S}{2h\nu^3/c^2}\right)\left[\bar{J} + \epsilon' B - (1+\epsilon')S\right]. \tag{9.86}$$

The stimulation factor $1 + [(g_\ell + g_u)/g_u]2h\nu^3/c^2$ can normally be omitted, leaving this useful equation:

$$\frac{dS}{dt} = A_{u\ell}\left[\bar{J} + \epsilon B - (1+\epsilon)S\right]. \tag{9.87}$$

In the following discussion we will first consider an example in which flight time is dominant and the spontaneous lifetime is neglected, and then one in which the reverse occurs. Let us return to the time-dependent transport equation for a line formed by a two-level atom with partial redistribution, and we will also include a background continuous opacity. The equation including time of flight is

$$\frac{1}{c}\frac{\partial I_\nu}{\partial t} + \mathbf{n}\cdot\nabla I_\nu = -(k_L\phi(\nu) + k_c)I_\nu \\ + (1-\epsilon)k_L \int d\nu' R(\nu,\nu')J_{\nu'} + \epsilon B\phi(\nu) + k_c S_c. \tag{9.88}$$

Now we form the Laplace transform, with $\tilde{I}_\nu(p) = \int_0^\infty dt\, \exp(-pt) I_\nu(t)$. The equation becomes

$$\mathbf{n}\cdot\nabla\tilde{I}_\nu = -\left[k_L\phi(\nu) + k_c + \frac{p}{c}\right]\tilde{I}_\nu \\ + (1-\epsilon)k_L \int d\nu' R(\nu,\nu')\tilde{J}_{\nu'} + \frac{1}{p}\epsilon B\phi(\nu) + \frac{1}{p}k_c S_c + \frac{1}{c}I_\nu(0). \tag{9.89}$$

What we notice about this equation is that it is identical to a steady-state line transfer equation with an extra contribution p/c in the continuous opacity, an effective continuous emissivity of $k_c S_c/p + I_\nu(0)/c$, and an effective Planck function of B/p.

We want to find out about the rate at which the intensity relaxes toward a steady state, assuming there is one. The steady-state intensity obeys this equation

$$\mathbf{n}\cdot\nabla I_\nu^s = -[k_L\phi(\nu) + k_c]I_\nu^s + (1-\epsilon)K_L \int d\nu' R(\nu,\nu')J_{\nu'}^s + \epsilon B\phi(\nu) + k_c S_c. \tag{9.90}$$

If we define ΔI_ν to be $I_\nu - I_\nu^s$, then its Laplace transform obeys this equation

$$\mathbf{n}\cdot\nabla(\Delta\tilde{I}_\nu) = -\left[k_L\phi(\nu) + k_c + \frac{p}{c}\right]\Delta\tilde{I}_\nu \\ + (1-\epsilon)k_L \int d\nu' R(\nu,\nu')\Delta\tilde{J}_\nu + \frac{1}{c}\Delta I_\nu(0). \tag{9.91}$$

A theorem on Laplace transforms tells us what the effect is of k_c on the time dependence: if $(\Delta I_\nu)_0(t)$ is the solution without continuous absorption, then $\exp(-k_c ct)(\Delta I_\nu)_0(t)$ is the solution including continuous absorption. So we can drop k_c at this point with no loss of generality, since it can always be put back later using this result.

Let us consider the case of complete redistribution in an infinite homogeneous medium with an initially white, isotropic departure $\Delta I_\nu(0)$ from the steady-state intensity. For convenience we convert the frequencies to Doppler units, and use the line profile function on the Doppler-width scale $\phi(x)$, and define a new variable $\beta = p\Delta\nu_D/(k_L c)$. The redistribution function becomes $R(x,x') = \phi(x)\phi(x')$.

The transform then obeys this equation

$$\Delta\tilde{I}_x[(\phi(x)+\beta] = (1-\epsilon)\phi(x)\tilde{\bar{J}} + \frac{\Delta\nu_D}{k_L c}\Delta I_\nu(0). \tag{9.92}$$

Multiplying this equation by $\phi(x)/[\phi(x)+\beta]$ and integrating over x leads to

$$\tilde{\bar{J}} = (1-\epsilon)\tilde{\bar{J}}\int_{-\infty}^{\infty} dx\,\frac{\phi^2(x)}{\phi(x)+\beta} + \frac{\Delta\nu_D}{k_L c}\Delta I_\nu(0)\int_{-\infty}^{\infty} dx\,\frac{\phi(x)}{\phi(x)+\beta}. \tag{9.93}$$

A function $F(\beta)$ that depends on the shape of the line profile, often called the curve-of-growth function, is defined by

$$F(\beta) = \int_{-\infty}^{\infty} dx\,\frac{\phi(x)}{\phi(x)+\beta}. \tag{9.94}$$

The first integral in (9.93) is seen to be $1-\beta F(\beta)$. Therefore the solution of (9.93) is

$$\tilde{\bar{J}} = \frac{\Delta\nu_D \Delta I_\nu(0)}{k_L c}\frac{F(\beta)}{\epsilon + (1-\epsilon)\beta F(\beta)}. \tag{9.95}$$

Our task now is to consider specific shapes of the line profile function, approximate $F(\beta)$ for each of these for $\beta \to 0$, and then, since $\beta \propto p$, use the approximation to get the inverse Laplace transform, and thereby find $\bar{J}$ for large t. For a square line profile, i.e., a "line" that has a constant opacity rather than a smooth shape, we use $\phi(x) = 1$ for $0 \le x \le 1$, with the result that $F(\beta) = 1/(1+\beta)$. Then we see that

$$\tilde{\bar{J}} = \frac{\Delta\nu_D \Delta I_\nu(0)}{k_L c}\frac{1}{\epsilon+\beta} = \Delta I_\nu(0)\frac{1}{p + \epsilon k_L c/\Delta\nu_D}. \tag{9.96}$$

The inverse Laplace transform gives $\bar{J} = \Delta I_\nu(0)\exp(-\epsilon k_L c t/\Delta\nu_D)$. In other words, the initial departure of the intensity from the steady-state value decays with a mean lifetime given by the time to make $1/\epsilon$ flights with a mean free path of $\Delta\nu_D/k_L$.

With Doppler broadening the curve of growth function turns out to be

$$F(\beta) \sim 2\sqrt{\ln\left(\frac{1}{\sqrt{\pi}\beta}\right)}, \tag{9.97}$$

when β is small. We will refer to the $\sqrt{\ln}$ expression as $C(\beta)$; it is a slowly-varying function, ranging from 1 to 3 over quite a large range in β. The transform of $\bar{J}$ can then be written as

$$\tilde{\bar{J}} \sim \frac{\Delta\nu_D \Delta I_\nu(0)}{k_L c}\frac{2C(\beta)}{\epsilon + (1-\epsilon)p\Delta\nu_D 2C(\beta)/(k_L c)}. \tag{9.98}$$

In the Laplace inversion, most of the contribution comes from values of p that make $\beta \approx \epsilon$. Making this substitution in $C(\beta)$ leads to

$$\bar{J} \sim \Delta I_\nu(0) \exp\left(-\frac{\epsilon k_L ct}{\Delta\nu_D 2C(\epsilon)}\right). \tag{9.99}$$

This is almost the same result, with the same mean decay time, as for the square profile, except that the mean free path and the mean free time are increased by the factor $2C(\epsilon)$.

For a Voigt profile it is sufficient to calculate $F(\beta)$ using the large-x asymptotic form of the profile, $\phi(x) \sim a/(\pi x^2)$. Then it is easy to see that $F(\beta) \sim \sqrt{\pi a/\beta}$. This will only apply for $\beta \ll a$. Then the transform is

$$\tilde{\bar{J}} \sim \frac{\Delta\nu_D \Delta I_\nu(0)}{k_L c} \frac{\sqrt{\pi a/\beta}}{\epsilon + \sqrt{\pi a \beta}} = \Delta I_\nu(0) \frac{1}{\sqrt{p}[\sqrt{p} + \epsilon\sqrt{k_L c/(\pi \Delta\nu_D a)}]}. \tag{9.100}$$

Finding the inverse Laplace transform in the tables leads to

$$\bar{J} \sim \Delta I_\nu(0) \exp\left(\frac{\epsilon^2 k_L ct}{\pi \Delta\nu_D a}\right) \operatorname{erfc}\left(\epsilon\sqrt{\frac{k_L ct}{\pi \Delta\nu_D a}}\right). \tag{9.101}$$

In this case the decay time is of order a/ϵ^2 times the flight time for a line-center mean free path.

In every case, the mean decay time is about equal to the flight time for the thermalization length corresponding to the given profile and value of ϵ. This is at first surprising since the flight time adds up all the flight distances rather than taking the net displacement. But we recall the longest-single-flight picture, which says that the sum of all the flights is just about the same as the longest one, and also about the same as the net displacement.

There are other cases to consider, such as PRD, and other problems to analyze, such as the time-dependent spreading of line radiation in an infinite medium with no quenching or absorption. We will round out the present discussion by treating that problem with CRD in the case that the spontaneous lifetime dominates the time dependence rather than the flight time.

The dynamical equation for the source function becomes

$$\frac{\partial S(\tau, t)}{\partial t} = A_{ul}\left[\int_{-\infty}^{\infty} d\tau' \, K_1(\tau - \tau')S(\tau') - S(\tau)\right]. \tag{9.102}$$

This becomes Holstein's (1947a,b) and Biberman's (1947) integro-differential equation for trapped resonance-line radiation when we recognize that S is proportional to the excited-state population. We suppose that at $t = 0$ there is a thin sheet of excited atoms at $\tau = 0$. We Fourier analyze this equation with respect to τ and obtain

$$\frac{\partial \tilde{S}}{\partial t} = A_{u\ell}[\tilde{K}_1(k) - 1]\tilde{S}. \tag{9.103}$$

At the initial time the Fourier transform of the delta function representing the sheet of excitation is $\tilde{S}(0) = 1$. And therefore the solution for the transform is

$$\tilde{S}(k, t) = \exp\left\{-A_{u\ell}[1 - \tilde{K}_1(k)]t\right\}. \tag{9.104}$$

In order to find how $S(t)$ varies with large t we have to investigate $\tilde{K}_1(k)$ for $k \to 0$. This is the work done by Ivanov and coworkers and reported in Ivanov (1973). The general expression is

$$\tilde{K}_1(k) = \int_{-\infty}^{\infty} dx \, \frac{\phi^2(x)}{k} \tan^{-1} \frac{k}{\phi(x)}. \tag{9.105}$$

The results obtained from the asymptotic evaluation of the integral are:

$$\tilde{K}_1(k) \sim \begin{cases} 1 - \dfrac{\pi}{4} \dfrac{|k|}{\sqrt{\ln(1/(\sqrt{\pi}|k|))}} & \text{Doppler} \\ 1 - \dfrac{\sqrt{2\pi a|k|}}{3} & \text{Voigt.} \\ 1 - \dfrac{\sqrt{2\pi|k|}}{3} & \text{Lorentz} \end{cases} \tag{9.106}$$

The inversion to find $S(\tau, t)$ then gives in the Doppler case

$$S(\tau, t) \sim \frac{W}{\pi} \frac{1}{\tau^2 + W^2}, \tag{9.107}$$

in which W is a Lorentzian width in optical depth given by

$$W = \frac{\pi A_{u\ell} t}{4\sqrt{\ln(A_{u\ell} t/\sqrt{\pi})}}. \tag{9.108}$$

The wave of excited atoms spreads linearly in time with an effective velocity

$$\frac{W \lambda_{\mathrm{mfp}}}{t} = \frac{\pi A_{u\ell} \lambda_{\mathrm{mfp}}}{4\sqrt{\ln(A_{u\ell} t/\sqrt{\pi})}}. \tag{9.109}$$

This is guaranteed to be small compared with c since the lifetime dominates the time of flight only if $A_{u\ell}\lambda_{\mathrm{mfp}} \ll c$.

In the Voigt case the form for $S(\tau, t)$ at large t is

$$S(\tau, t) \sim \frac{9}{(A_{u\ell} t)^3} z^3 g_F(z), \tag{9.110}$$

in terms of the variable $z = A_{u\ell}t/(3\sqrt{\tau})$ and the function $g_F(z)$ (not the same as the earlier $g(z)$) that is related to the Fresnel integrals, see Abramowitz and Stegun (1964), Section 7.3, (7.3.6). The range of τ that contains half the excited atoms is found to be $|\tau| \leq 0.86(A_{u\ell}t)^2$. Remarkably, the wave of excited atoms spreads at an accelerated rate as time goes on! The effective wave speed is of order $A_{u\ell}^2\lambda_{\rm mfp}t$, which is $A_{u\ell}t$ times larger than in the Doppler case. Thus after several scatterings the Voigt wave of excited atoms is moving much faster than the Doppler one. This model is only for CRD; with PRD the spread of excited atoms is significantly slower. We can estimate the mean number of scatterings by $N \approx A_{u\ell}t$. Then the width of the excited atom distribution would be approximately the thickness of the finite slab that would have that mean number of scatterings, in other words, $\tau \approx N \approx A_{u\ell}t$. This corresponds to the same wave speed as in the Doppler case.

Holstein (1947b) treats the homogeneous infinitely long cylinder in a calculation that was extended by Payne and Cook (1970). These authors were seeking the decay constant eigenvalue for the bounded medium; Payne and Cook also provide the spatial variation of S for large time. The scaling of characteristic size with time agrees with the results above; the shape functions are modified for cylindrical geometry vs slab geometry.

10
Refraction and polarized light

10.1 Refraction

Our picture of the radiation has to be modified when the atoms, ions, and electrons respond coherently to the oscillating electric field. This gives the medium a dielectric constant and therefore an index of refraction, and the transport of radiation is modified. The discussion we give here of refraction effects in radiation transport is derived from Cox and Giuli (1968).

The material in a volume element dV can have a mean electric dipole moment $\mathbf{p}$, induced by the local electric field $\mathcal{E}$, given by

$$\mathbf{p} = \chi \mathcal{E} dV, \tag{10.1}$$

where χ is the AC electric susceptibility of the material. This coherent oscillating dipole modifies the propagation of the electromagnetic waves, since now there is a dielectric constant

$$\epsilon = 1 + 4\pi\chi, \tag{10.2}$$

and the phase velocity of electromagnetic waves will be

$$v_p = \frac{c}{n}, \tag{10.3}$$

where $n = \sqrt{\epsilon}$ is the refractive index. (We will continue to use $\mathbf{n}$ for the direction vector of photon propagation. The scalar symbol will always be the refractive index, and the vector will refer to the photon direction.) The susceptibility, dielectric constant, and refractive index may all be complex, since the atomic dipoles do not necessarily oscillate just in phase with the local electric field.

This continuous-medium picture that leads to the dielectric constant is the result of another application of coarse graining. It does not make sense unless the wavelength of the electromagnetic wave is large compared with the particle spacing. When this is done more carefully, the result is found to depend on the two-particle correlation functions for the plasma species at the values of **k** and ω appropriate for the electromagnetic wave. We cannot go into that level of detail here.

The very simplest example of the plasma susceptibility is provided by the free electrons. The susceptibility per electron is found by solving an equation of motion which we write

$$m\ddot{\mathbf{r}} + m\nu_c\dot{\mathbf{r}} = -e\mathcal{E}, \tag{10.4}$$

in which $\mathcal{E}$ is presumed to oscillate as $\exp(-i\omega t)$. We have included a "drag" force $-m\nu_c\dot{\mathbf{r}}$ that represents heuristically the electron–ion collisions that produce relaxation of the electron velocities to the mean. This term would be the right-hand side $(\partial f/\partial t)_{\rm coll}$ term in the Boltzmann equation for the electron distribution function. The solution for the electron displacement is

$$\mathbf{r} = \frac{e\mathcal{E}}{m(\omega^2 + i\omega\nu_c)}, \tag{10.5}$$

so the susceptibility per electron is

$$\alpha = -\frac{e\mathbf{r}}{\mathcal{E}} = -\frac{e^2}{m}\frac{1}{\omega^2 + i\omega\nu_c}. \tag{10.6}$$

The complex dielectric constant becomes

$$\epsilon = 1 - \frac{\omega_{pe}^2}{\omega^2 + i\omega\nu_c}, \tag{10.7}$$

where ω_{pe} is the electron plasma frequency $\sqrt{4\pi N_e e^2/m}$. The collision frequency ν_c is generally quite small, and if it is neglected the dielectric constant is real and the refractive index is $n = \sqrt{1 - (\omega_{pe}/\omega)^2}$ provided that ω is above the plasma frequency. For frequencies below the plasma frequency electromagnetic waves cannot propagate in this picture.

Let us return to the case that ϵ and n have a small imaginary part and see what that does to the propagation of the wave. The wave vector is $\omega/v_p\hat{\mathbf{k}} = n\omega\hat{\mathbf{k}}/c$. The exponential $\exp(i\mathbf{k}\cdot\mathbf{r})$ contains a factor $\exp[-\Im(n)\omega\hat{\mathbf{k}}\cdot\mathbf{r}/c]$, which corresponds to an absorptivity, i.e., attenuation coefficient of the energy flux, of

$$k_\nu = \frac{2\omega}{c}\Im(n). \tag{10.8}$$

We can also express this as

$$k_\nu = \frac{\Im(\epsilon)\omega}{\Re(n)c} = \frac{\omega_{pe}^2\nu_c}{\Re(n)c(\omega^2 + \nu_c^2)}. \tag{10.9}$$

The collision frequency, for Coulomb collisions with ions Z, is a number of order $N_i(Ze^2/kT)^2\sqrt{kT/m}$, and when this and the definition of the plasma frequency are inserted, the absorptivity for $\omega \gg \nu_c$ is found to be of order

$$k_\nu \approx \frac{4\pi Z^2 e^6}{c(mkT)^{3/2}\omega^2 \Re(n)} N_e N_i. \tag{10.10}$$

This is nothing more or less than the free–free absorptivity, *after correction for stimulated emission*, that we discussed earlier (cf. (8.79)), apart from the numerical factors and the factor of the real refractive index in the denominator. We see that free–free absorption emerges from a completely classical plasma physics discussion, in which, if need be, the correlations of the plasma species can be included.

If the oscillating dipoles represented by all the electrons are added incoherently instead of coherently, then each one produces scattered radiation according to the Rayleigh-scattering formula $\sigma = (8\pi/3)k^4|\alpha|^2$, which becomes

$$\sigma = \frac{8\pi}{3}\left(\frac{e^2}{mc^2}\right)^2 \frac{\omega^2}{\omega^2 + \nu_c^2}. \tag{10.11}$$

This is exactly Thomson scattering except for the roll-off at very low frequency. Thus plasma dispersion and Thomson scattering are alternatives, and which occurs under a particular set of plasma conditions depends on the correlation functions. All three processes, Thomson scattering, free–free absorption, and plasma dispersion, must be treated in a unified way for dense plasmas.

Returning to the case of a real index of refraction, we inquire next what happens when we apply quantization to the electromagnetic field defined using the macroscopic fields that account for the dielectric contribution. We must understand that quanta of these fields do not represent just a disturbance of $\mathcal{E}$ and $\mathcal{H}$ traveling though space, but include the sympathetic response of the plasma as well. The energy in such waves is really partly in field energy proper, and partly in material energy. If we decline to use the macroscopic fields for our photon states, then we will find that the transport equation, which itself results from a coarse-graining procedure, will transport radiation at the wrong rate or in the wrong direction.

Our "dressed" photons, the quanta of the macroscopic electromagnetic field, still have the energy $\hbar\omega$, but their momentum is $\hbar\mathbf{k} = n\omega\hat{\mathbf{k}}/c$. The number of photon states of a particular (transverse) polarization in a volume V in a range dk of k and a solid angle $d\Omega$ is still

$$\frac{Vk^2 dk d\Omega}{(2\pi)^3}, \tag{10.12}$$

but when k is replaced by its value in terms of ω this becomes

$$\frac{n^2\omega^2 d\Omega d\omega}{(2\pi)^3 c^2 v_g}, \tag{10.13}$$

where v_g is the group velocity

$$v_g \equiv \frac{d\omega}{dk}. \tag{10.14}$$

The thermodynamic equilibrium intensity can be calculated by multiplying the number of photons per mode, the energy per photon, the number of modes per unit volume per unit frequency per unit solid angle and the group velocity of the photon. The Bose–Einstein value for the number of photons per mode in thermodynamic equilibrium at temperature T is unchanged, as is the energy of the photon, while the number of modes is multiplied by n^2c/v_g and the group velocity is multiplied by v_g/c; we conclude that the equilibrium intensity is altered by the dielectric according to

$$B_\nu \rightarrow n^2 B_\nu. \tag{10.15}$$

The modification of the expansion of the vector potential **A** in the creation and annihilation operators requires a careful calculation of the energy density including the material contribution for a dispersive medium, see Landau and Lifshitz (1960). This leads to a modification of the multiplicative factor $(2\pi\hbar c^2/\omega V)^{1/2}$ to $(2\pi\hbar c v_g/n\omega V)^{1/2}$. As a result the cross section for emission processes is modified by a factor n^2c/v_g from the density of states times a factor v_g/cn from the **A** expansion, or a factor n overall. The cross sections for photoabsorption are modified by a factor v_g/cn from the **A** expansion and a factor c/v_g from the division by the photon flux, or a factor $1/n$ overall. This gives the results quoted by Cox and Giuli (1968),

$$k_\nu \rightarrow \frac{k_\nu^0}{n}, \tag{10.16}$$

$$j_\nu \rightarrow n j_\nu^0, \tag{10.17}$$

where the quantities with superscript ^{0} are the ones given by the atomic physics in the absence of refraction. See also Mercier (1964) and Dawson and Oberman (1962). It is an interesting and unsettled question how general the relation is between the refracting absorptivity and a hypothetical nonrefracting absorptivity.

The transport equation including refraction is modified in three respects. First, the rays are bent, so the transport operator must correspond to differentiation along a curved path. Second, the absorption and emission must use the modified values of absorptivity and emissivity. The third correction is that owing to the refraction, which can produce focusing or defocusing of a beam, the intensity is not itself constant, even in the absence of absorption and emission. It is I_ν/n^2 that is constant. This can be shown either by calculating how a beam expands in solid angle following a cluster of rays, or by invoking the second law of thermodynamics and the relation (10.15). The transport equation as modified is

$$\frac{1}{c}\frac{\partial(I_\nu/n^2)}{\partial t} + \mathbf{n}\cdot\nabla\left(\frac{I_\nu}{n^2}\right) + \frac{d\mathbf{n}}{ds}\cdot\nabla_{\mathbf{n}}\left(\frac{I_\nu}{n^2}\right) = \frac{1}{n^2}(j_\nu - k_\nu I_\nu). \quad (10.18)$$

The vector $d\mathbf{n}/ds$ is the rate of bending of the ray direction per unit length measured along the ray. This comes from Snell's law, which takes this form for our problem:

$$\frac{d\mathbf{n}}{ds} = (\mathsf{I} - \mathbf{nn})\cdot\nabla\log n. \quad (10.19)$$

As discussed by Cox and Giuli (1968), the definitions of energy density and radiation pressure are modified from those for the nonrefracting medium. The three moments are now given by

$$E_\nu = \frac{1}{v_g}\int_{4\pi} I_\nu d\Omega, \quad (10.20)$$

$$\mathbf{F}_\nu = \int_{4\pi} \mathbf{n} I_\nu d\Omega, \quad (10.21)$$

and

$$\mathsf{P}_\nu = \frac{n}{c}\int_{4\pi} \mathbf{nn} I_\nu d\Omega. \quad (10.22)$$

The energy density has a division by the group velocity replacing the division by c, while the pressure has a division by the phase velocity c/n replacing the division by c.

The angle moments of (10.18) are complicated by the presence of the ray-bending term in the transport operator. Cox and Giuli (1968, Vol. 1, pp. 135–136, 173) carry out the calculations, from which we find the following:

$$\frac{\partial}{\partial t}\left(\frac{v_g}{c}E_\nu\right) + \nabla\cdot\mathbf{F}_\nu = 4\pi j_\nu - k_\nu v_g E_\nu, \quad (10.23)$$

$$\frac{\partial}{\partial t}\left(\frac{1}{n^2 c}\mathbf{F}_\nu\right) + \nabla\cdot\left(\frac{c}{n^3}\mathsf{P}_\nu\right) = -\frac{k_\nu}{n^2}\mathbf{F}_\nu \quad (10.24)$$

for the case of isotropic absorption, emission, and refraction. Equation (10.23) shows that our notions of energy and momentum conservation are modified by refraction. Equation (10.24) can also be written in steady state, as

$$\nabla \cdot \mathsf{P}_\nu = -\frac{n}{c} k_\nu \mathbf{F}_\nu + 3\mathsf{P}_\nu \cdot \nabla \log n. \tag{10.25}$$

The left-hand side is the actual rate at which the radiation field is gaining momentum; the first term on the right-hand side is the momentum imparted through absorption/emission processes, since the momentum flux is $k/\omega = n/c$ times the energy flux. The second term on the right is a ponderomotive force term.

In the diffusion limit the intensity goes to n^2 times the ordinary Planck function, and therefore E_ν goes to $n^2 c/v_g$ times $4\pi B_\nu/c$ and the pressure P_ν goes to n^3 times $4\pi B_\nu/3c$. As a result the diffusion expression for the flux is

$$\mathbf{F}_\nu = -\frac{4\pi n^2}{3k_\nu} \nabla B_\nu, \tag{10.26}$$

and if the absorptivity is put in terms of the nonrefracting value k_ν^0 this becomes

$$\mathbf{F}_\nu = -\frac{4\pi n^3}{3k_\nu^0} \nabla B_\nu. \tag{10.27}$$

We take the frequency integral of this and express the total flux in terms of a new Rosseland mean opacity:

$$\mathbf{F} = -\frac{16\sigma T^3}{3\kappa_R \rho} \nabla T, \tag{10.28}$$

with

$$\frac{1}{\kappa_R} = \frac{\displaystyle\int_0^\infty d\nu \frac{n^3}{\kappa_\nu^0} \frac{dB_\nu}{dT}}{\displaystyle\int_0^\infty d\nu \frac{dB_\nu}{dT}}, \tag{10.29}$$

Notice that the index of refraction is included (cubed) in the numerator, but not in the denominator. That is because we want the denominator to continue to be exactly $4\sigma T^3/\pi$, since this is used in obtaining (10.28). In the simple model that $n = [1 - (\omega_{pe}/\omega)^2]^{1/2}$ the numerator integral in (10.29) must be cut off below the plasma frequency ω_{pe}, and we can see that the integrand drops to zero as this limit is approached, so the contribution to $1/\kappa_R$ from frequencies between ω_{pe} and, say, $2\omega_{pe}$ is much reduced by the effect of refraction. If the density is high enough that $\hbar\omega_{pe} \approx kT$ there will be a significant increase in κ_R. This modification to the Rosseland mean opacity is also discussed by Cox and Giuli (1968).

The importance of refraction in most astrophysical problems is not great since the plasma frequency is given by

$$\hbar\omega_{pe} = 0.37133 \left(\frac{N_e}{10^{20}\,\mathrm{cm}^{-3}} \right)^{1/2} \quad \mathrm{eV}, \tag{10.30}$$

which puts it well below the optical frequency range at the densities likely to be encountered. It remains a significant issue at radio frequencies since, for example, even at the typical coronal density of $10^8\,\mathrm{cm}^{-3}$ the plasma frequency is 90 MHz. The correction to the Rosseland mean opacity indicated in (10.29) is often omitted even in the large opacity tabulations because in the conditions where this correction is significant there are uncertainties in the opacity calculation that are even larger.

10.2 Description of polarized light

We have so far assumed that the radiation field is equally strong in the two modes of polarization, and we have summed or averaged over the polarizations as appropriate, to get results for unpolarized light. This is really not correct in a number of circumstances, and in this section we will show what might be done about that. In truth, for most purposes we continue to use the results for unpolarized light, at the cost of some loss of accuracy, since the number of unknowns goes up by 2–4 times when polarization is considered, and therefore it is an issue of computational expense.

What is the mathematical description of polarized light? Here is the classical way of defining it, derived from Chandrasekhar (1960). Consider propagation in the z direction. The x component of the electric field is one possible random function of time and the y component is another. The mean square value of $\mathcal{E}_x$ is $4\pi/c$ times the intensity if all the radiation is propagating in the z direction and there is no y component, so we call $c\langle\mathcal{E}_x\rangle^2/4\pi$ I_x, the x intensity. Likewise for I_y. The sum $I_x + I_y$ is indeed the total intensity I. This would be the end of the story, and in fact it sometimes is, except for the possibility of correlation between $\mathcal{E}_x$ and $\mathcal{E}_y$. And it turns out that there are pieces of optical apparatus that can detect correlations between $\mathcal{E}_x$ and $\mathcal{E}_y$ when one of the two components has been shifted in phase by some controlled amount. Such an apparatus is the quarter-wave plate, for example, which introduces a shift of $\pi/2$. A mathematical approach that brings out the way that $\mathcal{E}_x$ and $\mathcal{E}_y$ may be correlated with a shift in phase is the following. We pick an interval of time $\mathcal{T}$ that is long enough for $1/\mathcal{T}$ to be smaller than the frequency resolution $\Delta\nu$ we are interested in. We expand $\mathcal{E}_x$ and $\mathcal{E}_y$ over this interval in complex exponential series:

$$\mathcal{E}_x(t) = \sum_k \left(a_{x,k} e^{-i\omega_k t} + a^*_{x,k} e^{+i\omega_k t} \right), \tag{10.31}$$

$$\mathcal{E}_y(t) = \sum_k \left(a_{y,k} e^{-i\omega_k t} + a^*_{y,k} e^{+i\omega_k t} \right). \tag{10.32}$$

If we now pick out one frequency of interest, say ω_k, the Fourier coefficients for the two field components are represented by two complex numbers, $a_{x,k}$ and $a_{y,k}$. From these we build a 2×2 Hermitian matrix $\boldsymbol{\rho}_k$ as follows:

$$\boldsymbol{\rho}_k = \begin{pmatrix} a_{x,k} \\ a_{y,k} \end{pmatrix} \begin{pmatrix} a^*_{x,k} & a^*_{y,k} \end{pmatrix} = \begin{pmatrix} |a_{x,k}|^2 & a_{x,k} a^*_{y,k} \\ a^*_{x,k} a_{y,k} & |a_{y,k}|^2 \end{pmatrix}. \tag{10.33}$$

The value of $\rho_{k,11}$ is the power spectrum of $\mathcal{E}_x$, and if this is summed over the frequencies then, by Parseval's theorem, we recover I_x. Thus this is proportional to the spectral intensity for the x polarization. Likewise for $\rho_{k,22}$, which gives the spectral intensity for the y polarization. The off-diagonal components come in when we introduce a funny polarizer that responds to

$$\left| \cos\alpha \, a_{x,k} + \sin\alpha \, e^{i\beta} a_{y,k} \right|^2, \tag{10.34}$$

where α and β are arbitrarily adjustable angles. The response is seen to be given by the expression

$$\begin{aligned} &\begin{pmatrix} \cos\alpha & \sin\alpha \, e^{i\beta} \end{pmatrix} \begin{pmatrix} |a_{x,k}|^2 & a_{x,k} a^*_{y,k} \\ a^*_{x,k} a_{y,k} & |a_{y,k}|^2 \end{pmatrix} \begin{pmatrix} \cos\alpha \\ \sin\alpha \, e^{-i\beta} \end{pmatrix} \\ &= \cos^2\alpha \, \rho_{k,11} + \sin^2\alpha \, \rho_{k,22} + \cos\alpha \, \sin\alpha (e^{i\beta} \rho_{k,21} + e^{-i\beta} \rho_{k,12}). \end{aligned} \tag{10.35}$$

The full matrix is needed to determine this response for all possible values of α and β.

The matrix $\boldsymbol{\rho}_k$ is often called the *coherency* matrix (cf., Born and Wolf (1989), Section 10.8.1). In quantum mechanical language it is a density matrix with respect to the polarization modes. The components have conventional designations as shown here:

$$\boldsymbol{\rho}_k = \frac{1}{2} \begin{pmatrix} I + Q & U + iV \\ U - iV & I - Q \end{pmatrix}, \tag{10.36}$$

where the quantities I, Q, U, and V are the *Stokes parameters*. The matrix is positive semi-definite, so while Q, U and V can each have either sign, I must be positive, and $I \geq \sqrt{Q^2 + U^2 + V^2}$.

The extension of our treatment to the quantized radiation field is fairly easy since expansions (10.31) and (10.32) are just like the representation of the fields in terms of creation and annihilation operators. The quantum equivalent to (10.33) is

$$\rho_k = \mathrm{Tr}\left[\rho\begin{pmatrix} a^\dagger_{\mathbf{k}x}a_{\mathbf{k}x} & a_{\mathbf{k}x}a^\dagger_{\mathbf{k}y} \\ a^\dagger_{\mathbf{k}x}a_{\mathbf{k}y} & a^\dagger_{\mathbf{k}y}a_{\mathbf{k}y}\end{pmatrix}\right], \tag{10.37}$$

in which ρ without a subscript is the overall density matrix for the radiation field, and the trace is taken over all the radiation states. This entails a summation over all occupation vectors $|n_{k\varpi}\rangle$ in which the ns take all possible values. In the case that the radiation field is relatively weak, the only nonvanishing density matrix values connect states with either no photons at all, or just one photon total in all the modes. In this case ρ_k reduces to just the part of ρ connecting the states that have one photon. That is, we can regard ρ_k itself as the radiation density matrix. The density matrix for situations with multiple photon occupancies is another interesting subject, but it takes us beyond what can be discussed in the intensity picture. For example, some measurements can distinguish between different probability distributions for the photon number that correspond to the same intensity. The intensity interferometer of Hanbury Brown and Twiss (Hanbury Brown and Twiss, 1954, 1956a,b; see also Baym, 1997) relies on such an effect, since photon statistics are not Poisson owing to the photons' boson nature.

So far, since we have chosen just one time interval $\mathcal{T}$, the ρ_k matrix is related precisely to one complex vector $(a_{x,k} \quad a_{y,k})$, therefore it is rank 1, and the Stokes parameters obey the relation $I = \sqrt{Q^2 + U^2 + V^2}$ precisely. This is not the general case. When the electric field contains noise, which is true when emission from thermal sources is considered, the time-dependent field components are stochastic processes and the Fourier coefficients over any finite time interval are random variables. In this situation the correct quantity to use to describe the radiation is the average of ρ_k over an infinite number of realizations of the time-dependent fields, or over an infinite number of different time windows $\mathcal{T}$. After averaging, ρ_k will still have the Hermiticity property, and it will still be semi-definite. In fact, it is exceedingly unlikely that $I = \sqrt{Q^2 + U^2 + V^2}$ will remain true, and therefore the matrix is highly likely to be positive definite. This means that every funny polarizer as described above will give some nonzero response.

Some general properties of the coherency matrix or the Stokes parameters are the following. When two beams of light are merged, supposing them to come from distinct sources so they are completely uncorrelated, then the coherency matrices or Stokes parameters can be added component by component. The case that $I = \sqrt{Q^2 + U^2 + V^2}$ is called elliptically polarized light; it contains just one, possibly complex, polarization mode. It is called elliptical because the points $(\mathcal{E}_x, \mathcal{E}_y)$ trace out an ellipse in the x–y plane. If $U = 0$ the semi-major and semi-minor axes of the ellipse are aligned with x and y. If $V = 0$ then the light is linearly polarized in general, and the plane of polarization depends on Q and U, the position angle with respect to the x-axis being $(1/2)\tan^{-1} U/Q$. If $U = 0$ then the light is x-polarized if $Q = I$ and y-polarized if $Q = -I$. In the $U = 0$ case, if also $Q = 0$, the light is circularly polarized. Circularly polarized light with $V = I$ has the relation that $a_{y,k} = \exp(-i\pi/2)a_{x,k}$ and therefore the y component of the field *leads* the x component of the field by a quarter period. Thus the point representing $(\mathcal{E}_x, \mathcal{E}_y)$ moves clockwise around a circle in the x–y diagram as t increases at a fixed z (assuming that the xyz coordinate system is right-handed). This is right-circular polarization. If a quarter-wave plate is used to *delay* $\mathcal{E}_y$ relative to $\mathcal{E}_x$ by $\pi/2$ for right-circularly polarized light, the result is light that is linearly polarized at $\theta = \pi/4$. In the case that $V = -I$ the motion is counter-clockwise around the circle, this is left-circular polarization, and the effect of the quarter-wave plate is to produce linearly polarized light at $\theta = 3\pi/4$.

Every Hermitian matrix can be diagonalized by a unitary transformation, which means that ρ_k can be expressed in terms of rotated, possibly complex, orthogonal polarization modes, with respect to which the two field components are completely uncorrelated. There are two physical pictures of a beam of arbitrarily polarized light that go with this mathematical statement. The first picture is to regard the beam as the sum of two parts, of which the first part is a mixture of equal amounts of the two modes, and is therefore unpolarized, and the second part is a residual amount of just one of the two modes, which is thus elliptically polarized. Thus any kind of light can be formed by the mixture of some unpolarized light with a beam of elliptically polarized light. The alternative physical picture is that any kind of light is the sum of suitable amounts of two particular orthogonal elliptical polarizations. Orthogonal elliptical polarizations, by the way, differ from each other by interchanging the major and minor axes, and reversing the sense of rotation. So linear x polarization is orthogonal to linear y polarization; linear polarization along $y = 2x$ is orthogonal to linear polarization along $x = -2y$. Right-circular polarization is orthogonal to left-circular polarization.

Elliptically polarized light that is normalized to unit intensity, so $Q^2 + U^2 + V^2 = 1$, corresponds to a point on the unit sphere in QUV space. This is called the Poincaré sphere. It gives an easy way of visualizing types of polarization, and the actions of various polarizers. (See Huard (1997).) The "north" pole of the Poincaré sphere is right-circularly polarized light and the "south" pole is left-circularly polarized. The equator contains the types of linear polarization. The longitude is twice the position angle of the major axis of the ellipse, and the latitude determines the ellipticity according to latitude $= 2\epsilon$ with $\tan\epsilon =$ minor axis/major axis. The northern hemisphere contains the right-handed polarizations, and the southern hemisphere the left-handed ones. Diametrically opposite points on the Poincaré sphere represent orthogonal polarizations. The action of a phase plate or compensator is visualized as a rotation of the sphere about a certain direction. For example, the quarter-wave plate produces a 90° rotation about the Q axis, so that U becomes V and V becomes $-U$. In general, the action of polarizers and compensators is represented in matrix language by $\mathbf{I}' = \mathsf{M}\mathbf{I}$, a transformation mapping the old Stokes vector $\mathbf{I}$ into a new one $\mathbf{I}'$. The 4×4 matrix M is called the Mueller matrix for the device.

If we change from one set of polarization basis vectors, for example the vectors $\mathbf{e}_x$ and $\mathbf{e}_y$ we have been discussing, to some other pair $\mathbf{e}_1$ and $\mathbf{e}_2$, where we will require that $\mathbf{e}_1$ and $\mathbf{e}_2$ are normalized and orthogonal: $\mathbf{e}_1^* \cdot \mathbf{e}_1 = \mathbf{e}_2^* \cdot \mathbf{e}_2 = 1$ and $\mathbf{e}_1^* \cdot \mathbf{e}_2 = 0$, what happens to the coherency matrix? Suppose that

$$\begin{pmatrix}\mathbf{e}_x & \mathbf{e}_y\end{pmatrix} = \begin{pmatrix}\mathbf{e}_1 & \mathbf{e}_2\end{pmatrix} M, \tag{10.38}$$

in terms of a 2×2 unitary matrix M. Then the vector field Fourier amplitude can be expressed in two equivalent ways,

$$\begin{pmatrix}\mathbf{e}_x & \mathbf{e}_y\end{pmatrix} \begin{pmatrix} a_{x,k} \\ a_{y,k} \end{pmatrix} \tag{10.39}$$

or

$$\begin{pmatrix}\mathbf{e}_1 & \mathbf{e}_2\end{pmatrix} \begin{pmatrix} a_{1,k} \\ a_{2,k} \end{pmatrix}. \tag{10.40}$$

Therefore the amplitudes are related by

$$\begin{pmatrix} a_{1,k} \\ a_{2,k} \end{pmatrix} = M \begin{pmatrix} a_{x,k} \\ a_{y,k} \end{pmatrix}. \tag{10.41}$$

When these amplitudes are used to construct the coherency matrix ρ_k' with respect to these new polarization modes, it turns out to be

$$\rho_k' = M \rho_k M^\dagger. \tag{10.42}$$

So the unitary transformations that diagonalize the coherency matrix do indeed just correspond to picking different polarization modes for the basis, as suggested above.

As an example of this transformation let us take the modes

$$\mathbf{e}_1 = \frac{-\mathbf{e}_x + i\mathbf{e}_y}{\sqrt{2}}, \tag{10.43}$$

$$\mathbf{e}_2 = \frac{\mathbf{e}_x + i\mathbf{e}_y}{\sqrt{2}}. \tag{10.44}$$

These are the circular polarization modes. The matrix M in this case is

$$M = \frac{1}{\sqrt{2}} \begin{pmatrix} -1 & -i \\ 1 & -i \end{pmatrix}. \tag{10.45}$$

Applying this to find the transformed coherency matrix in terms of the Stokes parameters gives

$$\rho'_k = \frac{1}{2} \begin{pmatrix} I + V & -Q - iU \\ -Q + iU & I - V \end{pmatrix}. \tag{10.46}$$

So as a result of the transformation V now plays the role that Q did before, $-Q$ plays the role that U did before, and $-U$ plays the role that V did before. We recognize $\mathbf{e}_1$ as the basis vector for the right-circular polarization mode, and $\mathbf{e}_2$ as the basis vector for the left-circular polarization.

It must be added here that, sadly, the conventions related to the Stokes parameters and the coherency matrix are not too well established. Some authors assume that the time-dependent exponential factor is $\exp(i\omega t)$ instead of $\exp(-i\omega t)$. This reverses the sign of V. Some authors define a positive V as referring to counter-clockwise rotation of the electric vector, which does the same thing. And there are other variations. In this presentation $\exp(-i\omega t)$ and clockwise rotation have been assumed, and in this and other respects the conventions described by Rees (1987) have been followed.

10.3 Transport equation for an isotropic medium

The transport equation for polarized light we shall consider first will include only the effect of scattering on the polarization components, and refraction and any dependence of the absorptivity on polarization mode will be put aside for now. We will deal with a vector-valued intensity based on the four Stokes components,

$$\mathbf{I} = \begin{pmatrix} I \\ Q \\ U \\ V \end{pmatrix}. \tag{10.47}$$

The absorptivity is (in our approximation) a scalar, but the emissivity is also a vector, since it is potentially different for the four components. Thus the transport equation is

$$\frac{1}{c}\frac{\partial \mathbf{I}_\nu}{\partial t} + \mathbf{n} \cdot \nabla \mathbf{I}_\nu = \mathbf{j}_\nu - k_\nu \mathbf{I}_\nu. \tag{10.48}$$

Incidentally, the additivity of Stokes vectors is what allows us to write a linear transport equation. There is an implicit assumption that the gains or losses of radiation in a volume element have a random phase compared with the radiation field itself. That will not be the case if the "scatterers" are spatially ordered, for example.

The discussion hinges on the form of the Stokes vector emissivity $\mathbf{j}_\nu$. Thermal emission is unpolarized, and therefore it corresponds to

$$\mathbf{j}_\nu = \begin{pmatrix} k_\nu B_\nu \\ 0 \\ 0 \\ 0 \end{pmatrix}. \tag{10.49}$$

If this is the only kind of emission, if any radiation incident at the boundaries is unpolarized, and if any radiation present at the initial time is unpolarized, then the transport equation shows that the radiation remains everywhere unpolarized at all time, a common-sense result. However, we shall see that scattering processes produce polarization when there is none present to begin with.

There is a simple result for the Stokes vector emissivity for scattering processes like Thomson and Rayleigh scattering. To see how to get there from the quantum mechanics, we need to begin with a density-matrix form of Fermi's Golden Rule, which is

$$\dot{\rho}_{f'f} = \frac{2\pi}{\hbar} \sum_{ii'} \langle f'|H_{\rm int}|i'\rangle \rho_{i'i} \langle i|H_{\rm int}|f\rangle \delta(E_f - E_i), \tag{10.50}$$

where the initial states i and i' are one degenerate set in energy, with populations and possible correlations as given by $\rho_{i'i}$, and the final set of states f and f' are another degenerate set, and furthermore there is overall conservation of energy. In the case of Rayleigh and Thomson scattering the indices on the initial density (polarization) matrix are ϖ_1' and ϖ_1, and we want to find the contribution to the final state density matrix with indices ϖ_2' and ϖ_2. The first matrix element of $H_{\rm int}$ contains a factor $\mathbf{e}^*_{\mathbf{k}_2\varpi_{2'}} \cdot \mathbf{e}_{\mathbf{k}_1\varpi_{1'}}$ and the second matrix element contains a factor $\mathbf{e}^*_{\mathbf{k}_1\varpi_1} \cdot \mathbf{e}_{\mathbf{k}_2\varpi_2}$. The product of these two factors is the dependence of the differential cross section for coherency matrix components on the polarization modes. If we define the direction cosine matrix C by $C_{\varpi_1\varpi_2} = \mathbf{e}^*_{\mathbf{k}_1\varpi_1} \cdot \mathbf{e}_{\mathbf{k}_2\varpi_2}$, then the polarization-matrix rate of emission due to scattering of photons in the solid angle $d\Omega_1$ can be written as the matrix product

$$N_e \left(\frac{e^2}{mc^2}\right)^2 C^\dagger \rho_k C d\Omega_1. \tag{10.51}$$

Obtaining the direction cosine matrix explicitly is an exercise in angular algebra; indeed the cleanest way of representing angular quantities such as $\boldsymbol{\rho}_k$ and the related emissivity function is by means of irreducible tensors, for which all the methods of Racah angular algebra may be invoked. We are not able to delve into that here.

The direction cosine matrix C is easy to give if the polarizations are referred to the plane containing the initial and final photon directions. Let polarization 1 be perpendicular to that plane and polarization 2 be parallel to it. Then the direction cosine matrix is

$$C = (\mathbf{e}^*_{\mathbf{k}_1\varpi_1} \cdot \mathbf{e}_{\mathbf{k}_2\varpi_2}) = \begin{pmatrix} 1 & 0 \\ 0 & \cos\Theta \end{pmatrix}, \tag{10.52}$$

where Θ is the scattering angle. Working out the matrix products in (10.51) using the definition (10.36) of $\boldsymbol{\rho}_k$ and then rearranging the polarization components as Stokes vector components gives

$$\frac{1}{2}N_e\left(\frac{e^2}{mc^2}\right)^2 \begin{pmatrix} 1+\cos^2\Theta & \sin^2\Theta & 0 & 0 \\ \sin^2\Theta & 1+\cos^2\Theta & 0 & 0 \\ 0 & 0 & 2\cos\Theta & 0 \\ 0 & 0 & 0 & 2\cos\Theta \end{pmatrix} \begin{pmatrix} I \\ Q \\ U \\ V \end{pmatrix} d\Omega_1 \tag{10.53}$$

for the contribution to the Stokes emissivity vector from this particular initial photon direction. However, before being used to actually solve the equation of transfer the matrix appearing here must still be transformed to account for the fact that the polarization modes must be referred to a consistent azimuthal angle of reference, since the scattering plane rotates around as the photon directions vary. Equation (10.53) as it stands is sufficient to show us that unpolarized radiation scattered through $\Theta = 90°$ becomes 100% linearly polarized perpendicular to the scattering plane.

10.4 Polarized light in an anisotropic medium

The equation of transfer in the preceding subsection assumes that the absorption coefficient is a scalar quantity; i.e., it is the same for the two modes of polarization. Another implicit assumption is that the index of refraction is the same for the two modes. Neither of these things is true for an anisotropic medium. This is a medium in which the dielectric tensor is not a scalar tensor, and in which, therefore, the speed of light is different in different directions. This is encountered in birefringent crystals, and in astrophysics in describing light propagation through magnetized plasmas. Four examples of the latter are: (1) the formation of Zeeman split or broadened spectral lines in the solar photosphere (Beckers, 1969), (2) remote probing of the earth's ionosphere using microwave emission of O_2 (Lenoir, 1968), (3) formation of the cyclotron line in the accretion column of a neutron star (Mészáros and Nagel, 1985), and (4) Faraday rotation of radio signals from distant pulsars by the interstellar medium.

The last of these examples points out the need to consider in the transfer equation not only the absorption of the Stokes components, but also the rotation – the exchange of energy *between* components. The theory of Faraday rotation has long included the rotation effect based on classical optics without incorporating it in the transfer equation, while the theory of spectral line formation in the magnetized solar photosphere initially ignored the rotation (Unno, 1956). A unified treatment including both effects has now been given for the radio astronomy case by Lenoir (1967) and for the solar Zeeman effect by Rachkovsky (1962), E. Landi Degl'Innocenti and M. Landi Degl'Innocenti (1972). A general formulation for the Zeeman effect problem based on QED is found in E. Landi Degl'Innocenti (1983). E. Landi Degl'Innocenti (1987) reviews the application of his methods to the formation of polarized spectral lines in the solar atmosphere.

Lenoir (1967) uses a semi-classical development based on Maxwell's equations to arrive at this form of the transfer equation expressed in terms of the coherency matrix $\boldsymbol{\rho}$

$$\frac{d\boldsymbol{\rho}}{ds} + \mathsf{G}\boldsymbol{\rho} + \boldsymbol{\rho}\mathsf{G}^{\dagger} = 2B_{\nu}\mathsf{A}, \tag{10.54}$$

in which G is, in effect, $-2\pi i/\lambda$ times the complex tensor index of refraction, $\mathsf{G}^{\dagger}$ is its Hermitian adjoint, and A is its Hermitian part, $(\mathsf{G} + \mathsf{G}^{\dagger})/2$. (Lenoir's phases have to be corrected for time dependence $\exp(-i\omega t)$.) In Lenoir's formulation he assumes the permittivity ϵ_0 of vacuum but includes a permeability tensor $\boldsymbol{\mu} = \mu_0(\mathsf{I} + \boldsymbol{\chi})$ that incorporates the plasma Zeeman effect, and it is then found to be sufficient to take $\mathsf{G} = -(i\pi/\lambda)\boldsymbol{\chi}_{\perp}$, where $\boldsymbol{\chi}_{\perp}$ stands for the 2×2 projection of $\boldsymbol{\chi}$ into the plane perpendicular to the direction of propagation. Alternatively, the ideal permeability μ_0 may be used, and the permittivity given by the general dielectric tensor $\boldsymbol{\epsilon}$. The result is the same, with $\boldsymbol{\epsilon}/\epsilon_0 - \mathsf{I}$ replacing $\boldsymbol{\chi}$.

Lenoir (1967) specifically considers LTE radiative transfer in the microwave region of the spectrum, thus his absorption coefficient implicitly contains a correction factor $1 - \exp(-h\nu/kT) \approx h\nu/kT$ for stimulated emission, and he may use the Rayleigh–Jeans approximation $B_{\nu} \approx 2kT\nu^2/c^2$. He therefore refers to brightness temperature rather than to the specific intensity and its coherence components. However, nothing in the formulation prevents using it, as quoted above, for the coherency matrix in intensity units, provided it is understood that G has been corrected for stimulated emission.

In the foregoing equations all the tensors are 2×2. If the polarization basis vectors are $\mathbf{e}_x$ and $\mathbf{e}_y$, corresponding to the relation (10.36) between the coherency matrix and the Stokes parameters, then the tensor $\boldsymbol{\chi}_\perp$, for example, is just

$$\boldsymbol{\chi}_\perp = \begin{pmatrix} \chi_{xx} & \chi_{xy} \\ \chi_{yx} & \chi_{yy} \end{pmatrix} \tag{10.55}$$

in the case that the radiation direction in question is the $+z$ direction. If, as above, a different set of polarization vectors is used, related to these by a unitary transformation with a matrix M, then the transformed coherency matrix $\boldsymbol{\rho}'$ of (10.42) will obey the transfer equation

$$\frac{d\boldsymbol{\rho}'}{ds} + M\mathsf{G}M^\dagger\boldsymbol{\rho}' + \boldsymbol{\rho}'M\mathsf{G}^\dagger M^\dagger = 2B_\nu M\mathsf{A}M^\dagger, \tag{10.56}$$

as we can see by multiplying (10.54) on the left by M and on the right by $M^\dagger$; the matrices pass through the derivative with respect to s if the polarization modes are space-independent. The form of the transfer equation is unchanged in this basis, and it is necessary just to transform the G and A tensors to the new basis.

The transfer equation for the coherency matrix $\boldsymbol{\rho}$ can be made into a useful transfer equation for the Stokes vector $\mathbf{I}$. If the G matrix is written as

$$\mathsf{G} = \begin{pmatrix} G_{11} & G_{12} \\ G_{21} & G_{22} \end{pmatrix}, \tag{10.57}$$

then working out the matrix products in (10.54) using (10.36) leads to

$$\frac{d\mathbf{I}}{ds} + \mathsf{K}\mathbf{I} = \mathbf{j} = \mathsf{K}\mathbf{S}, \tag{10.58}$$

with the absorption matrix K and the emissivity and source function vectors $\mathbf{j}$ and $\mathbf{S}$ defined by

$$\mathsf{K} = \begin{pmatrix} \phi_I & \phi_Q & \phi_U & \phi_V \\ \phi_Q & \phi_I & \phi'_V & -\phi'_U \\ \phi_U & -\phi'_V & \phi_I & \phi'_Q \\ \phi_V & \phi'_U & -\phi'_Q & \phi_I \end{pmatrix}, \quad \mathbf{j} = B_\nu \begin{pmatrix} \phi_I \\ \phi_Q \\ \phi_U \\ \phi_V \end{pmatrix}, \quad \mathbf{S} = B_\nu \begin{pmatrix} 1 \\ 0 \\ 0 \\ 0 \end{pmatrix}. \tag{10.59}$$

The seven parameters that appear in (10.59) are related to the G elements by

$$\begin{aligned}
\phi_I &= \Re(G_{11}) + \Re(G_{22}), \\
\phi_Q &= \Re(G_{11}) - \Re(G_{22}), \\
\phi_U &= \Re(G_{12}) + \Re(G_{21}), \\
\phi_V &= -\Im(G_{12}) + \Im(G_{21}), \\
\phi'_Q &= \Im(G_{11}) - \Im(G_{22}), \\
\phi'_U &= \Im(G_{12}) + \Im(G_{21}), \\
\phi'_V &= \Re(G_{12}) - \Re(G_{21}).
\end{aligned} \tag{10.60}$$

This notation follows Jefferies, Lites, and Skumanich (1989); the ϕs should not be confused either with angles or with values of line profile functions, although it turns out that for polarized line transfer the unprimed ϕs are various combinations of profile functions of the Zeeman components times geometrical factors, and the primed parameters contain the corresponding dispersion functions, which are similar to the derivatives of the profiles; see below.

The action of the ϕ' elements of K in (10.58) is worthy of note. These elements, by themselves, cause the Stokes parameters Q, U, V to rotate as one progresses along the ray. The axis of rotation is $(\phi'_Q, \phi'_U, \phi'_V)$, and the magnitude of this vector gives the angle of rotation per unit of path length. The effect can be visualized by considering a counter-clockwise rotation of the Poincaré sphere about this direction.

The diagonal ϕ_I elements of K produce uniform absorption of all the components of **I**, so that the percentage and type of polarization are unchanged as I is reduced. But the effect of ϕ_Q, ϕ_U, and ϕ_V is to selectively absorb one polarization and to absorb less the orthogonal one. Specifically, the polarization component along $(Q, U, V) = (\phi_Q, \phi_U, \phi_V)$ is absorbed at the increased rate $\phi_I + (\phi_Q^2 + \phi_U^2 + \phi_V^2)^{1/2}$, while the orthogonal polarization $(Q, U, V) = (-\phi_Q, -\phi_U, -\phi_V)$ has the decreased absorption coefficient $\phi_I - (\phi_Q^2 + \phi_U^2 + \phi_V^2)^{1/2}$. (The large and small absorption coefficients are the same as twice the eigenvalues of A, the Hermitian part of G. A condition of physical reasonableness is that A should be positive definite.) If the path length through this material is sufficiently large, then the attenuated emerging radiation is 100% polarized along $(Q, U, V) = (-\phi_Q, -\phi_U, -\phi_V)$, regardless of its initial polarization. So the recipe for making a polarizer that passes only linear polarization in the x direction ($Q > 0, U = V = 0$) is to provide a slab of material with large, negative ϕ_Q, $\phi_I \approx |\phi_Q|$, and negligible values of the other parameters.

A simple example of the transfer equation for anisotropic polarized light is provided by Faraday rotation of cosmic radio waves due to magnetized interstellar plasma. The dielectric tensor ϵ for this case is a generalization of (10.7) discussed earlier. Following Allis, Buchsbaum, and Bers (1963) and correcting the phases for time dependence $\exp(-i\omega t)$ leads to the dielectric tensor, for the case of a uniform static field $\mathbf{B}$ along the z axis, given by

$$\frac{\epsilon}{\epsilon_0} = \mathsf{I} - \frac{\omega_{pe}^2}{2\omega}\begin{pmatrix} \ell + r & -i(\ell - r) & 0 \\ i(\ell - r) & \ell + r & 0 \\ 0 & 0 & 2p \end{pmatrix}, \tag{10.61}$$

with the definitions

$$r = \frac{1}{\omega + \omega_B + i\nu_c}, \quad \ell = \frac{1}{\omega - \omega_B + i\nu_c}, \quad p = \frac{1}{\omega + i\nu_c}, \tag{10.62}$$

in which ω_B is the electron cyclotron frequency eB/mc and ω_{pe} and ν_c are, as before, the electron plasma frequency $\sqrt{4\pi N_e e^2/m}$ and the electron–ion collision frequency. Making the assumption that both ω_B and ν_c are much smaller than ω leads to

$$\frac{\epsilon}{\epsilon_0} \approx \left(1 - \frac{\omega_{pe}^2}{\omega^2} + i\frac{\omega_{pe}^2 \nu_c}{\omega^3}\right)\mathsf{I} + i\frac{\omega_{pe}^2 \omega_B}{\omega^3}\begin{pmatrix} 0 & 1 & 0 \\ -1 & 0 & 0 \\ 0 & 0 & 0 \end{pmatrix}. \tag{10.63}$$

The isotropic part of this tensor is what we found before; the second term is new. The components as written are for a coordinate system with the z axis along $\mathbf{B}$. But the matrix in the second term can also be written in tensor language as $b_{ij} = e_{ijk}B_k/B$ in terms of the anti-symmetric tensor e_{ijk} which is $+1$ if ijk is an even permutation, -1 if ijk is an odd permutation, and is otherwise 0. Since e_{ijk} and B_k transform like a good tensor and vector under rotations, the same formula can be used to express this matrix for a general orientation of $\mathbf{B}$, which gives

$$\frac{\epsilon}{\epsilon_0} \approx \left(1 - \frac{\omega_{pe}^2}{\omega^2} + i\frac{\omega_{pe}^2 \nu_c}{\omega^3}\right)\mathsf{I} + i\frac{\omega_{pe}^2 \omega_B}{\omega^3 B}\begin{pmatrix} 0 & B_z & -B_y \\ -B_z & 0 & B_x \\ B_y & -B_x & 0 \end{pmatrix}, \tag{10.64}$$

and therefore

$$\mathsf{G} = \left(\frac{i\omega_{pe}^2}{2\omega c} + \frac{\omega_{pe}^2 \nu_c}{2\omega^2 c}\right)\mathsf{I} + \frac{\omega_{pe}^2 \omega_{Bz}}{2\omega^2 c}\begin{pmatrix} 0 & 1 \\ -1 & 0 \end{pmatrix}, \tag{10.65}$$

where the identity matrix I is now 2×2, and ω_{Bz} is a new variable, $\omega_{Bz} = \omega_B B_z/B = eB_z/mc$. In contrast to ω_B, ω_{Bz} can have either sign, depending on the angle between $\mathbf{B}$ and the $+z$ axis. The Faraday rotation will apparently depend on only that component of $\mathbf{B}$ that is along the direction of propagation of the wave.

The scalar imaginary (dispersive) part of G disappears in the transfer equation, since it has no effect on the intensities. It is, however, responsible for pulsar dispersion measures. The real scalar part of G represents absorption due to inverse bremsstrahlung, as discussed before. Given this result for G, the elements of K can be read off: $\phi_I = \omega_{pe}^2 \nu_c/(\omega^2 c)$, $\phi_Q = 0$, $\phi_U = 0$, $\phi_V = 0$, $\phi'_Q = 0$, $\phi'_U = 0$ and $\phi'_V = \omega_{pe}^2 \omega_{Bz}/(\omega^2 c)$. Neglecting the emission by the cold interstellar material, the Stokes transfer equations become

$$\frac{dI}{ds} = -\frac{\omega_{pe}^2 \nu_c}{\omega^2 c} I, \tag{10.66}$$

$$\frac{dQ}{ds} = -\frac{\omega_{pe}^2 \nu_c}{\omega^2 c} Q - \frac{\omega_{pe}^2 \omega_{Bz}}{\omega^2 c} U, \tag{10.67}$$

$$\frac{dU}{ds} = -\frac{\omega_{pe}^2 \nu_c}{\omega^2 c} U + \frac{\omega_{pe}^2 \omega_{Bz}}{\omega^2 c} Q, \tag{10.68}$$

$$\frac{dV}{ds} = -\frac{\omega_{pe}^2 \nu_c}{\omega^2 c} V. \tag{10.69}$$

If the light from the distant source is partially linearly polarized, then at the source $V = 0$, and it remains so. We see that I, Q, and U are attenuated at the same rate by the inverse bremsstrahlung. But the coupling of Q and U produces rotation of the plane of polarization. The rate of rotation can be calculated by

$$\frac{d\theta}{ds} = \frac{d}{ds}\left(\frac{1}{2}\tan^{-1}\frac{U}{Q}\right) = \frac{1}{2}\frac{1}{Q^2 + U^2}\left(Q\frac{dU}{ds} - U\frac{dQ}{ds}\right) = \frac{\omega_{pe}^2 \omega_{Bz}}{2\omega^2 c}. \tag{10.70}$$

The rotation per unit path length is proportional to the electron density, the projection of **B** on the line of sight and inversely to the square of the frequency. If the component of **B** in the direction of propagation is positive, the sense of the rotation is counter-clockwise. It often happens that the total rotation angle as observed on the earth for a certain radio source is many times 2π at typical frequencies. What is observed in these cases is a linear variation of the position angle of linear polarization with λ^2 over a small range of λ. The coefficient is proportional to $\int N_e B_z \, ds$ over the line of sight to the source, called the rotation measure.

E. Landi Degl'Innocenti's (1983) form of the transfer equation for $\boldsymbol{\rho}$ is

$$\frac{d\boldsymbol{\rho}}{ds} = \frac{1}{2}\left(\mathsf{f} + \mathsf{f}^\dagger\right) - \left(\mathsf{g}\boldsymbol{\rho} + \boldsymbol{\rho}\mathsf{g}^\dagger\right) + \left(\mathsf{h}\boldsymbol{\rho} + \boldsymbol{\rho}\mathsf{h}^\dagger\right). \tag{10.71}$$

The three new matrices that appear here: f, g, and h, take the places of G and A, and are expressed by E. Landi Degl'Innocenti as specific sums of atomic density matrix elements multiplied by two factors of quantities like $\langle a|\mathbf{p}\cdot\mathbf{e}|b\rangle$ and by the complex line absorption profile function, about which more below. They represent spontaneous emission, absorption, and stimulated emission, respectively. The stimulated emission term, with h, enters the transfer equation as a subtraction from the absorption term, as expected. The spontaneous emission matrix f is the stimulated emission matrix multiplied by the factor $2h\nu^3/c^2$, also as expected. When the atomic density matrices reduce to simple populations, and furthermore are in LTE, then the h matrix becomes $\exp(-h\nu/kT)$ times the g matrix and we can denote g − h by G, with the result that f is $\mathsf{G}B_\nu$. This reduces E. Landi Degl'Innocenti's equation to Lenoir's form.

E. Landi Degl'Innocenti's result for the g matrix can be looked at in more detail in the simple case of absorption for a normal Zeeman triplet, for example in a ${}^1S_0$–${}^1P_1$ transition. The ${}^1P_1$ level is split by the Zeeman effect into the $M = -1$, $M = 0$, and $M = 1$ sublevels, and the line is split into the $0 \to -1$ σ_- component, the $0 \to 0$ π component and the $0 \to 1$ σ_+ component. The frequencies of the components turn out to be $\nu = \nu_0 - \nu_B$, ν_0, and $\nu_0 + \nu_B$, respectively, where ν_0 is the unperturbed frequency and $\nu_B = eB/4\pi mc$ is the normal Zeeman splitting. The expression for the components of the g matrix at frequency ν becomes

$$\begin{aligned} g_{\alpha\beta} = C\big[&\langle u, -1|\mathbf{p}\cdot\mathbf{e}_\alpha|\ell, 0\rangle\langle u, -1|\mathbf{p}\cdot\mathbf{e}_\beta|\ell, 0\rangle^*\,\Phi(\nu_0 - \nu_B - \nu) \\ &+ \langle u, 0|\mathbf{p}\cdot\mathbf{e}_\alpha|\ell, 0\rangle\langle u, 0|\mathbf{p}\cdot\mathbf{e}_\beta|\ell, 0\rangle^*\,\Phi(\nu_0 - \nu) \\ &+ \langle u, +1|\mathbf{p}\cdot\mathbf{e}_\alpha|\ell, 0\rangle\langle u, +1|\mathbf{p}\cdot\mathbf{e}_\beta|\ell, 0\rangle^*\,\Phi(\nu_0 + \nu_B - \nu)\big], \end{aligned} \tag{10.72}$$

in which C is a certain combination of atomic constants with the population density of atoms in the lower level. The complex function $\Phi(\Delta\nu)$ has a real part H, which is the Voigt profile function for a Doppler width $\Delta\nu_D$ and a Lorentzian width $\Gamma = a\Delta\nu_D$, and the imaginary part is $2F$, the dispersion function that can be derived from H using the Kramers–Kronig relation. If we define $v = \Delta\nu/\Delta\nu_D$, then the functions are given by

$$\begin{aligned} \Delta\nu_D\Phi(\Delta\nu) &= \frac{i}{\pi^{3/2}}\int_{-\infty}^{\infty}\frac{\exp(-t^2)\,dt}{v + ia - t} \\ &= \frac{1}{\sqrt{\pi}}H(a, v) + \frac{2i}{\sqrt{\pi}}F(a, v) \end{aligned} \tag{10.73}$$

with

$$\tag{10.74}$$

$$H(a, v) = \frac{a}{\pi}\int_{-\infty}^{\infty}\frac{\exp(-t^2)\,dt}{(v - t)^2 + a^2} \tag{10.75}$$

$$F(a, v) = \frac{1}{2\pi}\int_{-\infty}^{\infty}\frac{(v - t)\exp(-t^2)\,dt}{(v - t)^2 + a^2}, \tag{10.76}$$

also

$$\Delta\nu_D \Phi(\Delta\nu) = \frac{1}{\sqrt{\pi}} w(v + ia) \tag{10.77}$$
$$= \frac{1}{\sqrt{\pi}} \exp\left(-(v + ia)^2\right) \operatorname{erfc}(a - iv). \tag{10.78}$$

The last relation connects the complex profile shape with the complex error function (Abramowitz and Stegun, 1964).

The factors like $\langle u, -1|\mathbf{p}\cdot\mathbf{e}_\alpha|\ell, 0\rangle$ evaluate to physical constants times the reduced matrix element $\langle u||\mathbf{p}||\ell\rangle$ times the trigonometrical factor relating the direction of $\mathbf{e}_\alpha$ to the -1 spherical component of a unit vector along the **B** field. In fact, the vector **p** can be expressed as

$$\mathbf{p} = -p'_1\mathbf{e}'_{-1} + p'_0\mathbf{e}'_0 - p'_{-1}\mathbf{e}'_1 \tag{10.79}$$

in terms of its spherical components in an $x'y'z'$ coordinate system with the z' axis along **B**, and the spherical basis vectors in this coordinate system. An application of the Wigner–Eckart theorem to $\langle u, -1|\mathbf{p}|\ell, 0\rangle$ gives

$$\langle u, -1|\mathbf{p}|\ell, 0\rangle = -\langle u, -1|p'_{-1}|\ell, 0\rangle\mathbf{e}'_1$$
$$= -\langle u||\mathbf{p}||\ell\rangle \begin{pmatrix} 1 & 1 & 0 \\ 1 & -1 & 0 \end{pmatrix} \mathbf{e}'_1, \tag{10.80}$$

where the $3j$ symbol has the value $1/\sqrt{3}$. Expressing $\mathbf{e}'_1$ in xyz coordinates requires selecting a direction for the x' axis in the plane perpendicular to **B**; a different choice of this direction will multiply this matrix element by a complex phase factor $\exp(i\delta)$. Fortunately this will cancel out when $\langle u, -1|\mathbf{p}\cdot\mathbf{e}_\alpha|\ell, 0\rangle$ is combined with $\langle u, -1|\mathbf{p}\cdot\mathbf{e}_\beta|\ell, 0\rangle^*$. Arbitrarily selecting $\mathbf{e}'_x$ to lie in the x–y plane leads to

$$\begin{aligned} \mathbf{e}'_x &= -\sin\phi_B\mathbf{e}_x + \cos\phi_b\mathbf{e}_y \\ \mathbf{e}'_y &= -\cos\theta_B\cos\phi_B\mathbf{e}_x - \cos\theta_B\sin\phi_B\mathbf{e}_y + \sin\theta_B\mathbf{e}_z \\ \mathbf{e}'_z &= \sin\theta_B\cos\phi_B\mathbf{e}_x + \sin\theta_B\sin\phi_B\mathbf{e}_y + \cos\theta_B\mathbf{e}_z, \end{aligned} \tag{10.81}$$

where θ_B and ϕ_B are the polar and azimuthal angles, respectively, of **B** in the xyz coordinate system. The result is

$$\begin{aligned} \mathbf{e}'_1 = \frac{1}{\sqrt{2}}\big[&(-\sin\phi_b - i\cos\theta_B\cos\phi_B)\,\mathbf{e}_x \\ &+ (\cos\phi_B - i\cos\theta_B\sin\phi_B)\,\mathbf{e}_y + i\sin\theta_B\mathbf{e}_z\big], \end{aligned} \tag{10.82}$$

from which the contributions of just the σ_- Zeeman component to G are found to be

$$G_{11}(\sigma_-) = C'\left(\sin^2\phi_B + \cos^2\theta_B\cos^2\phi_B\right)\Phi(\nu_0 - \nu_B - \nu), \quad (10.83)$$

$$G_{12}(\sigma_-) = C'\left(-\sin^2\theta_B\sin\phi_B\cos\phi_B - i\cos\theta_B\right)\Phi(\nu_0 - \nu_B - \nu), \quad (10.84)$$

$$G_{21}(\sigma_-) = C'\left(-\sin^2\theta_B\sin\phi_B\cos\phi_B + i\cos\theta_B\right)\Phi(\nu_0 - \nu_B - \nu), \quad (10.85)$$

$$G_{22}(\sigma_-) = C'\left(\cos^2\phi_B + \cos^2\theta_B\sin^2\phi_B\right)\Phi(\nu_0 - \nu_B - \nu), \quad (10.86)$$

in which C' is the constant $C' = C[1 - \exp(-h\nu_0/kT)]|\langle u||\mathbf{p}||\ell\rangle|^2/6$.

The contributions of the σ_- component to the K matrix follow from (10.59) with

$$\begin{aligned}
\phi_I(\sigma_-) &= C''(1 + \cos^2\theta_B)\,H_r,\\
\phi_Q(\sigma_-) &= -C''\sin^2\theta_B\cos 2\phi_B\,H_r,\\
\phi_U(\sigma_-) &= -C''\sin^2\theta_B\sin 2\phi_B\,H_r,\\
\phi_V(\sigma_-) &= C''2\cos\theta_B\,H_r,\\
\phi'_Q(\sigma_-) &= -C''\sin^2\theta_B\cos 2\phi_B\,2F_r,\\
\phi'_U(\sigma_-) &= -C''\sin^2\theta_B\sin 2\phi_B\,2F_r,\\
\phi'_V(\sigma_-) &= C''2\cos\theta_B\,2F_r.
\end{aligned} \quad (10.87)$$

In these relations $C'' = C'/(\sqrt{\pi}\Delta\nu_D)$ and H_r stands for the *red* component (the σ_- component) profile function $H(a, v)$ with $v = (\nu_0 - \nu_B - \nu)/\Delta\nu_D$, and likewise for F_r.

In a similar way the contributions of the blue (σ_+) and unshifted (π) components, denoted by subscripts b and p, respectively, can be evaluated. The final result is represented by these relations:

$$\begin{aligned}
\phi_I &= C''\left[2\sin^2\theta_B\,H_p + (1 + \cos^2\theta_B)(H_b + H_r)\right],\\
\phi_Q &= C''\sin^2\theta_B\cos 2\phi_B\,(2H_p - H_b - H_r),\\
\phi_U &= C''\sin^2\theta_B\sin 2\phi_B\,(2H_p - H_b - H_r),\\
\phi_V &= C''2\cos\theta_B\,(H_r - H_b),\\
\phi'_Q &= C''\sin^2\theta_B\cos 2\phi_B\,(4F_p - 2F_b - 2F_r),\\
\phi'_U &= C''\sin^2\theta_B\sin 2\phi_B\,(4F_p - 2F_b - 2F_r),\\
\phi'_V &= C''2\cos\theta_B\,(2F_r - 2F_b).
\end{aligned} \quad (10.88)$$

This transfer equation is illustrated with a calculation, following Rees (1987), of the Stokes line profiles of a generic line formed in the magnetized solar atmosphere. The line formation model is the Milne–Eddington one, with the assumption

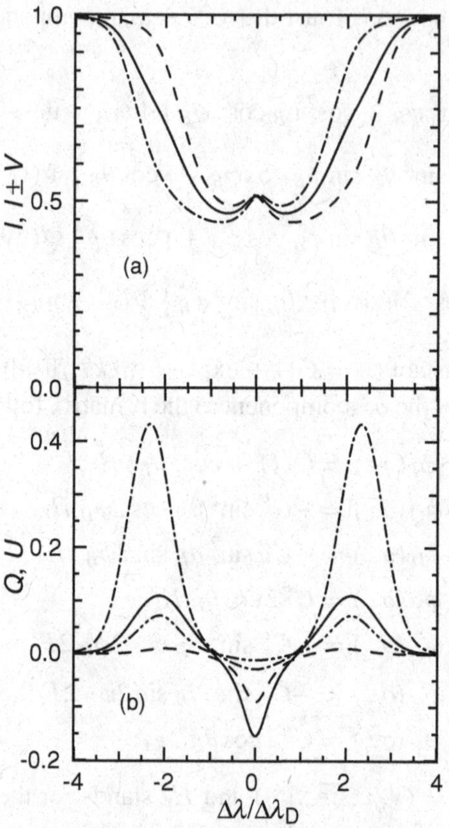

Fig. 10.1 Stokes parameter profiles are shown as functions of wavelength for a line formed in a sunspot. (a) The curves show the total intensity I (solid) and the circularly polarized components $I + V$ (dashed) and $I - V$ (dash-dotted). (b) The linear polarization parameters Q and U are shown both with and without Faraday depolarization. Solid curve: U with depolarization; dashed curve: Q with depolarization; dash-dotted curve: U without depolarization; dash-double-dotted curve: Q without depolarization.

of LTE, a ratio of line opacity to continuous opacity that is independent of depth, a linear variation of the Planck function with continuous optical depth, and a constant geometry of the magnetic field. For this model the line-center ratio of line to continuous opacity is 10, the Planck function is $B_\nu \propto 1 + (3/2)\tau_c$, the line wavelength is 4500 Å, the Voigt parameter is $a = 0.01$, the Zeeman shift is 1 Doppler width unit, the atomic mass is that of iron, 56, the temperature used to calculate the Doppler width is taken to be 5000 K and the magnetic field is inclined 60° to the vertical direction at an azimuth of 30° with respect to the x axis. The radiation is viewed along the vertical. The magnetic field implied by these numbers is close to 2000 G, which is reasonable for a sunspot.

The profiles of the Stokes components are illustrated in Figure 10.1. The upper portion shows the total intensity profile, and also shows the profiles for right-circular and left-circular polarization, i.e., of $I + V$ and $I - V$. At this high value of the magnetic field there is a very sizeable shift of the profile between the two circular modes. Indeed, the two σ components are almost resolved in the total intensity. The magnetometer, which looks for modulation of the intensity in the wing of the profile as the polarization is flipped between left and right circular, easily measures this large shift. The lower part of the figure shows the profiles of Q and U by themselves. Because the magnetic field is viewed obliquely, linear polarization is seen with a similar magnitude to circular polarization, except for one effect. That is, that the Faraday rotation produces depolarization of the linear components. This is indicated in the figure. The linear polarization is much larger when the Faraday rotation is absent. The degree of linear polarization peaks in the line wings near where the circular polarization is largest. By using models such as this the observers can infer the magnitude and direction of the transverse magnetic field component, albeit with some inaccuracy. At lower values of the field the contrast between the two circular polarizations becomes very slight, and careful signal processing is needed to extract the longitudinal field. The linear polarization effect is even smaller, so the derived transverse fields can never be as accurate as the longitudinal ones. The reason why the linear polarization is so small, besides Faraday depolarization, can be seen in (10.88): ϕ_Q and ϕ_U have the form of a second difference of the profile function H, and will be proportional to ω_B^2 for small ω_B, while ϕ_V is a first difference and will be proportional to ω_B to the first power.

This example has assumed LTE and a normal Zeeman triplet. The majority of the lines we might want to study have so-called anomalous Zeeman patterns, and the level populations may well be non-LTE. The complete formulation by E. Landi Degl'Innocenti (1983) encompasses these complications.

11
Numerical techniques for radiation transport

In this chapter we give brief discussions of the main solution algorithms for radiation hydrodynamics problems, some of which are very quick and approximate and some of which represent the best attempts at accuracy. It is unfortunately true that "you get what you pay for" in these calculations, and accuracy comes at a considerable cost. The earlier material in this book has brought out quite a few different processes that complicate the endeavor, such as the effects of fluid velocity on radiation quantities, the complicated spectral dependence of the opacity, non-LTE, refraction, and polarization. These effects are not too hard to include singly, although with some effort, but accounting for all of them has not seemed to be a practical objective up to the present time. And of course these difficulties are compounded many-fold in higher-dimensional geometries. Our discussion of algorithms will begin with the low-budget methods that may be priced just right for many purposes, after some preliminary observations about solution strategy. Some general references on this subject are the following: The conference volume *Astrophysical Radiation Hydrodynamics* (Winkler and Norman, 1982) is a good place to start. A meeting that included presentations on many of the current advanced hydrodynamics methods was the 12th Kingston Meeting on Theoretical Astrophysics held in Halifax in 1996 (Norman, 1996). Starting points for surveying the advanced numerical methods in radiation transport are the pair of books by Kalkofen (1984,1987) and the workshop proceedings *Stellar Atmospheres: Beyond Classical Models* (Crivellari, Hubeny, and Hummer, 1991). The most comprehensive review of the astrophysical methods to date is provided by the 2002 Tübingen workshop *Stellar Atmosphere Modeling* (Hubeny, Mihalas, and Werner, 2003).

11.1 Splitting hydrodynamics and radiation

Operator splitting is a time-honored method for calculating initial value problems that consist of different kinds of physics, of which at least some must be treated in an implicit fashion. Early descriptions of the application of this idea to radiation hydrodynamics are found in the cepheid and RR Lyrae stellar pulsation calculations by Christy (1966) and Cox *et al.* (1966). Another radiation hydrodynamic calculation from around the same time is the supernova model of Colgate and White (1966).

The idea is simple: advance physical process (A) as if it were the only activity during the time step, then use that result as the starting point and advance physics (B) for the same time step as if it were the only activity, and so on through all the processes. To fix some of these ideas, suppose we represent all the variables in our problem, represented in some discrete way in space, as a vector **X**, and suppose it satisfies a system of equations we write as

$$\frac{d\mathbf{X}}{dt} = \mathbf{A}[\mathbf{X}] + \mathbf{B}[\mathbf{X}] + \cdots, \tag{11.1}$$

where **A** and **B** are operators that perform different kinds of physics. We discretize time using the set $t^1, t^2, \ldots, t^n, \ldots$. The *explicit* method of time differencing is approximately this:

$$\frac{\mathbf{X}^{n+1} - \mathbf{X}^n}{\Delta t} = \mathbf{A}[\mathbf{X}^n] + \mathbf{B}[\mathbf{X}^n] + \cdots. \tag{11.2}$$

When we write $\mathbf{A}[\mathbf{X}^n]$ we mean that no information *later* in time than t^n is included. The information from several prior time steps may be combined to provided an estimate of the forward difference that is of higher order than the first in Δt. Including all the physical processes we want to is no problem in this approach since we need only keep track of the time derivative contributions from all the processes, calculated by taking spatial derivatives and doing the integrations, then add these up at the end to get the total amount by which to advance **X**. The failure mode of this approach is that it is almost always subject to a time-step constraint imposed either by the condition for numerical stability, or by accuracy considerations. This takes the form

$$\Delta t \left\| \frac{\delta \mathbf{A}}{\delta \mathbf{X}} \right\| < c, \tag{11.3}$$

where c is some numerical value rather smaller than 1, and likewise for the other operators.

In terms of the particular physical processes we need to consider, the stability limits that arise from this reasoning are the Courant limit, $c_s \Delta t / \Delta x < 1$, from the hydrodynamics, and a radiation Courant limit $c\Delta t/\Delta x < 1$ if we were so brave as to do radiation transport with explicit time differencing. If we use the radiation diffusion approximation, then the stability limit is $K_R \Delta t / (\rho C_v (\Delta x)^2) < 1$. We may or may not be able to live with the Courant limit; it depends on how long we want to evolve the problem compared with the hydrodynamic time scale. The radiation limit is usually the one that hurts. The radiation diffusion limit can be factored in this way:

$$\frac{K_R \Delta t}{\rho C_v (\Delta x)^2} = \frac{16\sigma T^4}{3\rho C_v T V} \frac{1}{\kappa_R \rho \Delta x} \frac{V \Delta t}{\Delta x}, \tag{11.4}$$

where we have introduced a typical flow speed V. The first factor on the right-hand side is, apart from the factor $16/3$, the inverse of the *Boltzmann number* for the flow. This factor can easily be of order 100. The second factor is the reciprocal of the optical depth of a zone. This optical depth certainly becomes as small as unity. The third factor is the time step compared with the flow time across a zone; we would hope that this would be about unity. So the inverse Boltzmann number makes this stability criterion too large by a factor that may be 100. Therefore explicit radiation diffusion is not a good idea. We reach a similar conclusion using the explicit radiation transport equation. In a non-LTE problem we face the stiffness of the kinetics equations, on which we have commented earlier.

There is a negative aspect of implicit time differencing. The rather large truncation error associated with using a time step large compared with the natural time scales based on $\|\delta \mathbf{A}/\delta \mathbf{X}\|$ makes the numerical representation quite dissipative, in the sense that noise is filtered out. This numerical dissipation can suppress real instabilities in the problem, and eliminate real high-frequency components that are physically significant. The validity checks that are applied as the problem runs smoothly along with giant time steps may fail to reveal that high-frequency modes would develop if they were allowed to. Of course, suppressing the high-frequency modes is precisely why one wants to use the implicit method in the first place, but there may not then be a way to verify that the significant results are being obtained correctly.

We turn then to implicit differencing. We try this:

$$\frac{\mathbf{X}^{n+1} - \mathbf{X}^n}{\Delta t} = \mathbf{A}[\mathbf{X}^{n+1}] + \mathbf{B}[\mathbf{X}^{n+1}] + \cdots . \tag{11.5}$$

Now we face a major computational challenge. The values we want to find, $\mathbf{X}^{n+1}$, appear in the (nonlinear) functions on the right-hand side, and to make matters worse, they appear in every term. Notwithstanding the obstacles, this approach has been used very successfully in a number of astrophysical problems such as stellar pulsation and protostar collapse. The attack on the problem is direct: set up all the equations as a nonlinear system for the unknowns $\mathbf{X}^{n+1}$ and apply the multivariate Newton–Raphson method. The initial guess for $\mathbf{X}^{n+1}$ might be taken to be $\mathbf{X}^n$. At each step the Jacobian matrix elements are calculated, the largest part of the cost, and the linear system for the next set of corrections to $\mathbf{X}^{n+1}$ is solved by direct elimination. The notable successes of this method have been in 1-D spherical geometry. Having only one spatial dimension is kind to direct elimination as a method of solving a banded linear system: the cost scales linearly with the number of zones and as the cube of the matrix bandwidth, i.e., of the number of variables per zone. Direct elimination is much more costly in two dimensions, and we begin to look for different solution methods. There are also problems with ensuring Newton–Raphson convergence; it may be necessary to severely restrict the time step to ensure rapid convergence.

Here then is operator splitting. If the time derivative of $\mathbf{X}$ is split into k pieces, then there are k partial time steps to advance from time t^n to time t^{n+1}:

$$\begin{aligned}
\frac{\mathbf{X}^{n+1/k} - \mathbf{X}^n}{\Delta t} &= \mathbf{A}[\mathbf{X}^{n+1/k}], \\
\frac{\mathbf{X}^{n+2/k} - \mathbf{X}^{n+1/k}}{\Delta t} &= \mathbf{B}[\mathbf{X}^{n+2/k}], \\
&\vdots \\
\frac{\mathbf{X}^{n+1} - \mathbf{X}^{n+(k-1)/k}}{\Delta t} &= \mathbf{F}[\mathbf{X}^{n+1}].
\end{aligned} \tag{11.6}$$

This does in fact converge to a solution of the differential equation as $\Delta t \to 0$, as we can see by adding the equations and Taylor-expanding the right-hand sides about $\mathbf{X}^n$. However, it is only first-order accurate. The order of accuracy is improved, when there are just two operators $\mathbf{A}$ and $\mathbf{B}$, by alternating cycles on which $\mathbf{A}$ is done first, then $\mathbf{B}$, with cycles that do the operators the other way around (called *Strang splitting*). With more than two operators the practice is to do $ABC\ldots$ on one cycle and $\ldots CBA$ on the next.

What are the advantages? Some parts of the physics may not be stiff at all, and those operators may be advanced using an explicit equation, leaving the implicit differencing for the parts that *are* stiff. When an implicit equation has to be solved for one variable in the splitting method the bandwidth of the linear system is much reduced. It is nine times faster to solve three linear systems with bandwidth one than to solve one linear system with bandwidth three. If you do not really need to solve two of the three systems in the split case, the gain is a factor 27. The lower dimensionality of the nonlinear system helps greatly with the robustness of the Newton–Raphson convergence. The disadvantage is that the error of the time differencing is increased, and it may not be very easy to estimate. Strang splitting helps with this, but there can still be the problem that $\mathbf{A}$ might move the solution in the wrong direction, creating an error that $\mathbf{B}$ has to correct. This problem is not usually *too* severe, but it must be watched for.

The stellar pulsation calculations for RR Lyrae stars by Christy (1966) and for cepheids by Cox *et al.* (1966) were made up of spherical Lagrangian zones with just three unknowns per zone: the radius, velocity, and temperature. The calculations proceeded in a staggered way with time, with the velocity being updated first, a half-time-step later the radii were updated, and finally the temperatures. Only the temperature equation was implicit. The Newton–Raphson method applied to the material energy equation led to a tri-diagonal system for the temperature corrections, which is about as easy as linear systems get.

With two or more space dimensions the choices become more painful. The considerations about time step limits still apply, so perhaps the hydrodynamics can be done explicitly, although some of the modern methods (e.g. Godunov's method, see Section 3.2.2) may still use splitting as a convenience. The radiation equations are a major problem now. The radiation diffusion equation has the character of an elliptic equation in space after the time differencing is done, and this is not at all as easy to solve as a scalar two-point boundary-value problem in one dimension. So even if the operator splitting is applied and the radiation equation is treated separately, the solution requires an iterative linear system solver such as the conjugate gradient (CG) or alternating-direction implicit (ADI) method. It will also be seen below that the adequacy of diffusion as a substitute for properly angle-dependent radiative transfer is more questionable in two and three dimensions than in one.

The coupling of the material temperature to the radiation field, through the material internal energy equation, has the helpful property that it involves only local quantities, apart from the advection term (which is lumped with the hydrodynamics processes in the splitting method), unless thermal conductivity must be considered. Often the conduction flux is negligible, so on the temperature coupling step the material temperature, or at least the Newton–Raphson correction to it, can be eliminated using a local equation so only the radiation field remains to be found from a large system of equations. This reduces the dimensionality by a factor 2, which is important when the solution cost varies as the cube of the number of unknowns per zone.

11.2 Thermal diffusion

Now we begin to walk through a hierarchy of increasing sophisticated and more costly, but not necessarily more accurate, algorithms for solving radiation hydrodynamics problems. We begin with the method Christy, Cox, Colgate, and others used, thermal diffusion, also called equilibrium diffusion. In this method the radiation field is removed from the problem and replaced using the relations derived for the diffusion limit, (6.66), (6.72), and (6.78), although the second order corrections in $E^{(0)}$ and $P^{(0)}$ are usually ignored, along with the relativistic corrections to $\mathbf{F}^{(0)}$. The combined energy equation for matter and radiation is used, which in effect adds aT^4/ρ to the internal energy, $aT^4/3$ to the pressure, and includes $\mathbf{F}^{(0)}$ as a flux. The advection parts of this having already been treated, what is left is an implicit equation for the temperatures at the advanced time step. The key part of making this equation implicit is using the advanced-time temperatures in the flux. That is, the equation looks something like this after discarding the advection flux and the work term:

$$\frac{\rho^{n+1}e^{n+1} + E^{n+1} - \rho^n e^n - E^n}{\Delta t} - \nabla \cdot \left[K_R(\rho^{n+1}, T^{n+1})\nabla T^{n+1}\right] = 0. \quad (11.7)$$

We repeat that this equation is not complete since the unnecessary terms for the present discussion have been dropped. The flux term in this equation is certainly not centered in time as it should be to make it second order accurate. That would be true if the flux in square brackets were replaced by the arithmetic average of the values at t^n and t^{n+1}, called Crank–Nicholson differencing. But that form, in the limit $K_R\Delta t/(\rho C_v(\Delta x)^2) \gg 1$, is susceptible to nonlinear numerical instabilities. Then we might try a weighted average, with a somewhat larger weight applied to the t^{n+1} flux. That does seem to solve the instability problem, but the solution remains noisier than if the fully backward-differenced form is used, as given first. This is another example of deliberately choosing the more dissipative numerical representation as a trade-off to obtain the highest possible time step.

Equation (11.7) is solved, as discussed earlier, using the Newton–Raphson method. The spatial derivative operators are first represented in whatever second order accurate form is permitted by the nature of the spatial zoning. In Eulerian calculations the ordinary centered second derivative formula can be used. There is some question about the proper spatial centering of the K_R factor, and different choices may be made. The Jacobian matrix that emerges when the equations are linearized is the matrix of the system that is to be solved. If the opacity-variation parts of the linearization could be ignored, then the system could be arranged to be symmetric and positive-definite, a very great advantage for the application of iterative solvers. Sometimes it is proposed to lag the opacities in time for just that reason. Good results have also been obtained including the opacity terms using nonsymmetric solvers, such as the direct elimination method in one dimension.

The failure mode of the thermal diffusion approximation is its poor performance in optically thin regions. Even the RR Lyrae pulsation calculations of 1965 revealed the shortcomings of the method, because the temperature throughout the atmosphere of the star was spuriously forced to a constant value by the use of a diffusion approximation in an optically thin region. In reality the temperature becomes decoupled from the radiation field, as discussed in Section 7.2.

11.3 Eddington approximation

The major objection just mentioned to thermal diffusion is removed if the assumptions (6.66), (6.72), and (6.78) are replaced with the simple closure relation $P = E/3$ and E and $\mathbf{F}$ are retained as variables. The equations determining them are (6.51) and (6.52), although the $1/c$ terms are usually dropped in the latter. The advection and work terms are also often dropped from the energy equation, but, as discussed earlier, this commits the errors of ignoring radiation energy density and work in the overall energy budget. The frequency integral of the opacity multiplied by the flux has to be approximated, since the spectral distribution of $\mathbf{F}_\nu$ is unknown; guided by thermal diffusion the Rosseland mean is used, leading to

$$\mathbf{F} = -\frac{c}{3\kappa_R \rho}\nabla E. \tag{11.8}$$

The energy coupling term on the right-hand side of (4.29) is approximated as in (8.89), or perhaps with the Rosseland mean here too.[1] The final result for the combined moment equations is this:

$$\rho\frac{D(E/\rho)}{Dt} + \frac{E}{3}\nabla\cdot\mathbf{u} - \nabla\cdot\left(\frac{c}{3\kappa_R\rho}\nabla E\right) = \kappa_P \rho c(aT^4 - E). \tag{11.9}$$

We need to stress at this point that the radiation quantities here are those in the comoving frame, even though the superscripts $^{(0)}$ have been dropped to reduce the clutter in the equations. Only by using comoving radiation are we permitted to evaluate the opacity and emissivity *sans* velocity effects.

We describe now how the temperature update proceeds when the (gray) Eddington approximation as just described is used. We repeat the internal energy equation given earlier,

$$\frac{\partial \rho e}{\partial t} + \nabla\cdot(\rho\mathbf{u}e) + p\nabla\cdot\mathbf{u} = -\kappa_P\rho c(aT^4 - E). \tag{11.10}$$

With operator-split hydrodynamics the advection and work terms in (11.10) have already been evaluated at the point the temperature update is being done. Everything else in this equation is local, so when the equation is linearized for the Newton–Raphson procedure there is just a simple linear equation to solve to obtain the correction to T in terms of that for E. Then this can be substituted into the linearized form of (11.9), which remains an elliptic equation for the corrections to E, at least provided the terms arising from the variation of the opacity with T are not too large. After the substitution of δT in terms of δE the structure of (11.9) is

[1] In the diffusion limit both E_ν and B_ν have the same spectral distribution, while the spectral dependence of $cE_\nu - 4\pi B_\nu$ is proportional to $(dB_\nu/dT)/\kappa_\nu$, which provides some justification for using the Rosseland mean in this place.

identical to the thermal diffusion equation apart from certain differences in the coefficients that correct the small-optical-depth errors in thermal diffusion. The cost to solve the elliptic equation is unchanged.

The significant technology issues connected with both thermal diffusion and Eddington-approximation calculations are: (1) making a finite-difference or finite-element representation of the partial differential equation, and (2) solving the resulting large sparse linear system of equations. Rapid progress has occurred in both areas and we will discuss this in Section 11.4. The computational techniques needed for radiation diffusion are not materially different from those applied to other engineering problems involving elliptic operators, such as heat conduction, electrostatics, and viscous incompressible fluid flow.

The boundary conditions required for (11.9), and also (11.7), come from reasoning similar to that leading to (5.33). We will repeat the argument in somewhat greater generality, to allow for the specification of an intensity of radiation that is incident on the problem at the boundary. We let I_B be this incident intensity, and if $\mathbf{n}_B$ is the unit outward-directed normal vector for a piece of the boundary, then I_B is defined for ray directions $\mathbf{n}$ that obey $\mathbf{n} \cdot \mathbf{n}_B < 0$, i.e., that point inward. Now, we do not know how much radiation will shine *out* of the problem at the boundary, but suppose we knew what the average cosine was for this *outward* intensity. That is, we think we are given everything we want to know about the *inward* intensity, but we make an *ansatz* about the *outward* intensity. The *ansatz* is

$$\frac{\int_{\mathbf{n}\cdot\mathbf{n}_B>0} d\Omega\, \mathbf{n}\cdot\mathbf{n}_B I(\mathbf{n})}{\int_{\mathbf{n}\cdot\mathbf{n}_B>0} d\Omega\, I(\mathbf{n})} = \langle\mu\rangle, \tag{11.11}$$

a value we think we know. If the integrals in the numerator had been over all solid angles instead of the outward hemisphere then the ratio would have been $\mathbf{n}_B \cdot \mathbf{F}/cE$. We add and subtract integrals over the inward hemisphere to make it look somewhat like that and find

$$\frac{\mathbf{n}_B \cdot \mathbf{F} + \int_{\mathbf{n}\cdot\mathbf{n}_B<0} d\Omega\, |\mathbf{n}\cdot\mathbf{n}_B| I_B(\mathbf{n})}{cE - \int_{\mathbf{n}\cdot\mathbf{n}_B<0} d\Omega\, I_B(\mathbf{n})} = \langle\mu\rangle. \tag{11.12}$$

This is the desired result. When it is rearranged it becomes a linear relation connecting the normal component of $\mathbf{F}$ with E at the boundary, possibly including an inhomogeneous term when there is incident radiation. It is a boundary condition of mixed type, i.e, neither Neumann ($\mathbf{n}_B \cdot \mathbf{F}$ specified) nor Dirichlet (E specified). We frequently want the boundary condition when there is vacuum or a "black absorber" outside so that $I_B = 0$. Then the relation is simply

$$\mathbf{n}_B \cdot \mathbf{F} = \langle\mu\rangle c E. \tag{11.13}$$

Oh yes, what do we take for $\langle\mu\rangle$? The most popular value is 1/2, and an argument for this was suggested in the earlier discussion of the exact Hopf function. Equation (11.13) with the choice $\langle\mu\rangle = 1/2$ is the Milne boundary condition. When (11.13) is combined with the Fick's law formula (11.8) for the flux, the vacuum boundary condition takes the form

$$E = -\frac{1}{3\langle\mu\rangle\kappa_R\rho}\frac{\partial E}{\partial n}, \tag{11.14}$$

in which $\partial E/\partial n$ is the normal derivative of E. A geometrical picture that goes with this equation is that a linear extrapolation of E outside the boundary reaches a value of zero at a distance $1/(3\langle\mu\rangle)$ mean free paths from the boundary; this is the *extrapolation length* implicit in the boundary condition. The extrapolation length is 2/3 of a mean free path if $\langle\mu\rangle$ is taken to be 1/2, and it is $1/\sqrt{3}$ mean free paths if $\langle\mu\rangle$ is taken to be $1/\sqrt{3}$. In a scattering-dominated diffusion problem the energy density in the interior, i.e., deeper within the medium than the boundary layer, is approximately proportional to the solution of Milne's first problem, which is $E \propto \tau + q(\tau) \approx \tau + q_\infty$, in which $q(\tau)$ is the Hopf function. If this relation is extrapolated to the place where $E = 0$, then the extrapolation length must be $q_\infty \approx 0.710\,45$ mean free paths. The most commonly used value for the extrapolation length is 2/3.

There is one more boundary condition that applies in other cases, and that is the reflection or symmetry condition. If a perfect mirror or perfect diffuse reflector is applied to the surface then the incoming intensity is forced to be exactly equal to the outgoing intensity and the flux vanishes. This also occurs when a piece of the boundary is part of a plane of reflection symmetry. Thus $\mathbf{n}_B \cdot \mathbf{F} = 0$ for those parts of the boundary. Notice that it is only the normal component of the flux that vanishes in this case. This boundary condition is exactly the Neumann type. Dirichlet boundary conditions do not seem to be quite physical. This is a statement that the full-sphere average of the intensity at a boundary point is a specified value. What is unphysical here is that it commits the problem to enter a conspiracy with the agents outside the boundary to make the average of their respective contributions to E come out to a given value. If the outside world is just a thermal bath, then the incoming intensity is the corresponding Planck function, but the emergent intensity is whatever it is, and the average will not necessarily be the same Planck function. The comparable specification of a nonzero value for the normal flux is more physical. The significance of this specification is that the agents outside have a battery that releases energy at a definite rate, and they capture whatever energy comes out through the boundary and give that back plus the energy released by their battery. When we put an inner boundary radius on a stellar atmosphere problem, and replace all of the star within that radius with a boundary condition, we are making an assumption like this. In this case the "battery" actually exists and is the nuclear energy source at the center of the star.

The Eddington approximation, unlike thermal diffusion, gives quite reasonable results in the optically-thin parts of the star. This is not to say that it is *accurate*, just that it is qualitatively correct. As we saw earlier, it gives an error of order 20% in the Eddington factor at $\tau = 0$ in the Milne problem. It yields a wave equation for light waves, which is qualitatively correct, for which the wave speed is $c/\sqrt{3}$, which is off by 42%. As Mihalas and Mihalas (1984, p. 518) say, in discussing radiative waves with thermal relaxation, "In our opinion the Eddington approximation should always yield results that are at least qualitatively correct."

One path toward making the Eddington approximation more accurate is to include an Eddington factor, which we discuss in Section 11.5.

11.4 Diffusion solvers

Solving a diffusion problem in one dimension that has been put into finite-difference form using a centered three-point formula such as this:

$$-A_i J_{i-1} + B_i J_i - C_i J_{i-1} = D_i, \tag{11.15}$$

is very simple indeed. The forward and back recursion scheme given by

$$E_i = \frac{C_i}{B_i - A_i E_{i-1}}, \tag{11.16}$$

$$F_i = \frac{D_i + A_i F_{i-1}}{B_i - A_i E_{i-1}}, \tag{11.17}$$

$$J_i = F_i + E_i J_{i+1} \tag{11.18}$$

is solved in the forward direction to obtain the Es and Fs, then a back substitution using the third equation gives the unknowns. If the tri-diagonal matrix $(-A_i \quad B_i \quad -C_i)$ is diagonally-dominant, so $B_i > |A_i| + |C_i|$, the recursion is guaranteed to be stable. This condition is almost always met with centered differencing of diffusion equations, so our problem is solved. The tri-diagonal recursion is so efficient that only a handful of floating-point operations are needed to obtain each of the unknowns we want; that is as good as it gets. So the 1-D problem is solved. Life is more difficult in two and three dimensions, and that is the topic of this section.

First, let us consider what *not* to do, if efficiency is the goal. A finite-difference formula representing a diffusion equation in two dimensions very often connects five or nine neighboring points on a more-or-less rectangular grid. The matrix of this system of linear equations has one row for each equation, and nonzero entries in that row in all the columns corresponding to mesh points that are coupled to the point in the middle. If the mesh points are ordered raster-fashion, going across in the x direction first, then up in y, the neighbor points in x to the middle point produce matrix entries immediately adjacent to the diagonal. But the neighbor points up or down in y produce matrix elements separated from the diagonal by about N_x columns, where N_x is the size of the mesh in the x direction. Both normal Gaussian elimination applied to this matrix and block-tri-diagonal elimination with $N_x \times N_x$ blocks lead to a solution cost of order $N_x^3 N_y$ operations. This is a cost that is N_x^2 operations per unknown, which is thousands of times worse than the 1-D case.

How much better can we do? By using iterative linear solution methods the cost can be brought down to something like $N_x^2 N_y$, or N_x operations per unknown. Some methods may do even better than this, but then it depends on how well-conditioned the matrix is. The (relatively) good news is that in three dimensions the scaling for the iterative methods is also of order N_x operations per unknown. Here is a laundry list of linear solver methods that we may want to discuss: conjugate gradient, conjugate gradient preconditioned by different methods, Chebyshev, ORTHOMIN, which is also known as the GMRES method, multigrid, and multigrid with a selection of preconditioners. These are all methods for solving large sparse linear systems. A system of nonlinear equations leads, by applying the Newton–Raphson method, to such a sparse linear system. But it may be that it is painful and expensive to actually compute and store the Jacobian matrix that is needed at each iteration. The Newton–Krylov method(s) are a way of carrying out the Newton iterations simultaneously with the GMRES or other linear solver iterations. All these methods will be discussed briefly in the remainder of this section. A study of a few promising candidate solvers for a radiation diffusion problem was reported by Baldwin *et al.* (1999).

We will follow the discussion by Saad (1996). Our goal is to solve a linear system of equations

$$Ax = b, \tag{11.19}$$

for a vector of unknowns x, which in most cases consist of one unknown per spatial cell in a 2-D or 3-D mesh. The cells, and the unknowns, are ordered in some way, such as in the raster scan.

11.4.1 Jacobi, Gauss–Seidel, and successive overrelaxation (SOR) method

These are the simplest, oldest, and poorest of the available methods. The idea is to separate A into its diagonal, subdiagonal and superdiagonal parts. That is,

$$A = -E + D - F, \tag{11.20}$$

in which D is the diagonal of A, $-E$ is the lower-triangular matrix that is the subdiagonal part of A, and $-F$ is the upper-triangular superdiagonal part of A. The E elements are the terms that couple a given cell to cells that come earlier in the raster scan, and F contains the couplings to cells that come later. In considering Jacobi iteration the linear system is written in this way:

$$Dx = b + Ex + Fx. \tag{11.21}$$

Then we solve by a process of iteration in which the current guess for x is put in on the right-hand side, and the diagonal system is solved for the next guess:

$$Dx^{k+1} = b + Ex^k + Fx^k. \tag{11.22}$$

With luck, this Jacobi iteration will converge. Clearly, if the E and F matrices are small in some sense compared with D, then there should be good convergence. More precisely, the method will converge if the largest, in magnitude, of the eigenvalues of the matrix $D^{-1}(E + F)$ is less than unity. Since for common finite-difference representations of the diffusion operator D is just *equal* to $E + F$ plus source-sink terms that may be small, this eigenvalue may be only slightly less than unity.

The Gauss–Seidel method is described by this equation:

$$-Ex^{k+1} + Dx^{k+1} = b + Fx^k. \tag{11.23}$$

So half of the off-diagonal part of A is kept on the left-hand side for the iteration. This is just about as easy to perform as the Jacobi iteration. For each iteration you scan through the mesh, updating the cells one at a time. When x is corrected in each cell, the new value replaces the old one, and the new value will be used for updating cells that come later in the scan. Only the cells that follow the given one will have just the prior iteration data available. In this case the convergence depends on the eigenvalues of $(D - E)^{-1}F$. Again, the eigenvalues are anticipated to be just slightly less than unity. It is found that they may be twice as far from unity as the eigenvalues for Jacobi iteration, which will cut the number of required iterations in half.

The thing that helps out the convergence of Gauss–Seidel (it would help for Jacobi too, but it is usually used with Gauss–Seidel) is SOR. For SOR, compared with Gauss–Seidel, some of the diagonal part D of A is put on the right-hand side of the equation along with the F part, and the rest of D and the E part are kept on the left:

$$\left(\frac{1}{\omega}D - E\right)x^{k+1} = b + \left(F - \frac{\omega - 1}{\omega}D\right)x^k. \tag{11.24}$$

With $\omega > 1$ this makes the corrections somewhat larger and accelerates the convergence; it can also make the corrections *too* large, and produce divergence (if $\omega > 2$). The optimum value for ω turns out to be

$$\omega = \frac{2}{1 + \sqrt{1 - \lambda^2}}, \tag{11.25}$$

in terms of the largest eigenvalue λ for Jacobi iteration. When ω has this optimum value the SOR eigenvalue becomes $\omega - 1$. SOR can yield a huge gain in convergence rate. When the Jacobi eigenvalue is 0.999, and the Gauss–Seidel eigenvalue is 0.998, then SOR has an eigenvalue of 0.914 provided ω is set to 1.914. That is a speed-up of 89 times. Empirically estimating the Jacobi eigenvalue and the optimum ω is not simple, however.

A useful extension of Jacobi iteration is *block Jacobi* iteration, for which the unknowns are partitioned into some number of groups, and for each iteration the equations belonging to each group are solved for the unknowns for that group using prior values of the unknowns in other groups. This is employed in parallel solution techniques for large systems for which the spatial domain is decomposed into subdomains, and each subdomain is given to a separate processor, or to a set of shared-memory processors. Block Jacobi iteration is by far the simplest method for solving the linear system in this case.

11.4.2 Alternating-direction implicit (ADI) method

In 2-D diffusion problems the matrix A often has the structure of a tri-diagonal matrix in the y direction combined with a tri-diagonal matrix in the z direction, as expected for an operator like

$$-\frac{\partial^2}{\partial y^2} - \frac{\partial^2}{\partial z^2}. \tag{11.26}$$

The matrix A also usually contains some diagonal pieces, such as source-sink terms for radiation diffusion. The essence of the ADI method is to perform two 1-D solutions per full iteration cycle, one in which the z operator is put onto the right-hand side, and a judiciously chosen diagonal component is added to both sides, leading to a tri-diagonal system in y on each z line. The second half of the iteration cycle is the reverse. Since the cost of a tri-diagonal solve is of the same order as the number of unknowns, one full ADI cycle has about the same cost as one SOR cycle. Saad (1996) mentions the result that the optimum convergence rate for ADI, with the best choice of that judiciously-chosen diagonal component, is the same as with a symmetrized SOR, in which the roles of E and F are switched on alternate iterations, using the optimum ω in that case.

It is interesting to consider what the convergence rates actually are, and how they depend on the mesh. For a simple Poisson equation problem with Dirichlet boundary conditions, on a $N \times N$ mesh, the Jacobi eigenvalue is about $\lambda = 1 - \pi^2/(2N^2)$. This means the optimum SOR eigenvalue is roughly $1 - 2\pi/N$. In order to reduce the initial error by a factor 10^6 the number of iterations will need to be $n_{\text{iter}} \approx 3\ln(10)N/\pi \approx 2.2N$. This is the basis for thinking that the iteration count may scale with the size of the mesh.

11.4.3 Krylov methods in general

A great many of the current iterative solution methods fall under the general description of "Keep multiplying the matrix into the current residual, and at each step combine all the vectors together in some way to get the next guess." This collection of vectors, which has the generic form $\{v_0, Av_0, A^2v_0, \ldots, A^{m-1}v_0\}$, spans what is called a Krylov subspace. What distinguishes the different methods in this class is the "combine all the vectors together" part. A recurrent theme is to choose the next iterate so that the error, or the residual, will be orthogonal to the Krylov subspace. Since the subspace becomes steadily larger as the iteration proceeds, the error can be quenched fast and faster.

11.4.4 CG method

The CG method of Hestenes and Stiefel (1952) and Lanczos (1952) is a very useful method for symmetric positive-definite matrices (all the eigenvalues are positive). Diffusion problems can in principle always lead to symmetric positive-definite systems of finite-difference equations, but in the application this is not always true. When it is, then the CG method is an excellent choice, usually with a suitable preconditioner. The CG algorithm, like the other Krylov methods, repeatedly corrects the current estimate of x by trying a displacement in the direction of a vector p, the search direction, which varies from iteration to iteration. The correct distance to move along the p direction for the next iterate is determined so that the new residual will be orthogonal to the Krylov subspace built up in steps from the initial residual. A wonderful property of the symmetric system is that if the new residual is just made orthogonal to the previous one, then orthogonality to the whole subspace is guaranteed. Then to find the new search direction the new residual is orthogonalized with respect to the previous search direction. This also guarantees orthogonality with all the previous ones. Given the new search direction, the next iteration can begin.

The mathematical expression of the algorithm is the following, where (f, g) is the notation for the vector inner product, which might also be written $f^{\mathrm{T}}g$, with ^{T}

standing for the transpose:

$$\alpha_j = \frac{(r_j, r_j)}{(Ap_j, p_j)}, \tag{11.27}$$

$$x_{j+1} = x_j + \alpha_j p_j, \tag{11.28}$$

$$r_{j+1} = r_j - \alpha_j A p_j, \tag{11.29}$$

$$\beta_j = \frac{(r_{j+1}, r_{j+1})}{(r_j, r_j)}, \tag{11.30}$$

$$p_{j+1} = r_{j+1} + \beta_j p_j. \tag{11.31}$$

The iteration begins with any good choice for x, and with $r = p = b - Ax$. At each step there is one matrix-vector multiplication required, and two inner products. The total number of floating-point operations is about equal to the number of nonzero elements in A.

The convergence rate varies as the iterations proceed, but the worst-case estimate depends on the condition number κ of A. The condition number is defined as $\kappa = \lambda_{\text{max}}/\lambda_{\text{min}}$ in terms of the largest and smallest eigenvalues of A. The eigenvalues are all real and positive. The error after many iterations is multiplied by the factor $(\sqrt{\kappa} - 1)/(\sqrt{\kappa} + 1)$ each iteration. For the Poisson problem mentioned above, the condition number of A is something like $\kappa = N^2/\pi^2$, with the result that the error amplification factor is $\approx 1 - 2\pi/N$, which is the same as for optimum SOR and about the same as optimum ADI. The advantage of CG is that there are no parameters to tune, the Achilles heel of the latter methods.

11.4.5 GMRES, ORTHOMIN, Ng, BCG, and Chebyshev methods

We turn to an algorithm that can be used for nonsymmetric matrices, which unfortunately often occur even when they ought not. This is the GMRES method. The idea of this method and a couple of its variants is that the m-dimensional Krylov subspace based on A and the initial residual vector r_0 is built up. The dimension m of the subspace may have to be chosen at the outset. Then a procedure (Gram–Schmidt or Householder's method (Saad, 1996)) is used to orthogonalize the vectors $r_0, Ar_0, \ldots, A^{m-1}r_0$ to form a set of basis vectors. Given this orthonormal set, it is easy to select a candidate solution x_m by adding to x_0 a linear combination of the basis vectors, where, in the case of GMRES, the L_2 norm of the residual $b - Ax_m$ is minimized. If this answer is not good enough, and it will not be if $m \ll N$, then the choices are: (1) start over again with x_m in place of x_0, or (2) keep on going, with the orthogonalization procedure applied to only the most recent k vectors. The latter modification is called quasi-GMRES or QGMRES by Saad (1996). The orthogonalization procedure in the basic method, and especially in QGMRES, becomes complex when the goals of well-conditioned numerical operations and storage minimization are taken into account.

There is a reorganized form of GMRES, called generalized conjugate residual (GCR), that recursively defines search direction vectors p_j such that all the vectors Ap_j are orthogonal. These take the place of the orthonormal basis in the GMRES method. The Krylov subspace is the same in the two cases, and both methods minimize the same norm of the residual, and therefore should be algebraically equivalent. The GCR algorithm is described by

$$\alpha_j = \frac{(r_j, Ap_j)}{(Ap_j, Ap_j)}, \tag{11.32}$$

$$x_{j+1} = x_j + \alpha_j p_j, \tag{11.33}$$

$$r_{j+1} = r_j - \alpha_j A p_j, \tag{11.34}$$

$$\beta_{ij} = -\frac{(Ar_{j+1}, Ap_i)}{(Ap_i, Ap_i)} \quad \text{for } i = 0, 1, \ldots, j, \tag{11.35}$$

$$p_{j+1} = r_{j+1} + \sum_{i=0}^{j} \beta_{ij} p_i. \tag{11.36}$$

The residual is available at each step, so it is easy to decide when to stop iterating. Unfortunately, unlike CG, the projection process involves more and more terms as j increases. The GCR algorithm, like GMRES, can either be stopped and restarted, or the projections can be limited to the most recent k vectors, *viz.*, the loop on i and the sum over i can be limited to $i = j - k + 1, \ldots, j$. The GCR algorithm is called ORTHOMIN(k) in that case. See Vinsome (1976).

A method due to Ng (1974) has been used by Olson, Auer, and Buchler (1986); it is described by Auer (1987) and compared by him with ORTHOMIN (Auer, 1991). It has very much the same flavor as ORTHOMIN, and is described as follows. A certain number k of simple relaxation iterations $x^{n+1} = x^n + b - Ax^n$ are performed, and the residuals $r^n = b - Ax^n$ are recorded. Then it is required that a new candidate for x given by

$$x = x_k - \sum_{p=0}^{k-1} \alpha_p (x^k - x^p) \tag{11.37}$$

should yield the minimum possible residual with respect to possible choices of the coefficients α_p. When this is worked out (see Auer (1991)) it implies that the final residual is r_k projected orthogonal to the space spanned by the vectors $r_k - r_p$, $p = 0, \ldots, k-1$. It turns out that Ng's method is identical to GMRES with $m = k$ and with a restart after each k iterations. The difference with ORTHOMIN(k) is that ORTHOMIN keeps on going without a restart, but uses a truncated orthogonalization.

Another wrinkle on Krylov-space methods is the bi-conjugate gradient (BCG) method. This uses a method due to Lanczos to develop two Krylov subspaces, one based on A and the other based on its transpose A^T. Sequences of basis vectors are chosen that are mutually orthogonal rather than orthogonal within each set. The logic is very similar to that of CG, but the solution at each step does not minimize the norm of a residual as in the CG case. Saad (1996) provides the details.

The final Krylov-type method we wish to discuss is the Chebyshev method described by Manteuffel (1977, 1978). The idea behind the Chebyshev method is that the residual at the nth step of the iteration is equal to the matrix $T_n[(d - A)/c]/T_n(d/c)$ multiplied by the initial residual. This matrix is a combination of powers of A up to the nth degree, so this is a Krylov-subspace method like the others we have discussed. The $T_n(z)$ are the complex Chebyshev polynomials and c and d are constants that are estimated based on knowledge of the eigenvalue spectrum of A. First of all, the method will not work unless all the eigenvalues have positive real part (A is positive definite). Then c and d should be such that an ellipse with its center at d and foci located at $d \pm c$ should be the smallest one possible that encloses all the eigenvalues. (If the major axis of the ellipse is aligned with the imaginary axis then c can be imaginary.) The Chebyshev polynomials have a maximal property, i.e., of the polynomials of a given degree that are bounded by 1 in the interval $-1 \leq z \leq 1$, they are the largest possible outside that range. This translates into making the residual as small as possible. The recurrence relation for Chebyshev polynomials leads to this setup of the iteration method:

$$x_n = x_{n-1} + dx_{n-1}, \tag{11.38}$$

$$r_n = b - Ax_n, \tag{11.39}$$

$$p_2^n = \frac{c^2 p_1^{n-1}}{4d - c^2 p_1^{n-1}}, \tag{11.40}$$

$$p_1^n = \frac{1 + p_2^n}{d}, \tag{11.41}$$

$$dx_n = p_2^n dx_{n-1} + p_1^n r_n. \tag{11.42}$$

The starting values are $p_1^0 = 2/d$, $p_2^0 = 0$ and $dx_0 = r_0/d$.

The estimation of c and d can be problematic. Calvetti, Golub, and Reichel (1994) provide an efficient algorithm for estimating the convex hull of the eigenvalues of A based on modified moments that are computed as the iteration proceeds. After a certain fixed number of iterations, or sooner if the residuals begin to increase, the convex hull estimate is updated, new values are derived for c and d, and the iteration is restarted.

The Chebyshev method has been successfully used in ALTAIR (Castor, Dykema, and Klein, 1992) for iterative solution of the system of kinetic equations for the atomic populations, and also to accelerate the net radiative bracket iterations.

11.4.6 Multigrid method

The multigrid method is not simply a method, it is a whole field of research. The reader is recommended to visit the web site `http://casper.cs.yale.edu/mgnet/www/mgnet.html` and consult the references listed there, such as the text by Wesseling (1992) and the tutorials of Henson (1987, 1999). The following discussion is aimed at merely giving the flavor of multigrid methods, and the literature must be consulted for the details. A system of linear equations $Ax = b$ may describe an elliptic PDE such as the radiation diffusion problem. The matrix A quite possibly has nice properties such as being symmetric and positive definite. Multigrid is the name for a method in which the solution $x = A^{-1}b$ is approximated in this way:

$$x \approx P A_{\text{coarse}}^{-1} R b, \tag{11.43}$$

in which A_{coarse} is a substantially smaller matrix than A, and P and R are rectangular matrices. The matrix R is called the restriction matrix, because it restricts or projects the vector it acts on to a smaller-dimensional subspace of the space containing x and b. The matrix P is called the prolongation matrix because it prolongs or interpolates the data from the subspace into the original larger space. In order to fix the ideas we can think of A as being a finite-difference operator on a fine mesh with a mesh spacing h, and A_{coarse} is the similar operator on the mesh with spacing $2h$, i.e., with every second mesh line omitted. In this picture P would be written as I_{2h}^{h} and R as I_{h}^{2h}. For a 2-D problem the dimension of A_{coarse} would be $1/4$ as large as that of A, and in three dimensions it would be$1/8$.

The coarse system, with mesh spacing $2h$, may still be too big to be solved readily. The coarsening process can then be applied recursively, so that

$$x \approx I_{2h}^{h} I_{4h}^{2h} \cdots I_{H}^{H/2} A_{H}^{-1} I_{H/2}^{H} \cdots I_{2h}^{4h} I_{h}^{2h} b. \tag{11.44}$$

The idea is that the mesh-space doubling proceeds to the point that the linear system on the coarsest mesh H is trivial to solve. The whole process of evaluating x using (11.44) is called a V-cycle; the picture is that the system size goes down, down, down as the restriction operators are applied to b, then the coarsest system is solved directly, after which the prolongation operators are applied up, up, up to give the answer on the finest mesh. The step-doubling and step-halving picture is only generic, of course. The restriction and prolongation operators can be anything that is convenient for the problem at hand, the only requirement being that they are readily applied and leave a small enough system at the coarsest level.

There are additional requirements if the matrices at every level are to preserve the symmetric positive-definite property of A itself. One of these is the symmetry condition

$$I_{2h}^{h} \propto (I_{h}^{2h})^{T} . \tag{11.45}$$

The replacement of the solution $x = A^{-1}b$ by the V-cycle does not solve the system exactly since the fine grid is finer than the coarse grid for a reason: the answer is more accurate. Thus it is still necessary to apply some relaxation of the solution on the finest grid. After one or more applications of the relaxation equation $x^{k+1} = x^{k} + r^{k}$, with $r^{k} = b - Ax_{k}$, there will be a final residual r. This is the quantity that should be used in place of b at the beginning of the V-cycle. At the end of the cycle, the result x is really the correction Δx that should be added to the solution on the finest scale. In fact, one relaxation operation can be applied at each level of refinement, during the down-down-down part of the V-cycle, so that the residual from that operation becomes the right-hand side for the system at the next-coarser level. Then on the up-up-up half of the V-cycle the prolonged corrections from the coarser level are added to the stored solution at that level to be passed on to the next finer level.

The multigrid methods succeed because relaxation at the finest scale quickly reduces the short-wavelength errors in the solution, while not affecting very much the long-wavelength error modes. But the hierarchical multigrid treatment extinguishes those long-wavelength errors. The cost of applying one V-cycle is just a modest factor larger than one relaxation cycle at the finest scale, so this improvement in reducing long-wavelength errors is almost free. Multigrid methods can have an iteration count that is well below the size N of the mesh; instead of being of order 100, iteration counts $\approx$10 are not unusual.

The multigrid application that is discussed by Baldwin *et al.* (1999), is more complicated than we have just described, and indicates how varied the multigrid concept can be. The multigrid approach in this case is called semicoarsening multigrid, or SMG. The "semi" in the name refers to the fact that in the 2-D problems considered only one direction, say x, is coarsened. The coarse operator at each level still includes the full fine-scale coupling in the other direction. The update operations at each level include a tri-diagonal solve in the y direction. Baldwin *et al.*, describe using SMG by itself, and also including one step of CG iteration before and after the V-cycle, i.e., using SMG as a preconditioner for CG. This turned out to be the most efficient of all the methods Baldwin *et al.*, compared on several of their test problems. Its competitors were simple SMG and one of the variations of preconditioned CG that will be described below.

11.4.7 Preconditioning

The topic of this subsection has already been mentioned several times, so we should find out what is meant by preconditioning. Recall that the CG method has an asymptotic convergence ratio given by $(\sqrt{\kappa}-1)/(\sqrt{\kappa}+1)$, where $\kappa = \lambda_{\max}/\lambda_{\min}$. The ratio is the condition number of A. A matrix A with a large value of κ is called ill-conditioned; a matrix with a small one (i.e., close to unity) is called well-conditioned. For an ill-conditioned matrix the convergence ratio is very close to unity, which means that CG, or any other iterative method, will be very slow to converge. In the other extreme, a matrix with a condition number of unity is already very close to a scalar times the identity matrix, which makes the convergence of the iterative methods immediate. Preconditioning is then the name for an operation that will take an ill-conditioned matrix and make it into a well-conditioned one, or at least better.

Consider again the simple relaxation method,

$$x^{k+1} = x^k + b - Ax^k = b + (I - A)x^k. \tag{11.46}$$

If A is very close to the identity matrix the convergence will be swift. Suppose A is not very close to the identity, but that another matrix M is at hand that is close enough to A that $M^{-1}A$ is reasonably close to the identity. Then if M^{-1} is applied to the linear system before setting up the relaxation equation, the iteration becomes

$$x^{k+1} = x^k + M^{-1}(b - Ax^k) = M^{-1}b + (I - M^{-1}A)x^k. \tag{11.47}$$

We now hope that the condition number of $M^{-1}A$ is much closer to unity than was the condition number of A, and that therefore the convergence ratio will be much less. In applying preconditioning we do not actually demonstrate the matrix M^{-1}, or multiply it by A. We just have to be able to solve a linear system with matrix M. The preconditioned iteration of whatever kind goes like this. First evaluate the residual in the accurate linear system using the current x:

$$r^k = b - Ax^k. \tag{11.48}$$

Then solve the following system for the preconditioned residual:

$$M\tilde{r}^k = r^k. \tag{11.49}$$

Now proceed with the chosen iteration method using $\tilde{r}^k$ where the residual would normally appear. Iteration methods like CG and GMRES require being able to multiply a vector p by A, and they generate the residual using a recursion relation. In this case, for the preconditioned system, p is first multiplied by A, and then the linear system $Mv = Ap$ is solved for the vector v that is put in the place where Ap should be.

In the previous subsection we discussed using multigrid as a preconditioner, and in the application by Baldwin *et al.* (1999) it was used to precondition CG. The second simplest of all preconditioners is the diagonal of the matrix, $M = D$. (The simplest is no preconditioning.) Diagonal preconditioning of simple relaxation is just Jacobi iteration. Diagonal preconditioning of CG may also be a good thing to do, as shown by Baldwin *et al.* CG is unaffected by a scale factor applied to the entire linear system, but diagonal scaling will balance the diagonal elements in different rows of the matrix, and according to Gershgorin's theorem[2] this should help balance the eigenvalues and make the condition number smaller. But there are better preconditioners for CG, as we see next.

Probably the most important preconditioner is incomplete LU factorization (ILU). We recall that LU factorization is the decomposition of A in this way: $A = LU$, in which L is a lower-triangular matrix with a unit diagonal, and U is an upper-triangular matrix. The factorization makes it very easy to then solve a linear system $Ax = b$, since $b = U^{-1}(L^{-1}b)$ can be evaluated by first doing $v = L^{-1}b$ recursively in the forward direction, then doing $U^{-1}v$ by recursion in the backward direction. If A is a sparse matrix, so that the nonzero elements in each row span quite a large number of columns on either side of the diagonal (for example, about N_x on each side for a 2-D diffusion problem with a 5-point or 9-point stencil), the LU decomposition will produce matrices L and U in which the intervening elements that are zero in A have become nonzero. That is, there is fill-in of the sparsity pattern. This is why the exact LU decomposition of A is expensive to do.

ILU is defined in this way. Perform a normal LU decomposition, except that at each point of the elimination process, if a nonzero value would be generated for some element of L or U where you do not want one, then discard that element and proceed. The "where you do not want one" part gives you some latitude. The commonly adopted choice is to throw away any elements that are places where the element of A vanishes. That is, "where you do not want one" is any place where $a_{ij} = 0$. Saad (1996) discusses the pseudo-code for achieving this. Saad also proves that the ILU factorization is a well-conditioned operation if A is an M-matrix.[3] Some of the iterative methods such as CG can be proved to be convergent only if A is an M-matrix. The result of ILU preconditioning of an M-matrix is also an M-matrix.

[2] The eigenvalues of a matrix lie in the union of the circles in the complex plane centered on each diagonal element with a radius equal to the sum of the absolute values of the remaining elements in the corresponding row; the same is true for columns.

[3] A matrix A is an M-matrix if it has a positive diagonal, negative off-diagonal elements, and the elements of A^{-1} are positive. The last condition can be replaced by the condition that the spectral radius of $I - D^{-1}A$ is less than unity, where D is the diagonal of A.

ILU is used as a preconditioner by letting the matrix M discussed earlier be LU, where L and U are the results of the ILU process. The ILU decomposition would be done once and for all and stored during the iterations. The solution of $Mv = Ap$ discussed above is evaluated as $v = U^{-1}(L^{-1}Ap)$. Saad provides examples of applying GMRES with ILU preconditioning as just described, and it shows big gains for several of the examples over simple GMRES.

Saad also discusses other variants of ILU. One of these is ILUT, which stands for incomplete LU factorization with thresholding. The ILUT algorithm takes two parameters. The first is a tolerance such that a fill-in element is discarded if its magnitude is less than the specified tolerance times the norm of the row. The second parameter is the maximum number of fill-in elements that will be kept, based on a list in which the elements are ordered by decreasing magnitude. Sample calculations by Saad show that ILUT is significantly more robust, and also faster, than simple ILU. For the test cases in Baldwin *et al.*, ILUT–GMRES performed fairly well, but was generally outperformed by preconditioned CG and multigrid. Of course, ILUT–GMRES is a method that works on nonsymmetric matrices for which CG variations do not, and many implementations of multigrid do not either.

For positive definite symmetric matrices the method of LU factorization can be modified somewhat to preserve the symmetry in the factors. It also allows only one of the factors to be stored. This is the Cholesky decomposition,

$$A = LL^T , \tag{11.50}$$

in which a single lower-triangular matrix L appears, and U has been replaced by the transpose of L. The algorithm for performing the Cholesky decomposition is almost the same as for LU except the square root of the diagonal element is extracted at each step and used to scale the column rather than using the diagonal element itself. The *incomplete Cholesky* (IC) preconditioner, introduced by Meijerink and van der Vorst (1977), is arrived at by performing a Cholesky factorization and discarding fill-in elements exactly as in ILU. Simple IC decomposition, with all fill-in elements discarded, is shown by Meijerink and van der Vorst to always succeed if A is an M-matrix. The implementation of IC-preconditioned conjugate gradient (ICCG) by Kershaw (1978) for radiation diffusion problems shows a great superiority over Gauss–Seidel, ADI and block-SOR. In Baldwin *et al.* a threshold variation ICT of IC preconditioning was applied with criteria the same as the ILUT factorization just described. In their sample calculations a threshold of 10^{-4} was allowed and a generous limit on the number of fill-in elements. The ICT-CG method turned out to be quite competitive with multigrid as the best method.

11.4.8 Nonlinear systems; Newton–Krylov method

The generic PDE for radiation diffusion is

$$\frac{\partial E}{\partial t} - \nabla \cdot [D(T)\nabla E] = \kappa\rho(4\pi B - cE), \tag{11.51}$$

where the diffusion coefficient D is a function of the material temperature T, which is determined by

$$\frac{\partial e(T)}{\partial t} = \kappa(cE - 4\pi B). \tag{11.52}$$

The dependences of D on T, of B on T, of κ on T, of e on T, and even of D on E and ∇E, are all quite nonlinear. The nonlinearity persists even if the equilibrium-diffusion assumption $cE = 4\pi B$ is made. To ensure stability this equation is discretized in time implicitly, as discussed earlier. That means that the time-advanced E^{n+1} will appear in the diffusion coefficient, as well as in the ∇E factor, as in

$$E^n - E^{n+1} + \Delta t \nabla \cdot \left[D\left(\frac{T^n + T^{n+1}}{2}\right) \nabla \left(\frac{E^n + E^{n+1}}{2}\right)\right] = 0, \tag{11.53}$$

in which T^{n+1} comes from E^{n+1} by an auxiliary calculation. This is to be solved for E^{n+1}. This is the prototype nonlinear diffusion problem.

The solution choices are: (1) lag D by using $D(T^n)$ instead of the time-centered form; (2) use the time-centered D but solve the equation in a Picard iteration in which D is updated after each solve;[4] (3) use Newton–Raphson iteration on E^{n+1}, including the variation of T^{n+1} and therefore D with E^{n+1}. Choice (1) is unacceptable because of inaccuracy and possible problems with thermal instability. For choice (2) the convergence is not nearly as good as with choice (3).

The Newton–Raphson method for a system of nonlinear equations $F(X) = 0$ is the iteration

$$J(X^n)X^{n+1} = J(X^n)X^n - F(X^n), \tag{11.54}$$

in which J stands for the Jacobian matrix $J = \partial F/\partial X$. The Jacobian of this system will be one of those large, sparse matrices, so this linear system that must be solved for each Newton iteration falls in the category we have been discussing. But an additional consideration is that the Jacobian matrix elements themselves may be costly to evaluate. In some other nonlinear problems, not radiation diffusion, it may be quite difficult to explicitly perform the differentiation for the Jacobian. This is where the Newton–Krylov method comes to the rescue.

[4] Picard iteration is an elegant way of saying "substitute the unknown back in and do it again."

Recall that all the Krylov-subspace methods for solving linear systems depend on the matrix A (here to be replaced by the Jacobian J) only through its products Av with specified vectors. But we have a way of doing Jv; it is

$$Jv = \lim_{\epsilon \to 0} \frac{1}{\epsilon}[F(X^n + \epsilon v) - F(X^n)]. \tag{11.55}$$

If we simply evaluate $[F(X^n + \epsilon v) - F(X^n)]/\epsilon$ with a suitable small ϵ, we should get an approximate value of Jv that is good enough for the purpose. The Newton–Krylov method, first applied to problems such as this by Brown and Saad (1990), is a double iteration with Newton iterations, and inner iterations using GMRES or another Krylov method, and using the relation just given to replace matrix-vector products.

The GMRES iterations may well need preconditioning to converge satisfactorily, but the preconditioners often explicitly use the matrix, e.g., as in ILU or IC preconditioning. What should be done about this? In a study of a preconditioned Newton–Krylov solution of equilibrium diffusion problems, Rider, Knoll, and Olson (1999) use one multigrid V-cycle based on the matrix A of the *linear* diffusion equation, i.e., the one with fully lagged coefficients, to precondition the GMRES iterations. It is not necessary to update A during the Newton–Raphson iterations. The amount of GMRES iteration within the Newton–Raphson loop is adjustable. It could fully converge the GMRES for every Newton iteration, or do only a single iteration, or something in between. Rider *et al.* choose to do a variable number of GMRES iterations, with the convergence tolerance tightening as the Newton iteration proceeds.

Jones and Woodward (2001) examine a problem of ground-water diffusion that is described using a nonlinear advection–diffusion equation. They apply the Newton–Krylov method using GMRES with two kinds of multigrid preconditioner: one uses pointwise coarsening and the other, the SMG from above, uses coarsening by planes through the mesh. The relaxation applied at each level is Gauss–Seidel. The GMRES convergence tolerance is adjusted dynamically as the Newton–Raphson converges. The results show that there is a trade-off between the simpler and cheaper pointwise multigrid and the more robust SMG.

The Chebyshev method can also serve as a nonlinear solver. Hyman and Manteuffel (1984) describe a nonlinear method that exploits the eventual linearity of Picard-type iteration $x^{n+1} = F(x^n)$ to apply ideas from the Chebyshev algorithm to estimate the optimum coefficients α and β in a relaxation equation

$$x^{n+1} = x^n + \alpha r^n + \beta(x^n - x^{n-1}), \tag{11.56}$$

where the residual is defined to be

$$r^n = x^n - F(x^n). \tag{11.57}$$

This performs quite well on a 3-D solution of Burgers's equation. Castor *et al.* (1992) use Chebyshev acceleration for the Picard iteration applied to the Net Radiative Brackets in the multilevel ETLA method (see Section 9.1.4) for non-LTE radiative transfer. The nonlinearity is severe in these problems, and the Jacobian is not readily obtained, nor is a preconditioner available. The Chebyshev acceleration method is the best that has been found for this problem.

Ng acceleration has also been used for nonlinear calculations. For example, the CONRAD code (MacFarlane, 1993) in use at the Fusion Technology Institute of the University of Wisconsin uses Ng acceleration for the level populations in a 1-D collisional-radiative equilibrium model based on single-flight escape probabilities.

11.5 Eddington factors and flux limiters

The Eddington-factor method, also called the variable Eddington factor (VEF) method, is simple in concept: if the precise ratio of the pressure tensor to the energy density were included as an *ad hoc* multiplier in the Eddington approximation equations, they would then become exact. This method of solving transport problems originated with the work of Gol'din (1964), under the name quasi-diffusion. The Eddington tensor nomenclature was introduced by Freeman *et al.* (1968).

Let the Eddington tensor T be defined by

$$\mathsf{P} \equiv \mathsf{T}E. \tag{11.58}$$

Substituting a relation like this into the monochromatic second moment equation in the comoving frame, (6.50), for example, leads to

$$\frac{1}{c}\frac{\partial \mathbf{F}_\nu}{\partial t} + \frac{1}{c}\nabla\cdot(\mathbf{uF}_\nu) + c\nabla\cdot(\mathsf{T}_\nu E_\nu) = -k_\nu \mathbf{F}_\nu. \tag{11.59}$$

If, as in Section 11.3, we drop the $1/c$ terms here and solve for $\mathbf{F}_\nu$ which is substituted into the first moment equation, we get

$$\rho\frac{D(E_\nu/\rho)}{Dt} + E_\nu \mathsf{T}_\nu{:}\nabla\mathbf{u} - \nabla\cdot\left[\frac{c}{\kappa_\nu \rho}\nabla\cdot(\mathsf{T}_\nu E_\nu)\right] = 4\pi j_\nu - \kappa_\nu \rho c E_\nu. \tag{11.60}$$

A frequency-averaged version of (11.60) takes the place of (11.9), and can be used in the same way. The solution cost is much the same, except that the partial differential equation is not self-adjoint in general, and therefore cannot, even in principle, be approximated by a difference representation with a symmetric matrix. Thus we are compelled to use nonsymmetric solvers from the outset.

Some other features of the VEF equation are seen by taking certain limiting cases. If the material terms and the velocity terms are dropped in (11.59) and also in (4.27), and then $\mathbf{F}_\nu$ is eliminated between them, the result is

$$\frac{\partial^2 E_\nu}{\partial t^2} - c^2 \nabla \cdot [\nabla \cdot (\mathsf{T}_\nu E_\nu)] = 0. \tag{11.61}$$

This equation has wave solutions that locally obey the dispersion relation

$$\frac{\omega^2}{c^2} = \mathbf{k} \cdot \mathsf{T}_\nu \cdot \mathbf{k}. \tag{11.62}$$

For a radiation front propagating in the direction $\mathbf{n}$ the Eddington tensor will be just $\mathbf{nn}$, which means the dispersion relation is $\omega = \mathbf{k} \cdot \mathbf{n}c$. This means the group velocity is exactly c in the direction $\mathbf{n}$. So the wave speed comes out right.

A similar limit that is informative is to consider a vacuum region in steady state that has a radiation field passing though it. Apparently this field must satisfy

$$\nabla \cdot (\mathsf{T}_\nu E_\nu) = 0. \tag{11.63}$$

If we expand the tensor divergence we find

$$E_\nu \nabla \cdot \mathsf{T}_\nu + \mathsf{T}_\nu \cdot \nabla E_\nu = 0, \tag{11.64}$$

and if we then multiply on the left by the inverse of the matrix T_ν and divide by E_ν we get

$$\mathsf{T}_\nu^{-1} \cdot \nabla \cdot \mathsf{T}_\nu = -\nabla \log E_\nu. \tag{11.65}$$

With a general tensor T_ν this equation will be inconsistent, because a condition for solubility is apparently

$$\nabla \times (\mathsf{T}_\nu^{-1} \cdot \nabla \cdot \mathsf{T}_\nu) = 0. \tag{11.66}$$

Putting this argument differently, we can say that since under the stated conditions there is certainly a solution for the radiation field, then (11.66) must be satisfied. But suppose that the tensor has been obtained by some approximation procedure and does not exactly obey (11.66), and suppose that the region is not exactly a vacuum, but there are small values of absorptivity and emissivity that we let tend to zero. Then what we expect is that this limit is a singular limit, and that the solution that is found in the limit does not have a finite value of the flux. This sad expectation is supported by some numerical experience.

This problem significantly impairs the robustness of the VEF method. A possible remedy is the following. Suppose there were an integrating factor q_ν such that

$$\nabla \cdot (\mathsf{T}_\nu E_\nu) = \frac{1}{q_\nu} \mathsf{T}_\nu \cdot \nabla (q_\nu E_\nu) \tag{11.67}$$

were true regardless of E_ν. Apparently the integrating factor would have to obey this equation

$$\mathsf{T}_\nu \cdot \nabla \log q_\nu = \nabla \cdot \mathsf{T}_\nu. \tag{11.68}$$

This does not seem like progress since the condition for this to be soluble is also (11.66). But now suppose that we extract a single scalar equation from this vector equation and obtain q_ν from it, then make the replacement (11.67) as an additional approximation. One possibility is to take the divergence of the equation, which gives

$$\nabla \cdot (\mathsf{T}_\nu \cdot \nabla \log q_\nu) = \nabla \cdot (\nabla \cdot \mathsf{T}_\nu), \tag{11.69}$$

as the equation from which q_ν is to be found. Given a numerical estimate of T_ν we would evaluate the second derivative on the right-hand side, then solve the Poisson-equation-like PDE for $\log q_\nu$. The additive constant is unimportant since a multiplicative factor in q_ν cancels out when the integrating factor is used. Because T_ν is symmetric and positive definite this self-adjoint elliptic equation is easy to solve using iterative methods. This approximation shows promise because the curl condition (11.66) is accurately obeyed in both the diffusion limit and the free-streaming limit.

If the formula for the flux is altered using (11.67) then the VEF equation becomes

$$\rho \frac{D(E_\nu/\rho)}{Dt} + E_\nu \mathsf{T}_\nu : \nabla \mathbf{u} - \nabla \cdot \left[\frac{c \mathsf{T}_\nu}{\kappa_\nu \rho q_\nu} \cdot \nabla (q_\nu E_\nu) \right] = 4\pi j_\nu - \kappa_\nu \rho c E_\nu, \tag{11.70}$$

which also now has the nice property of being self-adjoint. In fact, it is this property that makes the equation well-posed for any T_ν in the limit $k_\nu \to 0$.

This approach was introduced by Auer in spherical geometry, for which the curl condition is exactly satisfied, as a means of improving the conditioning of the VEF equation (11.60). We will return to that below in discussing spherical problems.

The question before us now is: where does the Eddington tensor come from? There are two general philosophies. The first is to use an analytic model based on the problem geometry that attempts to capture the main features of the tensor as they depend on that geometry. Let's take spherical geometry as the illustration. At a particular point we set up a local Cartesian coordinate system with the z axis in the radial direction, and x and y tangential. Axial symmetry about z means that the tensor is diagonal and the x and y diagonal elements are equal. Thus the tensor has the form

$$\mathsf{T} = \begin{pmatrix} \frac{1-f}{2} & 0 & 0 \\ 0 & \frac{1-f}{2} & 0 \\ 0 & 0 & f \end{pmatrix}, \tag{11.71}$$

since the trace must be unity. In other words, we let the rr component of the tensor be f, then the two transverse components are $(1 - f)/2$. We speak of f as *the* Eddington factor. This is defined in the same way as the Eddington factor in slab geometry discussed earlier. A simple analytic model for f of the type we are discussing is the formula that comes from assuming that the photosphere of a star radiates an equal intensity in all directions, and that the space above the photosphere is completely transparent. Thus at a point located above the photosphere the radiation field is constant within a certain cone and zero outside it. The half-angle of the cone is $\theta = \sin^{-1}(R_p/r)$ in terms of the photospheric radius R_p and the local radius r. The cosine of this angle is $\mu = \cos\theta = \sqrt{1 - (R_p/r)^2}$. Doing the integrations for P_{rr} and E leads to

$$f = \frac{1}{3}(1 + \mu + \mu^2). \tag{11.72}$$

As r approaches R_p the value of μ tends to zero, so the Eddington factor f tends to 1/3. So in this simple model we would adopt this formula for f in $r > R_p$ and set $f = 1/3$ in $r \leq R_p$. As a practical matter, it has been found that using such formulae for the Eddington factor ameliorates somewhat the errors in the Eddington approximation, but not enough.

The second general approach to the VEF method, now used exclusively, is to employ an auxiliary calculation that solves the radiative transfer equation as accurately as possible, with good resolution in angle space, and obtain the tensor point by point in space from the angle moments derived from this calculation. One may well ask, why bother with the VEF equation at all when an accurate angle-dependent transfer calculation will have to be done anyway? There are at least two reasons. One is that making the large set of radiation transport equations for many angles implicitly coupled through the material temperature leads to a system too costly to solve. For the auxiliary calculation the distribution of temperature is taken as given, which removes the implicitness and thus makes the transfer much cheaper. The second reason has to do with retardation, the presence of the time derivative term in the transport equation. The burden of carrying this term is severe – not in the cost of solving the spatial differencing with this small correction, but in the amount of storage required to carry all the intensities from one time step to the next. It may be true that dropping retardation still produces a tensor that is sufficiently accurate, even though the flux and energy density in the auxiliary calculation might have unacceptable systematic errors as a result. The case for the VEF method has pros and cons, and is by no means closed. We will return to this discussion in connection with approximate operator iteration methods (ALI), also called preconditioning.

There is another modification to the Eddington approximation that is somewhat related to the use of Eddington factors, and is an alternative to it. This is the *flux limiter*. The primary reference on flux limiters and their connection to the Eddington factor in one dimension is Pomraning (1982). The idea is to discard the $\partial \mathbf{F}/\partial t$ term in the flux moment equation and make the Eddington approximation $\mathsf{P}_\nu = (E_\nu/3)\mathsf{I}$, but compensate the errors of these approximations by including a correction factor in the diffusion coefficient:

$$\mathbf{F}_\nu = -\frac{c\mathsf{D}}{\kappa_\nu \rho} \cdot \nabla E_\nu. \tag{11.73}$$

The tensor (in general) D is the flux limiter. The only difference between D and T is which side of the divergence operator the tensor stands on; the inside for T and the outside for D. From this point on, the philosophies of flux limiters and Eddington factors begin to differ. There is no practical way to self-consistently calculate a flux limiter so as to produce agreement between solutions using (11.73) and accurate transport solutions. Instead, flux limiters are used in the way that analytically-based Eddington factors might be used but today are not. That is, a relatively simple formula is adopted for the flux limiter that captures some essential features of the problem, but which cannot be very accurate. Flux limiters are intended mainly to compensate for the omission of the $\partial \mathbf{F}/\partial t$ term. The raw Eddington approximation can give a flux that is arbitrarily large compared with cE if the gradient of E is large enough; this is something that can never happen if $\partial \mathbf{F}/\partial t$ is retained. This problem is corrected by making sure that D becomes small when ∇E is large, so the flux is indeed limited to be no larger than cE. The tensor D is invariably chosen to be a scalar tensor, i.e., just a scalar factor D, and this is considered to be a function of the dimensionless quantity

$$R = \frac{|\nabla E_\nu|}{\kappa_\nu \rho E_\nu}. \tag{11.74}$$

A small value of R means that the ordinary diffusion flux is small compared with cE_ν, and therefore no limiting should be necessary, and D should be $1/3$. A large value of R means that the physical limit on the flux is violated by the ordinary flux formula, and limiting is needed. The proper limit $\mathbf{F}_\nu \to cE_\nu$ is obtained if

$$D(R) \to \frac{1}{R} \quad \text{for} \quad R \to \infty. \tag{11.75}$$

The literature on flux limiters has become extensive, and we will quote three here:

$$D(R) = \begin{cases} \dfrac{1}{3+R} & \text{sum} \\ \dfrac{1}{\max(3, R)} & \text{max} \\ \dfrac{1}{R}\left(\coth R - \dfrac{1}{R}\right) & \text{Levermore} \end{cases} \tag{11.76}$$

The sum and the max flux limiters are just formulae chosen to have the right limits. Levermore's flux limiter is derived from an application of the Chapman–Enskog method of kinetic theory (Levermore, 1979, Levermore and Pomraning, 1981). Levermore's theory modifies the definitions given here by including a factor of the scattering albedo ϖ in the denominator of the definition of R, and the result for D then contains a factor ϖ in the denominator as well. The effect of the albedo is to leave both limits of the flux limiter unchanged, but to change the typical value of $E/|\nabla E|$ where D switches from one limit to the other, from one mean free path for the total absorptivity to one scattering mean free path. Thus in a problem with almost pure absorption the value of R would be large and the flux would be set to cE_ν even at large optical depth. This is appropriate if there is no internal source of radiation and only a beam incident at the boundary, but it is clearly wrong in the more usual case with an internal source.

There are some comments to be made about the application of flux limiting in an implicit radiation diffusion problem. It is evident that the global geometry of the problem, which determines the angular distribution of the radiation field and implicitly both the Eddington factor and the tensor D, cannot be encompassed by formulae like (11.76). That is, no matter how much skill is employed in selecting D, the error will none-the-less be similar to the 20% error of the Eddington approximation in general. For this reason there seems little to choose between the alternative expressions. The second point is that D is a nonlinear function of E_ν and its gradient, and this adds additional nonlinearity to (11.9) beyond that due to the temperature. Furthermore, in two or three dimensions the gradients in the different directions are combined in the diffusion coefficient since D depends on the norm of the gradient. The alternative of using a function for each coordinate direction that depends on that component of the gradient alone can lead to a flux vector that makes a large angle with ∇E_ν, and this is unphysical. The better approach is either to deal with the nonlinearities using Newton–Raphson or to lag the value of D in time.

11.6 Method of discrete ordinates

This method was originally introduced by Chandrasekhar (1960) to solve the standard problems of monochromatic or gray scattering in 1-D slab geometry, and it is particular to that geometry. We refer to Section 5.2 for the definition of the angle cosine μ and the formulation of the relation between the source function S and the mean intensity J, (5.25). The idea of the method of discrete ordinates is to choose a set of discrete values of $\mu = \{\pm\mu_i, i = 1, \ldots, n\}$ and a quadrature formula

$$\int_{-1}^{1} d\mu\, f(\mu) \to \sum_{i=1}^{n} w_i[f(-\mu_i) + f(\mu_i)]. \tag{11.77}$$

The quadrature will always be normalized so that

$$\sum_{i=1}^{n} w_i = 1. \tag{11.78}$$

Chandrasekhar confined himself to the even-order Gaussian quadratures on $[-1, 1]$ for which the values of μ_i are the positive zeroes of the Legendre polynomials $P_{2n}(\mu)$. The first approximation has a single value of μ_i which is the zero of $P_2(\mu)$, namely $1/\sqrt{3}$. Much more accurate results are obtained with other quadrature schemes, such as subdividing $[0, 1]$ into a large number of subintervals and applying three-point Gaussian quadrature on each subinterval. The Hopf function in Section 5.3 was obtained using twelve points chosen in this way with four subintervals; its accuracy is about 5.5 significant figures.

When this discrete set of angles is used for the transfer equation it takes this form

$$\pm\mu_i \frac{dI_i^{\pm}}{d\tau} = I_i^{\pm} - S. \tag{11.79}$$

For a fully computational approach to this problem this equation is then written in finite-difference form leading to a linear system of equations for $I_i^{\pm}$ at a set of discrete depths τ_j, which are solved in a standard way. Chandrasekhar's method does the analysis of the differential equations analytically. Equations (11.79) can be solved formally and substituted into the quadrature formula defining S. The result is exactly Milne's first integral equation, except that the E_1 kernel has been replaced according to

$$\frac{1}{2}E_1(|x|) \to \sum_i \frac{w_i}{\mu_i} \exp\left(-\frac{|x|}{\mu_i}\right). \tag{11.80}$$

This philosophy can be applied in a much more general way: If we are faced with any convolution-type integral equation on a full-space or a half-space, we can find approximations to the kernel in the form of a sum of exponentials and proceed exactly as for the Milne problem. For example, the non-LTE problem of line scattering with complete redistribution, which leads to an integral equation with the kernel $K_1(\tau)$ as described earlier, is treated in exactly this way. This is developed at some length in the book by Ivanov.

The next step in the analysis involves finding the function $T(z)$ that approximates $1 - \tilde{K}_1(i/z)$. Since the Fourier transform of $(a/2)\exp(-a|x|)$ is $1/(1 + k^2/a^2)$, the representation of $T(z)$ in the conservative scattering ($\varpi = 1$) case is

$$T(z) = 1 - \sum_{i=1}^{n} \frac{w_i}{1 - \mu_i^2/z^2} = \sum_{i=1}^{n} \frac{w_i \mu_i^2}{\mu_i^2 - z^2}. \tag{11.81}$$

We observe that $T(z)$ must be a rational function of z since it is a sum of rational functions. It has $2n$ poles, at the points $z = \pm\mu_i$. Since apparently $T(z) \propto 1/z^2$ for $z \to \infty$, it must have the form $R_{2n-2}(z)/S_{2n}(z)$, where R and S are polynomials of the indicated degrees. Therefore $T(z)$ should have $2n - 2$ zeroes $\pm z_\alpha$, $\alpha = 1, \ldots, n-1$. The computational work at this point is to actually use an algebraic root finder to evaluate (to high precision!) these roots. Once they are found $T(z)$ can be expressed in factored form as

$$T(z) = \frac{\prod_{\alpha=1}^{n-1} (1 - z^2/z_\alpha^2)}{\prod_{i=1}^{n} (1 - z^2/\mu_i^2)}, \tag{11.82}$$

where the multiplicative constant factor has been adjusted to satisfy $T(0) = 1$. Now it is simple to factor $T(z)$ into parts analytic in the left and right half-planes,

$$T(z) = \frac{1}{H(z)H(-z)}, \tag{11.83}$$

with

$$H(z) = \frac{\prod_{i=1}^{n} (1 + z/\mu_i)}{\prod_{\alpha=1}^{n-1} (1 + z/z_\alpha)}. \tag{11.84}$$

The problem is now essentially solved, because $H(\mu)$ is the angle dependence of the emergent intensity and $H(1/p)/p$ is the Laplace transform of the source function, for half-space problems. The Laplace inversion can be done easily with the method of residues since the poles of H, the points $-z_\alpha$, are already known.

There is a relation between all the roots and the quadrature points that comes from noting that $T(z) \sim -1/3z^2$ for $z \to \infty$ assuming that the quadrature formula (11.77) is accurate for the integral $\int \mu^2 d\mu = 1/3$. It is

$$\frac{\prod_{i=1}^{n} \mu_i}{\prod_{\alpha=1}^{n-1} z_\alpha} = \frac{1}{\sqrt{3}}. \tag{11.85}$$

Using this relation we see that the expansion of $H(z)$ for large z is

$$H(z) \sim \sqrt{3}(z + q_\infty), \tag{11.86}$$

where the constant q_∞ is given by

$$q_\infty = \sum_{i=1}^{n} \mu_i - \sum_{\alpha=1}^{n-1} z_\alpha. \tag{11.87}$$

Setting $p = 1/z$ thus gives the behavior of the Laplace transform of S at small p, namely

$$\tilde{S} \sim \sqrt{3}S(0)\left(\frac{1}{p^2} + \frac{q_\infty}{p}\right), \tag{11.88}$$

and therefore S for large τ is given by

$$S \sim \sqrt{3}S(0)(\tau + q_\infty). \tag{11.89}$$

In other words the large-τ value of the Hopf function comes out immediately from the solution of the characteristic equation $T(z) = 0$ for the roots z_α.

11.7 Spherical symmetry

Spherical radiative transfer is one step up in complexity from radiative transfer in slab geometry. Good reviews of spherical radiative transfer are found Mihalas (1978), p. 250ff, and Mihalas and Mihalas (1984), Section 83. For spherical problems all the scalars, such as opacities, temperature, source function, etc., are functions only of the radius (and perhaps time), but the intensities are functions of radius and μ, where μ is defined as the radial component of the direction vector **n**. However, spherical coordinates are curvilinear, which means that μ varies along a straight ray. We introduce coordinates based on the rays, which are the path length along the ray measured from the point of closest approach to the center,

$$z \equiv r\mu, \tag{11.90}$$

and the impact parameter of this ray relative to the center,

$$p \equiv r\sqrt{1-\mu^2}. \tag{11.91}$$

As a photon moves along the ray, p remains constant but r and μ vary as z increases by the distance traveled. The inverses of the relations giving z and p in terms of r and μ are

$$r = \sqrt{p^2+z^2} \tag{11.92}$$

and

$$\mu = \frac{z}{\sqrt{p^2+z^2}}. \tag{11.93}$$

The derivatives of these with respect to z at constant p give the variations of r and μ along the path:

$$\left(\frac{\partial r}{\partial z}\right)_p = \frac{z}{\sqrt{p^2+z^2}} = \mu, \tag{11.94}$$

$$\left(\frac{\partial \mu}{\partial z}\right)_p = \frac{p^2}{(p^2+z^2)^{3/2}} = \frac{1-\mu^2}{r}. \tag{11.95}$$

The correct form of the transport equation omitting velocities in (r, μ) coordinates is therefore

$$\frac{1}{c}\frac{\partial I_\nu}{\partial t} + \mu\frac{\partial I_\nu}{\partial r} + \frac{1-\mu^2}{r}\frac{\partial I_\nu}{\partial \mu} = j_\nu - k_\nu I_\nu. \tag{11.96}$$

Forming the first two moments of the transport equation is easy, since the μ-derivative term can be integrated by parts. We find

$$\frac{\partial E_\nu}{\partial t} + \frac{\partial F_\nu}{\partial r} + \frac{2F_\nu}{r} = 4\pi j_\nu - cE_\nu, \tag{11.97}$$

$$\frac{1}{c}\frac{\partial F_\nu}{\partial t} + c\frac{\partial P_\nu}{\partial r} + c\frac{3P_\nu - E_\nu}{r} = -k_\nu F_\nu, \tag{11.98}$$

which we would have expected from the general geometry relations given earlier if we were familiar with the form of a tensor divergence in spherical symmetry.

Much work has been done with radiative transfer in spherical symmetry in the steady-state case. We will pursue this briefly by dropping the time-derivative terms.

The moment equations then become

$$\frac{1}{r^2}\frac{\partial r^2 F_\nu}{\partial r} = 4\pi j_\nu - cE_\nu \tag{11.99}$$

and

$$c\frac{\partial P_\nu}{\partial r} + c\frac{3P_\nu - E_\nu}{r} = -k_\nu F_\nu . \tag{11.100}$$

Here is where the Eddington factor

$$f_\nu \equiv \frac{P_\nu}{E_\nu} \tag{11.101}$$

can be introduced, as well as Auer's integrating factor q_ν defined by

$$\log q_\nu = \int \frac{dr}{r}\left(3 - \frac{1}{f_\nu}\right). \tag{11.102}$$

This allows the flux to be calculated from E_ν using the divergence-like formula

$$F_\nu = -\frac{c}{k_\nu q_\nu}\frac{\partial}{\partial r}\left(q_\nu f_\nu E_\nu\right). \tag{11.103}$$

The spherical version of the VEF method then proceeds using this formula for F_ν and either (11.97) or (11.99). This part of the problem is then no more costly than slab geometry.

The question remains of how to calculate the Eddington factor. One approach is to select an angle mesh μ_i as well as a radius mesh r_j and convert (11.96) or its steady-state equivalent into a set of finite difference equations. By choosing an upwind form of differencing these equations can be solved in a single sweep from the upwind side in the downwind direction. (This idea of sweeping in the upwind-to-downwind direction will be explained more below, in connection with S_N methods.) We will not discuss this more here, because the accuracy turns out to be bad, and a better method is available.

The better method is to use the (p, z) variables as the coordinates. The mesh is actually constructed by finding the intersection points of the circles $r = r_j$ with the rays $p = p_i$. The z values come out to be $z_{i,j} = \pm\sqrt{r_j^2 - p_i^2}$. The two possible signs correspond to the two directions of propagation on the ray at a given radius, but we can also think of a long ray that starts outside the star at negative z, comes inward as z increases through negative values, reaches the point of closest approach at $z = 0$, then passes out again as z increases through the positive values. Such a mesh is illustrated in Figure 11.1. The transfer equation in these coordinates is

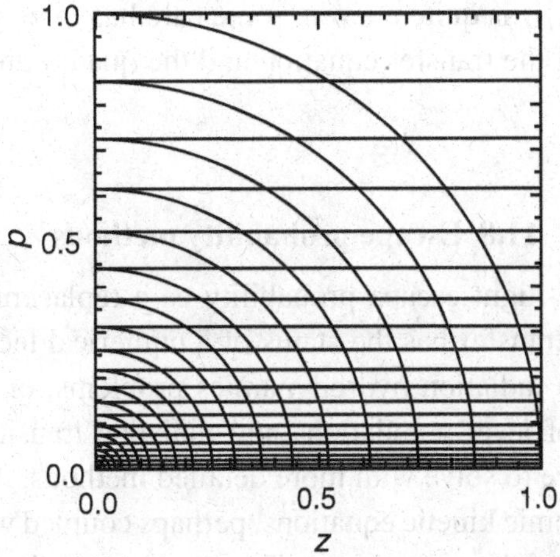

Fig. 11.1 Representation of the p, z mesh for spherical problems. Only the hemisphere $z \geq 0$ is shown.

simply

$$\left(\frac{\partial I_\nu}{\partial z}\right)_p = j_\nu - k_\nu I_\nu. \tag{11.104}$$

This is used in the auxiliary calculation for the VEF method to find I_ν given values of k_ν and j_ν. Sometimes this step is called the *formal solution*. The integration can be done by marching along each ray in the direction of increasing z. Alternatively, the Feautrier variables (see Section 5.6) $\mathcal{j}_\nu$ and $\mathcal{h}_\nu$ may be used, so that $\mathcal{j}_\nu$ obeys a two-point boundary-value problem,

$$\left(\frac{\partial}{\partial z}\right)_p \left[\frac{1}{k_\nu}\left(\frac{\partial \mathcal{j}_\nu}{\partial z}\right)_p\right] = k_\nu \mathcal{j}_\nu - j_\nu. \tag{11.105}$$

The boundary conditions are given on the symmetry plane at $z = 0$ – where $\mathcal{h}_\nu$ vanishes – and the outside radius. If there is an inner-boundary radius, then a condition there may replace the $z = 0$ condition for some rays.

However the intensity in the p, z coordinates is obtained, the angle moments are then calculated by

$$E_\nu(r_i) = \frac{2\pi}{cr_i}\int_{-r_i}^{r_i} dz_{i,j}\, I(p_i, z_{i,j}), \tag{11.106}$$

$$F_\nu(r_i) = \frac{2\pi}{r_i^2}\int_{-r_i}^{r_i} dz_{i,j}\, z_{i,j} I(p_i, z_{i,j}), \tag{11.107}$$

$$P_\nu(r_i) = \frac{2\pi}{cr_i^3}\int_{-r_i}^{r_i} dz_{i,j}\, z_{i,j}^2 I(p_i, z_{i,j}). \tag{11.108}$$

Because the mesh in $z_{i,j}$ is quite uneven, some care has to be given to making the spatial differencing of the transfer equation and the quadrature over z sufficiently accurate.

11.8 Escape probability methods

The use of the single-flight escape probability as a replacement for solving the equation of radiative transfer has the status of a numerical technique, since it enables solving coupled radiation hydrodynamics problems, or non-LTE problems with large numbers of level populations and radiative transitions that would be prohibitively expensive to solve with more detailed methods. At the simplest level these might be 0-D atomic kinetic equations, perhaps coupled with an energy equation for the evolution of the temperature. The "golden rule" of using single-flight escape probabilities in these situations was described earlier (see Section 9.3), and justified by Irons's theorem:

$$z \rightarrow p_{\text{esc}}. \tag{11.109}$$

The quantity z is the net radiative bracket, $z = 1 - \bar{J}/S$, and p_{esc} is the two-sided single-flight escape probability. For a spectral line p_{esc} is given by

$$p_{\text{esc}} = \frac{1}{2}\left[K_2(\tau) + K_2(\tau_0 - \tau)\right] \tag{11.110}$$

in terms of the kernel function defined by (9.18). Irons's theorem says that z and p_{esc} are equal in the mean sense. So the results should not be *too* bad in a 0-D or "one-zone" model if the mean value of p_{esc} is used to make the replacement $\bar{J} \rightarrow (1 - p_{\text{esc}})S$ in the kinetic equations, and also to approximate the term in the material energy equation

$$4\pi \int_0^\infty d\nu\, k_\nu (J_\nu - S_\nu) \approx -4\pi \sum_{\text{lines}} k_L\, p_{\text{esc}} S. \tag{11.111}$$

The results are much less satisfactory if z is replaced point-by-point with p_{esc} in a spatially-distributed model in a hydrodynamic simulation, for example. Even in the 0-D case, the approximation of the radiation field in photoionization continua using escape probabilities is not as accurate as for lines, and far from satisfactory.

An exception to the statement about escape probabilities not being accurate point-by-point may be high-velocity flows for which the Sobolev approximation (see Section 6.8) is valid. The Sobolev approximation may give results at the 10% level of accuracy while for static media the accuracy of the normal escape probability approximation is a factor 2 in good cases. However, Hummer and

Rybicki (1982) have given a rather pessimistic assessment of the accuracy of the simple Sobolev approximation in certain cases.

Another class of methods that has been proposed by Athay (1972b), Frisch and Frisch (1975) and Canfield, Puetter, and Ricchiazzi (1981), may fill the gap where point-by-point results of reasonable accuracy are needed but it is prohibitively expensive to solve all the transfer equations. It is reviewed by Rybicki (1984), and also by Athay, Frisch, and Canfield in the same volume. Athay calls his method *probabilistic radiative transfer*, the term also used by Canfield, while Rybicki prefers to call the method *second order escape probability*. Second order in Rybicki's nomenclature means that the method is derived from a quadratic integral formula rather than from a linear one, not that it is the second member in a systematic expansion. Neither term quite suggests what the method is without further explanation. The essence of the method is to obtain a first order differential equation for $\bar{J}$ (or z) as a function of optical depth in which the single-flight escape probability appears as a coefficient. This can be solved much more cheaply than doing detailed transfer calculations because there are no frequencies or angles to consider, and the equation is first order not second.

The derivation by Canfield *et al.* is the following. There is a quadratic exact integral of the Milne equation that was suggested by Frisch and Frisch; it is

$$\int_\sigma^\infty d\tau \, \frac{\partial \bar{J}}{\partial \tau} S(\tau) = \int_\sigma^\infty d\tau \, S(\tau) \frac{\partial}{\partial \tau} \int_\sigma^\infty d\tau' \, K_1(|\tau - \tau'|) S(\tau') + \frac{1}{2} S_\infty^2. \tag{11.112}$$

On the approximation that $S(\tau)$ is slowly varying on the scale of the width of the kernel K_1, the two factors of S can be factored out of the integrals on the right-hand side. What results is

$$\int_\sigma^\infty d\tau \, \frac{\partial \bar{J}}{\partial \tau} S(\tau) = -\frac{1}{2} S(\sigma)^2 [1 - K_2(\sigma)] + \frac{1}{2} S_\infty^2. \tag{11.113}$$

The K_2 function can be identified as $2p_{\text{esc}}$ for the semi-infinite medium on the basis of (11.110). Notice that this escape probability is two-sided. Differentiating the equation and dividing by S leads to

$$\frac{\partial \bar{J}}{\partial \tau} = \frac{\partial S}{\partial \tau} - \frac{\partial p_{\text{esc}}}{\partial \tau} S - 2 p_{\text{esc}} \frac{\partial S}{\partial \tau}. \tag{11.114}$$

This is the basic equation of the probabilistic radiative transfer/second order escape probability method. An equivalent equation was given by Frisch and Frisch (1975). Athay's (1972a) form is somewhat different. As noted by Rybicki, the integral of this relation, assuming $\bar{J}(\infty) = S_\infty$, gives

$$\bar{J}(\tau) = (1 - p_{\text{esc}}) S(\tau) + \int_\tau^\infty d\tau' \, p_{\text{esc}} \frac{\partial S}{\partial \tau'}. \tag{11.115}$$

Another way of writing the same equation is

$$z = p_{\mathrm{esc}} - \frac{1}{S}\int_{\tau}^{\infty} d\tau' \, p_{\mathrm{esc}} \frac{\partial S}{\partial \tau'}. \tag{11.116}$$

The first term in this relation is the ordinary escape-probability approximation. The second term is responsible for the improved accuracy of the second-order approximation.

If the relation $S = (1-\epsilon)\bar{J} + \epsilon B$ for the two-level atom is used to eliminate $\bar{J}$ from (11.114), assuming B is constant, then S must obey this equation:

$$(1-\epsilon)^{-1}\frac{\partial S}{\partial \tau} = \frac{\partial S}{\partial \tau} - \frac{\partial p_{\mathrm{esc}}}{\partial \tau} S - 2p_{\mathrm{esc}}\frac{\partial S}{\partial \tau}. \tag{11.117}$$

The integral of this equation that gives $S = B$ for $p_{\mathrm{esc}} = 0$ is

$$S = \frac{\sqrt{\epsilon}B}{\sqrt{\epsilon + (1-\epsilon)2p_{\mathrm{esc}}}}, \tag{11.118}$$

in exact agreement with Ivanov's approximation (9.26) for a semi-infinite slab ($\tau_0 \to \infty$).

This method has proved to be very useful for things like modeling quasar broad-emission-line spectra (Canfield *et al.*,1981). Canfield, McClymont, and Puetter (1984) describe applications of the method as well as an extension to finite slabs.

11.9 S_N methods

At this point we would like to distinguish long-characteristic methods from short-characteristic methods for solving the transfer equation. A long-characteristic method means that we march along a single straight ray to solve the equation, although the ray direction may be changing in terms of components of **n** along the local coordinate directions. An example of this was just seen in the p, z coordinates for spherical-symmetry problems. A short characteristic method is one in which a bundle of rays is created at each mesh point, each one of which goes in the direction of a certain **n** with respect to the local coordinates. The rays in this bundle are extended in the upwind direction only as far as the next spatial cell. Each spatial node has its own bundle, and these do not connect from one node to the next for two reasons: the ray directions are not parallel with the vectors joining neighboring nodes, and the ray directions at one node are not necessarily parallel to those at the neighboring node. The name S_N is often used for short-characteristic methods. The slab geometry case is an illustration of both long- and short-characteristic methods, because in this case the ray segments from all the nodes do join into long rays. For other geometries this is not true.

The S_N method is developing very rapidly at this time. Pomraning (1973) describes this method at an early state. The problems associated with the unhappy choice between inaccuracy (step differencing) and negative solutions (typified by the diamond-difference method) have vanished today, through the use of the discontinuous finite-element method (e.g., Dykema, Klein, and Castor (1996)) and the new corner-balance method of Adams (1997) and Castrianni and Adams (1998). The state of the S_N methods in 2002, including fast iterative solutions of implicit scattering problems, is reviewed by Adams and Larsen (2002).

One example of short characteristics has been encountered already, the spherical equation in r, μ coordinates. The S_N approach regards this as an advection problem in r, μ space. The equation can actually be cast into conservative form as

$$\frac{1}{r^2}\frac{\partial}{\partial r}\left(r^2\mu I_\nu\right)+\frac{\partial}{\partial\mu}\left(\frac{1-\mu^2}{r}I_\nu\right)=j_\nu-k_\nu I_\nu. \tag{11.119}$$

One approach to forming the S_N equations is to integrate this equation over a radial volume element $[r_i, r_{i+1}]$ and an angle element $[\mu_j, \mu_{j+1}]$. The "surface fluxes", $r^2\mu I$ in radius and $(1-\mu^2)I/r$ in μ, for that radius–angle cell are represented as interpolated values along the mesh lines, where preference is given to the values on the side of the mesh line corresponding to the cell from which the radiation is flowing. This is the "upwind differencing" idea. It should be familiar from the discussion of cell-centered advection in the Eulerian hydrodynamics methods, Section 3.2. The fluxes can be made first-order accurate, which makes the method second-order accurate, by doing the interpolation appropriately.

Another way to do it is to expand the intensity in a set of basis functions for each r, μ cell, such as the four functions needed to represent I with bilinear interpolation. Substituting this expansion for I into the transport equation yields a residual function that ideally would be zero everywhere, but of course will not be in practice. By taking projections of the residual on the basis functions (or perhaps another set) we derive enough equations to determine the unknown expansion coefficients. This is the finite element method. By allowing every cell to have its own set of expansion coefficients independent of its neighbors, i.e., by not enforcing continuity of I between neighbor cells, the number of degrees of freedom is increased and with this the ability to represent sharp changes in the intensity is improved. The "upwinding" enters in this version of the finite element method when the surface terms that arise from the integration over a cell are systematically evaluated using the variables on the upwind side of the cell boundary. This is the discontinuous finite element method.

Assuming that upwind differencing has been applied, the equation(s) for a given r, μ cell couple in the values for the neighbor cell at smaller r (if $\mu > 0$) or larger r (if $\mu < 0$) and at smaller μ. That means that it is possible to do a raster scan of the whole mesh in the proper order and find that all the neighbor-cell data that are needed at each point have already been computed. Thus one scan through the mesh is sufficient to evaluate all the intensities. This is what we mean by "sweeping" the mesh.

The radiation transport method in ZEUS-2D, an axial-symmetry 2-D Eulerian radiation hydrodynamics code (Stone and Norman, 1992a,b; Stone, Mihalas, and Norman, 1992), illustrates some of the short-characteristics ideas in two space dimensions. The characteristics of the transport equation in rz geometry are hyperbolae opening in the r direction, which means that either the angle advection is treated separately from spatial differencing, as just discussed, or the curved paths must be tracked. In ZEUS-2D the problem is solved by using the axial-symmetry extension of the pz coordinate system used for spherical geometry, as in Section 11.7. Tangent planes parallel to the z axis take the place of the rays with impact parameter p in the spherical case. There are as many tangent planes as there are zones in the r direction in the mesh. The slices by these tangent planes through the hydrodynamic mesh define the transport mesh in each plane. The number of cells in the z direction is the same as in the hydrodynamic mesh, while the number in the lateral direction varies from twice the number of radial zones to two, depending on the distance between the slice and the z axis. The transport on each tangent plane is computed for several values of n_z, the direction cosine of the ray with respect to the z axis. So using this transformation reduces the axial symmetry problem to one of transport in xy geometry with a Cartesian mesh.

The short characteristics method of Stone *et al.* (1992) applied in ZEUS-2D has features in common with the Mihalas, Auer, and Mihalas (1978) (MAM) method, and especially with the method of Kunasz and Auer (1988). The salient features of MAM are the following. The intensity is described by pointwise values located at the mesh nodes. For each of the chosen angles a ray is drawn through a given node O in that direction, both upwind and downwind, and it intersects the far sides of the neighboring zones at an upwind point M and a downwind point P. The transport equation is written in Feautrier form, see Section 5.6, along this characteristic, and Auer's (1976) Hermite differencing of the Feautrier equation is used, which gives a fourth-order accurate relation connecting the Feautrier values j and the source functions S at the three points M, O, P. Points M and P are not at nodes, so both j and S at these points are represented by quadratic interpolation of the values on the appropriate sides of the nine-node stencil surrounding O. The result is a nine-point differencing of the Feautrier form of the transport equation for this ray direction. Taken all together, for a given direction, the result is a block-tri-diagonal linear system for the unknowns j. Since the source functions may depend on all the js through the scattering term $\bar{J}$, the Rybicki elimination method (Rybicki, 1971) is applied. The downsides of this approach are two: (1) The block-tri-diagonal system is very expensive to solve. This is unavoidable with the Feautrier form. (2) The quadratic interpolations will produce ringing and can, and often do, yield negative intensities.

Kunasz and Auer (1988) depart from MAM by using the first order form of the equation of transfer – rather than the Feautrier form used by MAM – which is integrated (exactly) with the relation

$$I_O = I_M \exp(-\Delta\tau) + \int_0^{\Delta\tau} d\tau' \, S(\tau') \exp[-(\Delta\tau - \tau')], \qquad (11.120)$$

in which $S(\tau')$ must be represented by an interpolation function. They consider the alternatives of using linear interpolation of S between points M and O, or using parabolic interpolation based on the three points M, O, P. The values of S at points M and P may be given by linear or parabolic interpolation also. The parabolic interpolations for the upwind point M will sometimes make use of data one point beyond the nine-point stencil on the upwind face of the box defined by the nine points. This results in a total of 13 points on which data may be used to obtain I_O. Unlike the case with Feautrier differencing, the intensities can be calculated in a downwind sweep in the Kunasz and Auer method. This makes the operation count scale linearly with the product $N_x N_z$, where N_x and N_z are the number of mesh lines in those directions, whereas the MAM method scales as $N_x^3 N_z^2$. There is a heavy penalty for using the Feautrier variables in two or three dimension. The computational results in Kunasz and Auer (1988) indicate an unfortunate trade-off between accuracy and positivity in problems with discontinuous data, such as the searchlight beam. They point out that more typical problems are forgiving in this respect.

The differences in ZEUS-2D with respect to Kunasz and Auer (1988) are: linear interpolation replaces quadratic interpolation for the values at point M, and S is represented by linear interpolation along the ray. The solution of the transfer equation for the segment MO, given by (11.120), becomes

$$I_O = I_M \exp(-\Delta\tau) + (S_O - S_M)\exp(-\Delta\tau) + \frac{S_O - S_M}{\Delta\tau}[1 - \exp(-\Delta\tau)]. \qquad (11.121)$$

This differencing is first-order accurate, and it is not consistent with the diffusion limit. That is, the relation $J_\nu \approx S_\nu - \nabla \cdot (\nabla S_\nu / k_\nu)/(3k_\nu)$ will not be obeyed in the optically thick limit, although $\mathbf{F}_\nu \approx -\nabla S_\nu/(3k_\nu)$ may be. The scheme is also not conservative. These objections are addressed by using the transport solution only for the purpose of obtaining the Eddington tensor, see Section 11.5. The Eddington tensor is incorporated in a conservative cell-centered differencing of the radiation energy equation.

Two approaches to S_N radiation transport that do not use short-characteristic finite differences are the upstream corner-balance method (UCB) described by Adams (1997) and the nonlinear corner-balance method (NLCB) of Castrianni and Adams (1998). The Adams (1997) method is quite similar to the bilinear discontinuous finite element (BLD) method used by Dykema *et al.* (1996); see also Castor, Dykema, and Klein (1991). Adams (1997) reviews a variety of different methods. A general feature of these methods is to retain second-order accuracy while preserving positivity as much as possible. A very general result is that second-order accuracy and strict positivity (I can never be negative whatever the source function is, provided $S \geq 0$) are mutually exclusive in a linear algorithm. Nonlinear algorithms can have simultaneous second order accuracy and positivity, and indeed NLCB does.

A sample of how the BLD method works is the 1-D problem. Let us say we want to solve

$$\frac{\partial I}{\partial \tau} = -I + S \tag{11.122}$$

on a mesh with nodes $\tau_{1/2}, \tau_{3/2}, \ldots$, so that zone i is bounded by $\tau_{i-1/2}$ and $\tau_{i+1/2}$. We focus on zone i, and define $x = (\tau - \tau_{i-1/2})/(\tau_{i+1/2} - \tau_{i-1/2})$. For the linear discontinuous method the intensity in zone i is represented by

$$I = I_i^- + (I_i^+ - I_i^-)x. \tag{11.123}$$

That is, the value at $\tau_{i-1/2}$ is I_i^- and the value at $\tau_{i+1/2}$ is I_i^+. These variables are all independent, so at each node $i + 1/2$ the intensity is double-valued, having the value I_i^+ on the left and the value I_{i+1}^- on the right. The interpolation for I can be described by saying that we expand in a set of two basis functions; the basis function associated with the left-hand node is $w_0 = 1 - x$, and the function associated with the right-hand node is $w_1 = x$. The Galerkin prescription for the finite-element method is to substitute (11.123) into (11.122) and then form the projections on the two basis functions. There is upwinding built into this procedure: the integration over the interval $[\tau_{i-1/2}, \tau_{i+1/2}]$ is extended at its lower limit (upwind side) infinitesimally into the interval $[\tau_{i-3/2}, \tau_{i-1/2}]$. This brings in the value I_{i-1}^+ when $\partial I/\partial \tau$ is integrated. Carrying out these operations leads to the following equation, which is repeated for $k = 0, 1$:

$$\frac{1}{\Delta \tau}\left[(I_i^- - I_{i-1}^+)w_k(0) + (I_i^+ - I_i^-)\int_0^1 w_k(x)\,dx\right]$$
$$= (S_i^- - I_i^-)\int_0^1 w_k(x)(1-x)\,dx + (S_i^+ - I_i^+)\int_0^1 w_k(x)x\,dx. \tag{11.124}$$

It is perfectly possible to use this pair of equations as-is, but stability and positivity are improved by making a modification that is called "mass lumping". This consists of replacing $S_i^+ - I_i^+$ on the right-hand side of the $k = 0$ equation by $S_i^- - I_i^-$, and replacing $S_i^- - I_i^-$ on the right-hand side of the $k = 1$ equation by $S_i^+ - I_i^+$. The final result is

$$\frac{1}{\Delta\tau}\left[I_i^- - I_{i-1}^+ + \frac{1}{2}(I_i^+ - I_i^-)\right] = \frac{1}{2}(S_i^- - I_i^-), \tag{11.125}$$

$$\frac{1}{\Delta\tau}\left[\frac{1}{2}(I_i^+ - I_i^-)\right] = \frac{1}{2}(S_i^+ - I_i^+). \tag{11.126}$$

The solution of this pair of equations is

$$I_i^- = \left(1 + \Delta\tau + \frac{1}{2}\Delta\tau^2\right)^{-1}\left[I_{i-1}^+ + \Delta\tau\left(I_{i-1}^+ + \frac{1}{2}(S_i^- - S_i^+)\right) + \frac{1}{2}\Delta\tau^2 S_i^-\right], \tag{11.127}$$

$$I_i^+ = \left(1 + \Delta\tau + \frac{1}{2}\Delta\tau^2\right)^{-1}\left[I_{i-1}^+ + \frac{1}{2}\Delta\tau(S_i^- + S_i^+) + \frac{1}{2}\Delta\tau^2 S_i^+\right]. \tag{11.128}$$

The second equation can be solved recursively for all the I_i^+, from which the I_i^- follow using the first equation. We can see from the form of the equations that it is quite possible that this linear discontinuous method is second order accurate, and indeed it is. The average $I_i = (I_{i-1}^+ + I_i^-)/2$ is the adopted nodal value. This quantity is not guaranteed to be positive. If the source function increases dramatically in zone i, then I_i^- can become negative, and therefore so can the nodal average.

Adams (1997) describes first the simple corner balance (SCB) method. This introduces the idea of a "corner", which is a subdivision of a mesh cell obtained in this way: Define the center of the cell in some way, such as by averaging the coordinates of the vertices. Then in two dimensions draw lines from the center to the midpoints of all the sides of the cell, which can be an arbitrary polygon. These lines then divide the cell into corners, with each corner containing one of the vertices. In three dimensions the cell is sliced up by planes that connect the midpoints of two edges and the center of the cell, but these may be further divided into tetrahedra, depending on the method. In one dimension there are two "corners" per cell, the left half and the right half. With quadrilateral cells in two dimensions there are four corners per cell, and in a hexahedral mesh in three dimensions there are eight corners per cell. There can be 48 tetrahedra per hex, which is quite an obstacle to using tets, despite their attractive simplicity. The concept underlying the corner-balance methods is that the intensity is regarded as constant in each corner. The transport equation is treated by applying conservative finite-volume differencing on each corner. The node- or edge- or face-centered fluxes must be specified, and the usual S_N choices are: (1) diamond, which means that the edge flux is derived from the average intensity of the cells on each side; and (2) step, which means that the upwind intensity is used. Diamond is second order with ringing and the possibility of negative intensities, while step is first order and positive. In the corner-balance method the edges or faces of a corner can be either exterior, meaning that the adjacent corner belongs to another cell, or interior, meaning that the adjacent corner is in the same cell. The SCB method applies step to exterior edges and diamond to interior edges.

As pointed out by Adams (1997), for slabs and for Cartesian meshes in two dimensions, the equations of the SCB method are identical to the fully mass-lumped linear discontinuous method, i.e., to (11.125) and (11.126) in one dimension. There are some drawbacks of the SCB method. The first is that if the mesh is distorted in two or three dimensions (and SCB is no longer the same as lumped BLD in that case) then the wrong effective diffusion coefficient is obtained in the optically-thick limit. The boundary condition in the optically-thick limit may not be the proper diffusion boundary condition, depending on the cell geometry. The final drawback is that a linear system solve is required to obtain the corner intensities for all the corners in the cell; this is 2×2 in one dimension, see (11.125) and (11.126), but becomes a much more costly 8×8 in three dimensions. Adams addresses these problems with his UCB method.

The essence of Adams's UCB modification is to replace the diamond choice for the interior faces of the corners with an upwind expression that is not step, but a form that is itself derived by considering the optically-thick limit. Since it is upwind, this means that all the corners in the cell are upwind-differenced and can be solved in sequence, thus avoiding the linear system for all the corners of SCB. The numerical results shown by Adams (1997) illustrate the robustness but poor performance in some cases of SCB, the accuracy in most cases of non-mass-lumped BLD but the negativity it gives in a problem with complicated geometry, and the just-right behavior of UCB.

In both spherical geometry and 2-D axial-symmetry geometry the angle-derivative question must be faced, as discussed above. In spherical geometry the polar angle $\cos^{-1}\mu$ decreases along the ray in the direction of propagation. In axial-symmetry geometry the inclination of the ray to the symmetry axis z is constant, but the azimuthal angle of the ray with respect to the radial direction decreases along the ray. The relations are formally the same in the two cases, apart from a factor $(1 - n_z^2)^{1/2}$ to project path length into the x–y plane in the cylindrical case. The current S_N methods such as UCB and BLD do not apply a finite-volume method in angle, as described earlier, but represent the angle derivative in finite-difference form. Some of the background of this question is found in Chapter 9 of Richtmyer and Morton (1967).

This is illustrated for axial symmetry as follows. Let the direction vector have a component μ in the $+z$ direction, and let the angle between the ray's projection on the x–y plane and the radial direction be ϕ. (In much of the transport literature ϕ is used for the scalar flux, i.e., our $4\pi \bar{J}$, and ψ is used for the angular flux, our I, but that should not be a confusion here.) The impact parameter of the ray, a constant quantity, is $p = \sqrt{1-\mu^2}r \sin\phi$. Therefore $d\phi/dr$ along the ray is $d\phi/dr = -\tan\phi/r$. This means the transfer equation becomes

$$\mu \frac{\partial I}{\partial z} + \sqrt{1-\mu^2}\cos\phi \frac{\partial I}{\partial r} - \frac{\sqrt{1-\mu^2}}{r}\sin\phi \frac{\partial I}{\partial \phi} = -kI + kS. \quad (11.129)$$

This can also be written in conservative form as

$$\frac{\partial}{\partial z}(\mu I) + \frac{1}{r}\frac{\partial}{\partial r}\left(\sqrt{1-\mu^2}r\cos\phi I\right) - \frac{1}{r}\frac{\partial}{\partial \phi}\left(\sqrt{1-\mu^2}\sin\phi I\right) = -kI + kS. \quad (11.130)$$

It is the differencing of the last term on the left-hand side that is the question. In S_N schemes μ and ϕ are assigned specific values, and there will be, in general, several direction vectors with different values of ϕ for each value of μ; see below. If these ϕ values for the discrete directions are denoted by $0 < \phi_1 < \phi_2 < \cdots < \phi_K < \pi$, then we can define a mesh of cells in ϕ by $\phi_{k+1/2} \approx (\phi_k + \phi_{k+1})/2$ plus $\phi_{1/2} = 0$ and $\phi_{K+1/2} = \pi$. The average of the term in question in the transfer equation over the interval $[\phi_{k-1/2}, \phi_{k+1/2}]$ is given by

$$\frac{\sqrt{1-\mu^2}}{r\Delta\phi_k}\left[(\sin\phi I)_{k-1/2} - (\sin\phi I)_{k+1/2}\right], \quad (11.131)$$

with $\Delta\phi_k = \phi_{k+1/2} - \phi_{k-1/2}$. The flux-like quantity $(\sin\phi I)_{k+1/2}$ apparently vanishes at the ends, $k = 0$ and $k = K$. The question is, what value to assign points that are interior to the range? As is frequently the case in transport theory, the two common choices are step and diamond. With diamond differencing, the cell-center intensity is assumed to be the average of the cell-edge values on either side, $I_k = (I_{k-1/2} + I_{k+1/2})/2$; this is written as $I_{k-1/2} = 2I_k - I_{k+1/2}$ and used to evaluate the cell-edge flux at $k - 1/2$. The calculation procedure begins with a special starting calculation at $\phi = \pi$, for which the ϕ-flux vanishes, and this is used to provide $I_{K+1/2}$. The intensities at smaller values of k and ϕ are then found recursively. With step differencing the replacement $I_{k+1/2} = I_{k+1}$ is made. Diamond has been the choice in the traditional S_N method, *viz.*, Carlson (1963). Further information is available in Lewis and Miller (1984).

The average of the second term on the left-hand side of the transfer equation over $[\phi_{k-1/2}, \phi_{k+1/2}]$ leads to a coefficient $\sqrt{1-\mu^2}\langle\cos\phi\rangle_k$ that is identified with one of the specified direction cosines of the angle set; likewise, $\Delta\phi_k$ must be the azimuthal angle factor in the angular quadrature weight. Both these quantities are determined for a given quadrature set (see the next section). In the case that the intensity is precisely uniform and isotropic the curvature effects that arise from the second and third terms in the transfer equation must exactly cancel, which means that the effective coefficients $(\sin\phi)_{k+1/2}$ must satisfy this recursion relation:

$$(\sin\phi)_{k+1/2} - (\sin\phi)_{k-1/2} = \Delta\phi_k \langle\cos\phi\rangle_k. \tag{11.132}$$

This also is discussed by Lewis and Miller (1984). Morel and Montry (1984) point out that the curvature terms do not quite cancel out in forming the diffusion limit of the S_N equations with either step or diamond differencing, and this causes anomalous dips in the solution for $r \to 0$. They propose a weighted-diamond differencing in which instead of a 50–50 average $I_k = (I_{k-1/2} + I_{k+1/2})/2$, a weighted average is used with weights that depend on the mis-centering of $\cos\phi_k$ in $[\langle\cos\phi_{k-1/2}\rangle, \langle\cos\phi_{k+1/2}\rangle]$. This has proved to be quite successful.

It is often required to solve the S_N equations with a source function that includes scattering, either true scattering or with an effective source term in which thermal emission has been linearized and expressed in terms of the absorbed radiation. Source iteration (lambda iteration, Jacobi iteration) involves estimating the source function, solving the S_N equations for the intensities, using these to evaluate the scattering term(s) and thus obtain a new source function, and then repeating to convergence. As we have said before and will repeat again, this kind of iteration can be very slow to converge. The preconditioning methods that are often employed to speed up convergence are: (1) Eddington tensor (see Section 11.5), (2) diffusion synthetic acceleration, and (3) transport synthetic acceleration (TSA). The Eddington tensor method is not exactly an acceleration scheme, since in this method the material coupling is to the radiation moments obtained from the moment-equation solution, not to the S_N intensities. The Eddington tensor method, coupled with a BLD S_N scheme to obtain the tensor, is described in Klein *et al.* (1989). The coupled tensor diffusion equations describing a line scattering problem are solved by Klein *et al.* using another iterative method: the double splitting iteration. In this method two preconditioners are used, one after the other, on each iteration. The first preconditioner includes all nonlocal spatial coupling, but lags the frequency coupling; this is equivalent to Lambda iteration. The second preconditioner does the reverse; it includes all the frequency coupling but lags the spatial coupling. This is Jacobi iteration in the spatial sense. The double splitting iteration is found to perform well. Solving the 2-D tensor moment problem posed by the first preconditioner requires one of the solvers discussed in Section 11.4. Klein *et al.* use ORTHOMIN. The DSA method is based in the work of Kopp (1963) and later developed by Reed (1971) and especially Alcouffe (1976). In DSA a suitably defined diffusion operator is used to pre-condition the scattering terms in S_N, in just the way to be described in Section 11.11. It is found with some of the S_N schemes, particularly if the mesh is distorted, that the diffusion operator fails as a preconditioner; i.e., the spectral radius of the preconditioned iteration matrix is too large, and the iterations converge poorly or diverge. In such cases a helpful approach is to use a consistently-differenced S_N method of much lower order, perhaps S_2, as a preconditioner. This leads to the TSA method described by Ramoné, Adams, and Nowak (1997). In some perverse cases even this method may fail without user adjustments. This may be expected to be the case if the problem at hand is afflicted with ray effects (see the next section). In such cases the available angle sets are simply inadequate to describe the solution, and different angle sets will differ markedly from each other; it is no surprise if the effectiveness of the acceleration is poor when this happens.

It should be obvious that there is a lot of computational engineering that goes into doing these calculations, certainly in two and three dimensions, and the interested reader should consult the literature to learn the current state of the art.

11.10 What are the angles? The bad news

The nomenclature of S_N and in some cases the actual values of the angles come from the early neutron transport work, summarized by Carlson (1963), with later improvements summarized in Carlson (1970). In this work Carlson explains how the direction vectors, which correspond to certain points on the unit sphere, can be chosen to obey some important integral constraints, and especially the symmetry requirement of being unchanged under permutations of the three coordinate directions X, Y, Z. This symmetry requirement means that the pattern of the directions in the first octant, $x > 0$, $y > 0$, $z > 0$, must have three-fold symmetry: a rotation about the direction $(1, 1, 1)$ by $120°$ should take the pattern into itself. The additional assumption that the directions should be arranged in rows at constant latitude, i.e., of constant direction cosine with respect to any one of the three coordinate axes, means that in the simplest case the directions should resemble the graph of a triangle number. The triangle numbers are $1, 3, 6, 10, 15, \ldots, k(k+1)/2, \ldots$. These correspond to: rows of one direction; rows of one and two directions; rows of one, two, and three directions; and so on. For the kth triangle array there will be k different positive direction cosines with respect to one of the axes in the first octant. Taking into account the reflected directions that have negative values of the direction cosine means that the kth triangle array gives $N = 2k$ different direction cosines. This is the definition of N. The number of directions in one octant is therefore $k(k+1)/2 = N(N+2)/8$. We observe that N is always an even number in this nomenclature; N is the number of rows (i.e., the number of polar angles) in two octants. The triangle-number angle sets are Carlson's Set A. He also describes Set B, which differs from Set A in that the three vertices of the triangle are clipped off. With $N/2$ still taken for the number of surviving rows, the total number of directions per octant becomes $(N+8)(N-2)/8$. For a given $N \geq 6$ there are somewhat more angles in Set B than in Set A. The results for Sets A and B are tabulated by Carlson for N up to 8, and with some difficulty the general relations can be worked out.

In later work, Lathrop and Carlson (1965) and Carlson (1970) derived other quadrature sets that, unlike Sets A and B, exactly integrate certain higher-degree polynomials in the components of the direction vector. This is important when calculating transport using a highly-anisotropic scattering phase function, which is very often expanded in Legendre polynomials of the cosine of the scattering angle. The Carlson (1970) sets are called by the name "level-symmetric quadrature," and denoted symbolically by LQ_N. These have the triangle-number shape, with $N(N+2)/8$ directions per octant. The LQ_N quadrature integrates exactly all even polynomials in n_x, n_y, n_z up to degree $N-2$. The Lathrop and Carlson (1965) quadratures make the additional assumption that the quadrature weight for a given direction is the sum of three weights associated with the three direction cosines (as in Carlson (1963)) with the additional requirement that even polynomials in n_x up to the Nth degree be integrated exactly.

How many octants are needed to describe the radiation in a given problem? The symmetry of the radiation field, at a general location in space, is not as high as the spatial symmetry of the problem itself, since a symmetry element of the spatial structure does not correspond to a symmetry element of the radiation field at this particular location unless the location lies *on* the element. So, for example, a spherically-symmetric spatial structure has symmetry elements of the full rotation group, but only the rotations about an axis that passes through the center of symmetry and the point of observation are symmetries of the radiation field. The symmetry group of the radiation field for a spherical problem turns out to be $C_{\infty v}$ in Schoenflies notation. Thus the radiation field is axially symmetric, but the full range of polar angles from $-\pi$ to π is required. The same group and the same angles apply to 1-D slab geometry. But the symmetry is less for 1-D cylindrical geometry, the geometry of an infinite cylinder, since in that case the only symmetry elements of the spatial structure that leave the point of observation fixed are the reflections in a horizontal plane and a vertical plane, and a rotation about the radial direction by π, plus combinations of these. In other words, the group of the radiation field is C_{2v}. The radiation field is 2-D, but only two octants are needed to describe it: one with $n_r > 0$ and one with $n_r < 0$. That means a total of $N(N+2)/4$ directions for the typical S_N angle set. In the 2-D spatial geometries, xy and rz, there is only a single symmetry element that leaves the point of observation fixed, namely a reflection, corresponding to the group $\mathfrak{C}_{1h}$. The symmetry plane is the x–y plane in xy geometry, and the plane through the z-axis and the observation point in rz geometry. Thus four of the eight octants are required to describe the radiation field in either of the 2-D geometries, with $N(N+2)/2$ directions in all. If the spatial structure of the problem has any less symmetry than the ones so far mentioned, then the radiation field has no symmetries whatsoever at the general point of observation, and all eight octants are essential, with $N(N+2)$ directions.

The problem of finding an adequate angle set in two or three dimensions is much more severe than in one dimension. Recall that by using a total of 36 directions it was possible to get the 1-D Hopf function to something like six significant figures. The sad truth is that it is not at all hard to imagine a 2-D problem for which 36 directions will not even give one significant figure in the result.

Imagine that we are calculating the transport of hydrogen line and free–free emission in the solar corona, and that the source is a solar flare, where a large of amount of thermal energy has been deposited in a region perhaps only tens of kilometers in size. We want to calculate the UV radiation field that results on the opposite side of a supergranulation cell that is 30 000 km across. The clever solar physicist would do a hand calculation to get this answer, but what happens if the code is asked to do it?

The angle subtended by the flare-heated region at the observation point is less than (100 km)/(3×10^4 km), or 1/300 rad = 0.2 degree. If our trusty code is using S_N and has chosen an angle set that can accommodate the unexpected flare wherever it might occur, then the number of angle points it will have to use must be $4\pi/(1/300)^2$, which is 10^6! This might seem ridiculous, but it is true. If, say, a mere few hundred directions are used, which are spaced by about 9 degrees on the sphere, then as we move away from the flare the local flux as we compute it will fall off as $1/r^2$ until we are about 700 km away, and after that the results rapidly become worse. If we decide to walk away along one of the ray directions of our discrete set then we will find that the flux does not fall off very rapidly at all beyond 700 km. If, on the other hand, we walk away in a direction that falls between the rays the flux falls off much faster than $1/r^2$. By the time we are 3×10^4 km away there is a disagreement of several orders of magnitude between the flux in the ray directions and the directions in between.

The general name for this difficulty with S_N calculations is called "ray effects." There is no easy fix for this problem. It goes away once the angle set is sufficiently dense to resolve all the features that the solution contains.

A concept that could help is the idea of adaptive directions. Suppose the angle set at each spatial point is made dynamic, and at each time it moves toward the directions that would resolve the "important" features in the radiation field. This might do it, but it is not very easy to implement. How do you define "important"? Is this based on the gross magnitude of the intensity? What happens if you are interested in a critical feature that is not very large in magnitude? What happens when there are events in opposite hemispheres, both of which call for attention? What happens when there is a sudden onset, as in a flare? The directions may start wheeling toward the proper direction, but not get to where they are useful in resolving the flare until it is over. Or if the dynamic response is made brisk, the directions may chatter in a dreadful way and destroy the accuracy of the calculation. Finally, adaptive directions can end up making it impossible to perform the S_N sweep just discussed, which implies a large penalty in computing cost.

The summary of the angle crisis is as follows. Determine the smallest linear size of the important spatial inhomogeneities of absorptivity or emissivity for your problem. You would like to be able to resolve in angle the radiation produced by this object. Then decide on the largest viewing distance you really care about; this is no larger than the diameter of the problem, but it is also no larger than the longest mean free path for radiation you care about. If you view from a greater distance than that, you cannot see the object anyway. Divide the size by the distance and convert that to an angle. If your angle set cannot resolve that, then you are fooling yourself when you say that you are calculating radiation transport. You might just as well use diffusion; the answers would be no worse, and much cheaper to compute.

11.11 Implicit solutions – acceleration

The objective in this section is to describe preconditioning methods for radiative transfer. A general name that applies to many of these is accelerated lambda iteration (ALI), and this will be defined below. But first let us review the list of variables and equations we need to solve for the radiation hydrodynamics problem. There are the mass and momentum density and the corresponding conservation equations, which we may or may not choose to treat by operator splitting. There is the material temperature and its corresponding internal energy equation (or perhaps the total energy equation), which we are rather sure we cannot split. There are all the intensities that are needed to describe the radiation field for us. Depending on our needs these may be as few as a single frequency-integrated energy density, or as many as there are combinations of dozens to hundreds of frequencies with tens to thousands of angles. The intensities are surely coupled implicitly to the material temperature. If the problem is non-LTE then the level populations are another large set of unknowns that are locally coupled to the radiation field. That is, they enter the problem in a way similar to the temperature. For all except the smallest of these problems the Jacobian matrix of the set of nonlinear equations is too large for a direct solution. The point of the acceleration methods is to solve the linearized system iteratively, and the heart of the iterative methods is the way of preconditioning the iteration.

To fix the ideas consider a steady-state scattering problem with gray radiation and a constant albedo $\varpi < 1$, in other words the second Milne problem. The language introduced many years ago for the operation of acting upon the source function with $E_1(|\tau' - \tau|)/2$ is "applying the lambda operator". (See Kourganoff (1963).) In other words we define a linear operator Λ in this way:

$$\Lambda_\tau[S] \equiv \frac{1}{2}\int_0^\infty d\tau' E_1(|\tau' - \tau|)S(\tau'). \tag{11.133}$$

In terms of Λ, Milne's second problem is

$$S = (1 - \varpi)B + \varpi\Lambda[S]. \tag{11.134}$$

The name lambda has become attached to various iterative methods for solving the Milne problem. The first to consider is lambda iteration. This is the following operation applied repeatedly to convergence:

$$S^{n+1} = (1 - \varpi)B + \varpi\Lambda[S^n]. \tag{11.135}$$

This will indeed converge since $\varpi < 1$ and also the smallest eigenvalue λ of Λ is unity (for a full-space or half-space problem) or larger. The eigenvalue is defined as a value for which the equation

$$u = \lambda\Lambda[u] \tag{11.136}$$

has an admissible solution. This can take a very large number of iterations to converge. A physical picture of the iteration count is the following. Each step of the iteration is like letting a photon have one flight. Imagine releasing photons throughout the problem and tracking them through flight after flight until they all have either been quenched, through the destruction probability $1 - \varpi$, or have escaped. This number of flights will be roughly whichever is less of $1/(1 - \varpi)$ and $\mathcal{T}^2$, where $\mathcal{T}^2$ is the total optical thickness of the atmosphere. This is usually unacceptably large.

In linear algebra language lambda iteration is called Jacobi iteration. We can generalize our thinking about Milne's second problem by saying that S stands for any of the variables that describe the material, such as density, velocity, temperature or level populations, and the relation that gives S in terms of J stands for the hydrodynamic conservation laws and atomic kinetics equations that determine these material variables in terms of the radiation field. The lambda operator stands for the transport equation that determines the radiation field in terms of the material variables. Milne's second equation is thus a stand-in for the statement that all the material properties are self-consistent with the radiation they determine.

The answer to the question, what do you do when Jacobi iteration is too slow, is precondition. Here is how to precondition. We use our inventiveness and find a cheap but accurate operator Λ^* to approximate Λ. We suppose that after n steps we have an approximation S^n, but we are going to get the exact answer on the $(n+1)$th step. That will be the case if

$$\begin{aligned}S^{n+1} - S^n &= (1-\varpi)B + \varpi\Lambda[S^{n+1}] - S^n \\ &= (1-\varpi)B + \varpi\Lambda[S^{n+1} - S^n] + \varpi\Lambda[S^n] - S^n \\ &= \varpi\Lambda^*[S^{n+1} - S^n] + (1-\varpi)B + \varpi\Lambda[S^n] - S^n \\ &\quad + \varpi(\Lambda - \Lambda^*)[S^{n+1} - S^n]. \end{aligned} \tag{11.137}$$

We get our preconditioned iteration by neglecting the last term in the last equality on the grounds that it is a small operator applied to a small difference. Thus the accelerated iteration is

$$(1-\varpi\Lambda^*)[S^{n+1} - S^n] = (1-\varpi)B + \varpi\Lambda[S^n] - S^n. \tag{11.138}$$

The quantity on the right-hand side is the residual in the Milne equation after n iterations. The corresponding formula for $\bar{J}$ itself is

$$\bar{J}^{n+1} = \Lambda^*[S^{n+1}] + \bar{J}^n - \Lambda^*[S^n]. \tag{11.139}$$

Without the factor involving Λ^* on the left-hand side, the correction to S^n is just set equal to the residual, which is Jacobi iteration. Thus this factor is what provides the acceleration. As we see from the derivation, if the approximate lambda operator is accurate, then convergence is immediate. The name for methods like this is ALI or approximate operator iteration (AOI), depending on the author. The ALI methods have been described in many places. A place to start is Kalkofen (1987).

The amplification factor for this iteration, i.e., the factor by which the error is multiplied each time, depends on the eigenvalues of this operator:

$$(1-\varpi\Lambda^*)^{-1}\varpi(\Lambda - \Lambda^*). \tag{11.140}$$

We call the acceleration gentle if the operator $\varpi\Lambda^*$ is small in some sense, and aggressive if it is close to the unit operator. Since the operator $\varpi\Lambda$ itself is close to the unit operator in those situations in which we most need acceleration, the acceleration has to be aggressive to do any good. Too aggressive is bad, however. If $\varpi\Lambda^*$ is more than half-way from $\varpi\Lambda$ to the unit operator the accelerated iteration diverges.

The origin of the methods in this class is the work of Cannon (1973). From 1981–1986 this was picked up and extended by many others, including:

Scharmer (1981); Scharmer and Carlsson (1985); Werner and Husfeld (1985); Olson *et al.* (1986); Hamann (1985, 1986); Rybicki (1984); and Olson and Kunasz (1987). The variations are considerable, but a common theme is to let Λ^* be either a diagonal operator or one that couples nearest neighbors. The most "aggressive" diagonal that does not produce instability turns out to be approximately this:

$$\Lambda^* \approx 1 - \bar{p}_{\rm esc}(\text{zone}), \tag{11.141}$$

where $\bar{p}_{\rm esc}(\text{zone})$ is the zone-average single-flight escape probability from the given zone to any of its neighbors. The escape probability can either be calculated from expressions involving E_2 functions, integrals over line profiles corrected for the velocity field, etc., or obtained by just computing the lambda matrix in detail using the S_N equations or what you will, and discarding everything but the diagonal of it. In fact, the diagonal can be found by doing only a few local calculations so the wasted effort on the off-diagonal parts need not be done. (See Rybicki and Hummer (1991), Appendix B.) If $\bar{p}_{\rm esc}(\text{zone})$ is overestimated compared with the ideal value then the iteration becomes sluggish. If it is underestimated by a factor 2 the iteration will diverge. Thus a somewhat careful calculation of it is indicated; see Olson and Kunasz (1987).

The approximation (11.141) has an interesting consequence when it is used in the non-LTE rate equations. Equation (11.139) for $\bar{J}$, which determines the photoabsorption rate, becomes

$$\bar{J}^{n+1} \approx [1 - \bar{p}_{\rm esc}(\text{zone})]S^{n+1} + \bar{J}^n - [1 - \bar{p}_{\rm esc}(\text{zone})]S^n. \tag{11.142}$$

The net photoabsorption rate is calculated from

$$\mathcal{R}_{21} = N_2(A_{21} + B_{21}\bar{J}^{n+1}) - N_1 B_{12}\bar{J}^{n+1} \tag{11.143}$$

but we can simplify this expression by identifying S^{n+1} with the value computed from N_1 and N_2 using (9.6). Doing this yields

$$\mathcal{R}_{21} = N_2 A_{21}\bar{p}_{\rm esc}(\text{zone}) - (N_1 B_{12} - N_2 B_{21})\{\bar{J}^n - [1 - \bar{p}_{\rm esc}(\text{zone})]S^n\}. \tag{11.144}$$

We can pick off the coefficients of N_1 and N_2 on the right-hand side to be the effective radiative rate coefficients that are put into the kinetic equations. This approach has been used by Rybicki and Hummer (1991, 1992), and non-LTE calculations of supernova spectra using their method have been made by Hauschildt and colleagues (Hauschildt, Storzer, and Baron, 1994; Hauschildt, Baron, and Allard, 1997; Baron and Hauschildt, 1998).

The several non-LTE stellar atmosphere codes built on the ALI principle that were in existence in 1990 were reviewed by Hummer and Hubeny (1991). There is a technical point about these codes that is of interest in the present context. It is that many of the codes substitute the ALI approximation (11.139) into the rate equations, but then take into account the dependence of the coefficients in Λ^* on the level populations, thus forming a nonlinear system for the populations. In these methods (e.g., Werner (1987) and Carlsson (1986)), there is a double iteration loop, with outer ALI iterations and inner Newton–Raphson iterations. The Rybicki and Hummer (1991, 1992) modification lags the Λ^* coefficients, which makes the rate equations linear and therefore no Newton–Raphson iteration is necessary. The cost is additional ALI iterations, which in Rybicki and Hummer (1991, 1992) are minimized by using Ng (1974) acceleration. Dreizler and Werner (1991) describe a double-loop method that minimizes Newton–Raphson cost by using a quasi-Newton method due to Broyden (1965), in which the inverse Jacobian $J^{-1} = (\partial F/\partial x)^{-1}$ is approximated by a matrix B^{-1} that is formed recursively by adding on at each iteration a rank-1 matrix derived from Δx, ΔF, and the previous B^{-1}. This quasi-Newton method then has the flavor of the Newton–Krylov method described earlier. The TLUSTY code of Hubeny and Lanz (1995) represents a hybrid of ALI and traditional linearization, in that the radiative transitions can be treated with or without ALI at the user's option. A double iteration with inner Newton–Raphson iterations is used. These authors remark that they have found the approach of lagging the approximate lambda operator to fail for some problems

The VEF method (see Section 11.5) is another acceleration technique. In this case the approximate operator is the tensor diffusion operator, which should be quite accurate except that it may not reflect the changes that occur during the time step. We may suppose that one application of the tensor diffusion operator is sufficient to correct the estimated mean intensity instead of the tens of iterations that may be required with a diagonal approximate operator; however, the cost of a diffusion solution is much greater. The trade-off between cost per iteration and the iteration count may favor one method or the other in different problems. Alcouffe's diffusion synthetic acceleration method is a variation of this, but here the tensor is omitted and the unmodified Eddington operator is used as the accelerator (Alcouffe, 1976).

The implicit coupling of multifrequency radiation transport to the material energy equation is included within the general ALI framework as was hinted earlier. Equation (11.142) can be put not only into the kinetic equations, but into the internal energy equation as well, and used to derive the correction to the material temperature. No matrices dimensioned by frequency need to be solved in this procedure. Other types of acceleration can be used for the material energy equation, however. The multifrequency gray method is one such. The frequency-integrated implicit coupling equations are used with mean opacities based on the accurate multifrequency spectral distributions instead of the Planck and Rosseland means. The spectral distributions and the mean opacities are updated in the formal transfer part of the iteration cycle. This approach was used by Castor (1974a), and is discussed briefly by Pinto and Eastman (2000).

In general, there is now a well-filled storehouse of preconditioning methods to allow the solution of radiation hydrodynamic problems with large dimensions without directly solving any huge linear systems.

11.12 Monte Carlo methods

The Monte Carlo method for solving a linear transport equation like

$$\frac{1}{c}\frac{\partial I_\nu}{\partial t} + \mathbf{n}\cdot\nabla I_\nu = j_\nu - k_\nu I_\nu \qquad (11.145)$$

is really quite simple. We sample the distribution j_ν to create new particles in various zones with various frequencies and directions. We may also sample the boundary sources, if there are any. Then each particle is tracked through the problem until it leaves across a boundary or is destroyed. Every time step of the hydrodynamic problem each of the particles is tracked from zone to zone until the time for that step is used up, assuming the particle survives that long. When the particle is tracked through a particular zone, the optical thickness across the zone along the track is computed and used to sample whether the particle is destroyed in crossing the zone or not. At the end of the cycle the count of particles in each zone is an estimator of the energy density for that zone. Another, possibly less noisy, estimator is the sum of the track lengths for all the particles that crossed that zone during the time step. The effects of the fluid velocity, i.e., the Doppler and aberration effects, are easily taken into account. The particles are tagged with their fixed-frame frequency, but when a particle is tracked in a given zone it is transformed into the fluid frame for that zone to compute the probability of material interactions. When the emissivity is sampled, the fluid frame emissivity is used to get the frequency, and the direction is sampled, then the transformations are applied to get the fixed-frame values that the particle will carry. Compton scattering is included in full generality by sampling a relativistic electron velocity distribution and the Klein–Nishina cross section to determine whether a scattering event will occur in a given zone or not, and if it does, then the relativistic kinematics of the scattering process can be applied to find the new frequency and direction after scattering. In short, all these awkward processes can be accounted for in full, accurate detail.

A bit of concern arises with the need to implicitly couple the Monte Carlo radiation to the material temperature. The trick that was introduced by Fleck (see Fleck and Cummings (1971) and Fleck and Canfield (1984)) is to linearize the material energy equation and eliminate the temperature perturbation between that linearized equation and the linearization of the emissivity function. This produces an effective emissivity that is linear in the photoabsorption rate, much like the scattering source function in Milne's second problem or in the discussion of ALI methods. This is the "effective scattering" concept: absorption followed by thermal emission is treated like a scattering event. The frequency after emission is changed, however; it is resampled fom the thermal emission spectrum. It has been found by Larsen and Mercier (1987) that the Fleck–Cummings effective scattering formulation is inaccurate when the time step is longer than the radiative cooling time, which is when the implicit method is most needed. Some steps toward correcting this problem have been made by Ahrens and Larsen (2000), and by Martin and Brown (2001).

Can the "effective scattering" process be applied to non-LTE problems? This works perfectly well for resonance line scattering. Indeed, some of the most complete studies of line scattering with angle-dependent partial redistribution combined with fluid flow have been made this way. The failure point is when the particles must interact with excited atoms whose absorbing population is itself sensitive to the Monte Carlo estimate of a resonance line intensity. The combination of the noise and the nonlinear process produces large errors.

There are so many positives about the Monte Carlo method that it might seem surprising that any other method is used. The answer, of course, is the wildly exorbitant cost of doing such calculations. Here is one naive estimate of that cost. Let us say that we will be happy with our statistics if we can bin all the particles present on a given time step according to the zone they are in, the frequency they have, and the direction they are going, with perhaps 10^8 bins in all, not to be too greedy. For 1% statistics in every bin we would need 10^4 particles per bin, or 10^{12} in all. Now every particle crosses quite a few zones in a time step, and some dozens of floating point operations are needed for each zone that is crossed. So let us say 10^2–10^3 operations are needed per particle per time step. That means about 10^{14}–10^{15} operations in all per time step. What would it cost to do this calculation deterministically, using S_N, for instance? We have one intensity variable per photon bin, and we have to perform a few operations per variable per iteration cycle per time step. If 10^2 of the implicit coupling iterations are needed, then the work is 10^2–10^3 operations per variable per time step, or something like 10^{10}–10^{11} total operations per time step. That is very roughly 10^4 times less work than for Monte Carlo. But, say the Monte Carlo folks, requiring 10^4 particles per bin in every single bin is vastly more than you need if the number you want to estimate is one global quantity, such as the total power out of the problem. Sure, you need 10^4 particles in a bin whose contents are an observable on which you are focusing your attention, but who cares about all those other bins that you will not examine in detail. However, radiation hydrodynamics in a nonlinear problem, and what assurance is there that extremely poor statistics on large parts of the radiation field will not cause a serious bias in the estimate of even that one global quantity you want? This is a difficult question, and knowledge in this area is mostly empirical. Monte Carlo still lives with the mantle of being a very expensive method.

Monte Carlo methods cannot be done justice in a few lines. Some of the astrophysical applications have been to Comptonization in x-ray sources (Pozdnyakov, Sobol', and Syunyaev, 1979) and to resonance line transport in the presence of a velocity gradient (Caroff, Noerdlinger, and Scargle, 1972).

12
Examples

In this chapter we illustrate some of the ideas of radiation transport and hydrodynamics coupled with radiation transport by means of a small selection of examples. As described in the introduction, the challenging applications of the theory are left for the technical literature, and the problems presented here have been chosen for their simplicity or pedagogical value.

12.1 Marshak wave and evaporation fronts

The classic example of nonlinear radiation diffusion is the Marshak wave, first discussed by Marshak (1958). It is a self-similar thermal wave, treated with the thermal diffusion approximation, for a material with a constant specific heat and for which the Rosseland mean opacity varies as a power of the temperature. Hydrodynamic motion is ignored. This assumption is unrealistic, but is made for simplicity. This "thermal wave" is not a wave in the sense we used earlier; it does not come from a hyperbolic system of PDEs, and the dispersion relation does not yield wave speeds ω/k, etc. It is a wave in the sense that there is a characteristic structure, in this case a sharp temperature front, that moves through the material in the course of time, of which the shape remains fairly constant. The propagation law is not distance $\propto$ time, as expected for a hyperbolic system, but distance $\propto$ time$^{1/2}$ instead, owing to its diffusion nature.

A thorough discussion of how the thermal diffusion solution to this problem compares with transport solutions is given in Mihalas and Mihalas (1984), Section 103. Zel'dovich and Raizer (1967) devote Chapter 10 to thermal waves in general, and Section 7 to the Marshak problem. Pomraning gives an analytic solution in a linear case with $\kappa =$ constant and $C_v \propto T^3$ using nonequilibrium diffusion (Eddington approximation) and transport (Pomraning, 1979; Ganapol and Pomraning, 1983). Larsen and Pomraning (1980) use asymptotic analysis to obtain a system one order more accurate than thermal diffusion, using which it can be explained why the diffusion Marshak front speed is too great. Su and Olson (1996) present accurate solutions to Pomraning's linear problem in the Eddington approximation.

The geometry of the problem is a slab of material located in $z \geq 0$, and it is initially at zero temperature. Starting at time zero, radiation is applied at the $z = 0$ interface that is a blackbody at $T = T_0$, and this remains constant thereafter. This is one of those instances where the somewhat unphysical boundary condition is applied that the total energy density is specified, $E = aT_0^4$ in this case. It would be more sensible to require that the *incoming* radiation be the hemisphere flux σT_0^4, since the flux that comes back out from the problem is less, but that boundary condition makes the problem nonself-similar. We proceed with the simple boundary condition, and after all, the example is intended only to "guide the insight."

We let ρ be the material density and C_v be the specific heat, both constant. The thermal diffusion formula is used, so the flux is

$$F = -\frac{16\sigma T^3}{3\kappa_R \rho}\frac{\partial T}{\partial z}. \tag{12.1}$$

The opacity κ_R is assumed to follow a power law:

$$\kappa_R = \kappa_R(T_0)\left(\frac{T}{T_0}\right)^{-n}. \tag{12.2}$$

The exponent n will be set to either $n = 0$, which describes electron scattering, or to $n = 3$, which is representative of bound–free and free–free absorption. One thing to notice is that it is the opacity for temperatures near T_0 that is being represented in this way, not the opacity of the cold material in the slab at the start. When the formula for the opacity is put into (12.1) it becomes

$$F = -\frac{16\sigma}{3(n+4)T_0^n \kappa_R(T_0)\rho}\frac{\partial T^{n+4}}{\partial z}, \tag{12.3}$$

where the powers of T in the diffusivity have been combined with the T inside the gradient. The diffusion equation for the temperature follows directly from this, and is

$$\frac{\partial T}{\partial t} = \frac{1}{\rho C_v}\frac{\partial}{\partial z}\left(\frac{16\sigma}{3(n+4)T_0^n \kappa_R(T_0)\rho}\frac{\partial T^{n+4}}{\partial z}\right). \tag{12.4}$$

The temperature scaled by T_0 is the self-similar dependent variable for this problem,

$$g \equiv \frac{T}{T_0}, \tag{12.5}$$

and an inspection of the equation suggests that

$$\xi = \frac{K}{\sqrt{t}} z \tag{12.6}$$

be the scaled independent variable, with a suitable constant factor K. Collecting the constants in the equation gives this result

$$-\xi \frac{dg}{d\xi} = \frac{d^2 g^{n+4}}{d\xi^2}, \tag{12.7}$$

provided K is defined to be

$$K = \left(\frac{3(n+4)\kappa_R(T_0)\rho^2 C_v}{32\sigma T_0^3} \right)^{1/2}. \tag{12.8}$$

Equation (12.7) is to be solved with the boundary condition that $g = 1$ at $\xi = 0$, and another condition that is consistent with the material ahead of the thermal wave being at zero temperature. It will be seen that the solutions of (12.7) go to exactly zero at a finite value of ξ, which we will call ξ_{max}. There are an infinite number of solutions that go to zero at a particular ξ_{max}, but for all but one of these the flux tends to a nonzero value in the limit $\xi \to \xi_{max}$. That it should be zero seems obvious physically, but it also follows from (12.7) if we integrate from a value $\xi < \xi_{max}$ to a value of ξ located in the zero-temperature region ahead of the front. An integration by parts of the left-hand side leads to

$$\xi g + \int_\xi^\infty g \, d\xi = -\frac{dg^{n+4}}{d\xi}. \tag{12.9}$$

The left-hand side clearly tends to zero as $\xi \to \xi_{max}$, so the right-hand side, which is proportional to the flux, must tend to zero also. We can determine the behavior of g near the front by approximating the left-hand side of (12.9) with $\xi_{max} g$, which leads to the form

$$g \sim \left[\frac{(n+3)\xi_{max}}{n+4} (\xi_{max} - \xi) \right]^{1/(n+3)}. \tag{12.10}$$

This relation is the actual boundary condition at $\xi = \xi_{max}$, and used in conjunction with $g = 1$ at $\xi = 0$ it determines a unique solution. The value of ξ_{max} must be adjusted by trial and error until an integration from ξ_{max} to 0, started with the correct limiting form at the front, gives $g = 1$ at $\xi = 0$.

The values found for ξ_{max} are 1.232 for $n = 0$ and 1.121 for $n = 3$. The scaled temperature distributions for these two cases are shown in Figure 12.1. We see that for the higher value of n the temperature profile is closer to a square shape. In fact, the degree of squareness is remarkable. Almost all the material that has

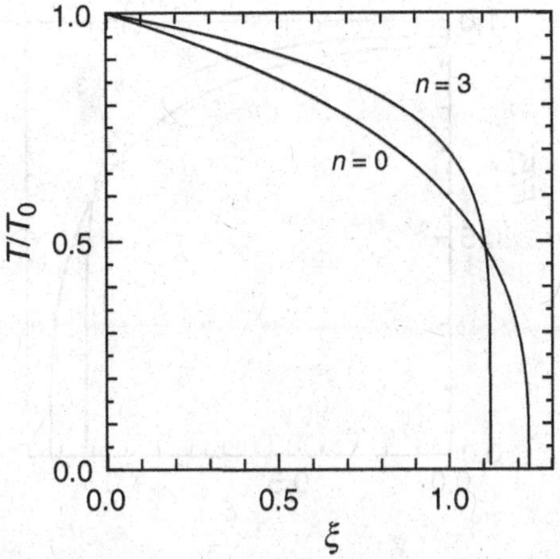

Fig. 12.1 Self-similar temperature distributions for Marshak waves with $\kappa_R \propto T^{-n}$. Abscissa: scaled distance from boundary. Ordinate: scaled temperature g.

been heated by the wave is at a temperature close to T_0, and the drop occurs very abruptly just at the front. This is a consequence of the "bleaching" associated with the wave. The material, very opaque to start with, becomes transparent as it is heated. Not totally so, but sufficiently transparent to allow the flow of radiation to equalize the temperature.

The flux is recovered from the solution by substituting the similarity variables into (12.1). The result is

$$F = -F_1 \frac{dg^{n+4}}{d\xi}, \tag{12.11}$$

where F_1 is the scaling value of the flux,

$$F_1 = \left[\frac{8(n+4)\sigma T_0^4 \rho C_v T_0}{3\kappa_R(T_0)\rho t}\right]^{1/2}. \tag{12.12}$$

The numerical solutions for the flux in units of F_1 for the $n = 0$ and $n = 3$ cases are shown in Figure 12.2. The other scale for flux we think about is $F_0 = cE_0 = 4\sigma T_0^4$, and if the flux is scaled to that we find

$$\frac{F}{cE_0} = -\left[\frac{(n+4)\rho C_v T_0 \lambda_R(T_0)}{6\sigma T_0^4 t}\right]^{1/2} \frac{dg^{n+4}}{d\xi}, \tag{12.13}$$

The ratio $\rho C_v T_0 \lambda_R(T_0)/\sigma T_0^4 t$ that appears here has a physical interpretation: it is the comparison of the heat content of a layer one Rosseland mean free path thick (at T_0) with the energy received from a blackbody at T_0 in the time t. If the

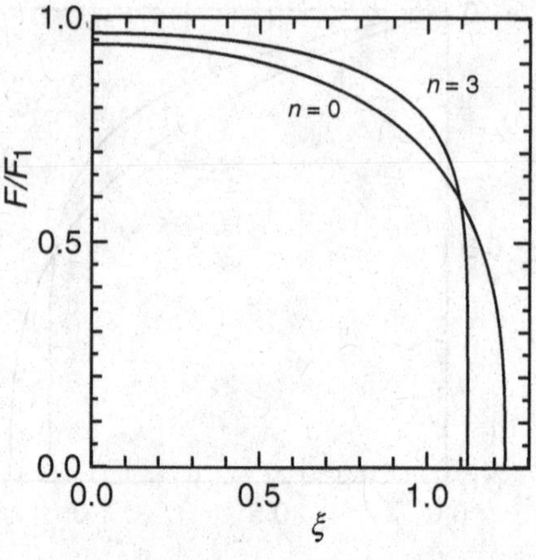

Fig. 12.2 Self-similar flux distributions for Marshak waves with $\kappa_R \propto T^{-n}$. Abscissa: scaled distance from boundary. Ordinate: flux in similarity unit F_1 defined in the text.

ratio is small, enough heat has been absorbed to warm a layer many mean free paths thick. This means that the absorption of heat slows down and the net flux across the $z = 0$ boundary is diminished. If the ratio is large then presumably the front cannot have penetrated even one mean free path. This means that the thermal diffusion approximation is poor. We see that in values of F/cE_0 that are larger than unity. In other words,

$$t > \frac{(n+4)\rho C_v T_0 \lambda_R(T_0)}{6\sigma T_0^4} \tag{12.14}$$

is a condition for this model to be valid. We also see the validity condition by substituting K back into the relation giving the scaling of z:

$$z = \lambda_R(T_0)\left[\frac{32\sigma T_0^4 t}{3(n+4)\rho C_v T_0 \lambda_R(T_0)}\right]^{1/2} \xi. \tag{12.15}$$

Thus the front will have penetrated several mean free paths only if the time obeys the condition (12.14).

Even when the time is late enough to obey condition (12.14) the flux can still violate $F < cE$ because F decreases toward the front much less steeply that $E \propto T^4$ does. The self-similar profile of F/cE is shown in Figure 12.3. This scaling function reaches values of 10–20 before the front is approached even moderately closely. As a result condition (12.14) would have to be obeyed by a large factor, perhaps 100 or more, for $F < cE$ to be satisfied over most of the heated region. Mihalas and Mihalas (1984) illustrate other calculations of the Marshak problem

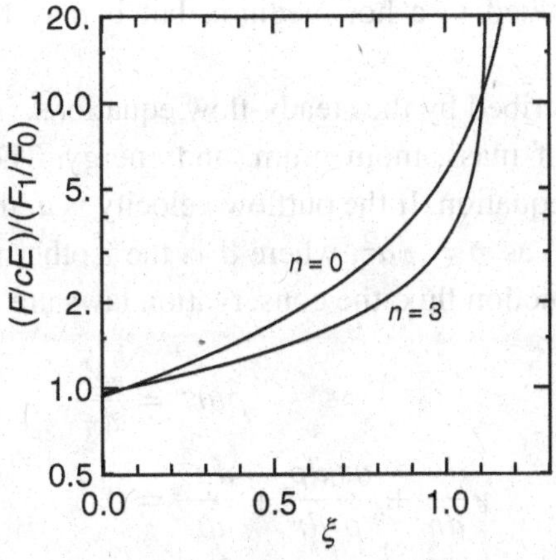

Fig. 12.3 The similarity function giving the profile of F/cE vs scaled distance.

for which the radiation transport approximation has been improved, either by using a flux limiter or by doing accurate angle-dependent transport. These results support the factor 100 suggested here. In particular applications it is possible that the condition is adequately satisfied, but that certainly should be checked before making rough estimates based on Marshak scaling.

The simple Marshak theory presented here is extended in another way by Hammer and Rosen (2003); rather than considering transport and multifrequency modifications to the theory, they present a simple asymptotic method that provides an analytic solution and that can be extended to consider subsonic thermal waves and materials with real physical properties – specific heat and opacity – in place of the power-law relations we have considered.

A problem that is related in some ways to the Marshak problem is evaporation of a cold spherical cloud that is immersed in a hot surrounding medium. This problem was originally treated by Cowie and McKee (1977) and is a central part of the multiphase model of the interstellar medium. A subsequent paper (Mckee and Cowie, 1977) incorporates the cooling effect of (optically thin) radiation. In the Cowie and McKee model the heat flux is caused by thermal conduction, not by radiation diffusion, but the analysis is just the same. The interior of the cloud is treated as cold and dense, and the evaporation front progresses into the cloud so slowly that a reasonable approximation is a steady outflow of the heated material. This flow begins near the front at a negligible speed and accelerates outward and in fact becomes transonic. It reaches quite high velocities far from the cloud. The temperature starts from some low value at the front and also increases outward, but it levels off far from the cloud at the temperature of the surrounding medium. This set of assumptions is not appropriate for all conditions when a cold cloud is embedded in a hot medium, but it may have some domain of applicability.

The problem is described by the steady-flow equations, in spherical symmetry, for the conservation of mass, momentum, and energy. The thermal conduction enters only in the last equation. If the outflow velocity is u, the density is ρ and the pressure is represented as $p = \rho a^2$, where a is the isothermal sound speed, then taking F_c for the conduction flux, the conservation laws are

$$\rho u r^2 = \frac{\dot{\mathcal{M}}}{4\pi}, \tag{12.16}$$

$$u\frac{du}{dr} + \frac{a^2}{\rho}\frac{d\rho}{dr} + \frac{da^2}{dr} = 0, \tag{12.17}$$

$$\frac{1}{2}u^2 + \frac{5}{2}a^2 + \frac{4\pi r^2 F_c}{\dot{\mathcal{M}}} = 0. \tag{12.18}$$

The conduction flux is obtained using a law like Spitzer's (Spitzer and Härm, 1953), but neglecting the variation of the Coulomb logarithm:

$$F_c = -K_S T^{5/2}\frac{dT}{dr}, \tag{12.19}$$

which we put in terms of the sound speed as

$$F_c = -K_S \mathcal{R}^{-7/2}(a^2)^{5/2}\frac{da^2}{dr}. \tag{12.20}$$

(Here $\mathcal{R}$ is the gas constant divided by the mean atomic weight.) Using this relation we can solve (12.18) for da^2/dr and obtain

$$\frac{da^2}{dr} = \frac{A}{(a^2)^{5/2}r^2}(u^2 + 5a^2), \tag{12.21}$$

where A is the collection of constants

$$A = \frac{\dot{\mathcal{M}}\mathcal{R}^{7/2}}{8\pi K_S}. \tag{12.22}$$

Eliminating ρ from the momentum equation (12.17) using the mass equation (12.16) leads to this form of the acceleration equation, very familiar from stellar wind theory:

$$\frac{1}{2}\left(1 - \frac{a^2}{u^2}\right)\frac{du^2}{dr} - \frac{2a^2}{r} + \frac{da^2}{dr} = 0. \tag{12.23}$$

Our strategy now is the following. What we know is that the velocity and temperature go to negligible values as $r \to R_0$, which is the radius for the cold cloud, and which we assume is given. We also know that the temperature goes to a value

T_∞ as $r \to \infty$, which means that $a \to a_\infty$. We do *not* know the evaporation rate $\dot{\mathcal{M}}$, and that is one major thing we want to find out, which means that A is an eigenvalue for the problem.

The initial attack is to make scale transformations of radius and velocity so that after scaling the sonic point, where $u = a$, is at unit radius, and both u and a are unity there. We forthwith adopt that scaling without changing the labels for the variables. We will put the proper units back in later. The first thing we observe is that since r, u, and a are all one at the sonic point, then if du^2/dr is to be finite there, as it must be if there *is* a transonic flow, then it must be true that

$$\left(\frac{da^2}{dr}\right)_s = 2 \tag{12.24}$$

for the sonic point value. That immediately determines A in the scaled units (but *not* in natural units),

$$A = \frac{1}{3}, \tag{12.25}$$

which means that the energy equation takes this form with no unknown coefficients:

$$\frac{da^2}{dr} = \frac{1}{3(a^2)^{5/2}r^2}(u^2 + 5a^2). \tag{12.26}$$

The rest of the solution method consists of solving (12.23) and (12.26) as a system, beginning with the sonic point conditions and integrating each way in radius. The sonic point is a singular point, of course, and the integration subroutines blow up if the integration is actually begun *at* the sonic point. It is necessary to take a small step away in the direction the integration should go. But what slope should be assumed for u^2? This requires an application of L'Hospital's rule:

$$\left(\frac{du^2}{dr}\right)_s = 2\lim_{r\to 1} \frac{2a^2/r - da^2/dr}{1 - a^2/u^2} = 2\frac{d(2a^2/r - da^2/dr)/dr}{d(1 - a^2/u^2)/dr}. \tag{12.27}$$

The derivatives are carried out using (12.26) to obtain the second derivative of a^2 and leaving the first derivative of u^2 as an unknown. What is found is

$$(u^2)' = 2\frac{38/3 - (u^2)'/3}{(u^2)' - 2}, \tag{12.28}$$

a quadratic equation that has the roots

$$(u^2)' = \frac{2}{3}(1 + \sqrt{58}), \qquad \frac{2}{3}(1 - \sqrt{58}), \tag{12.29}$$

from which we have to select the positive root for an accelerated flow.

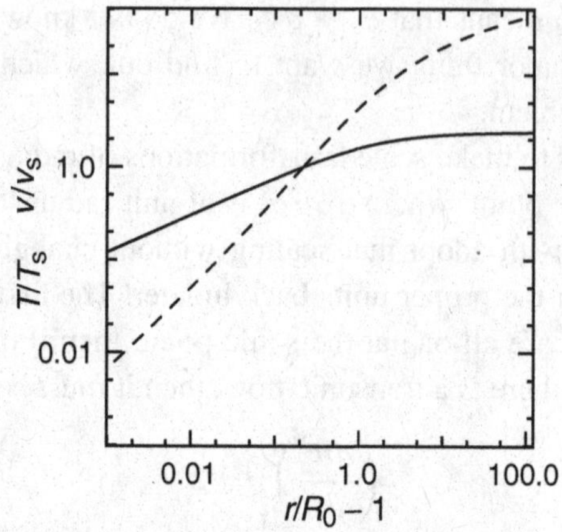

Fig. 12.4 Scaled temperature (solid curve) and velocity (dashed curve) vs radius for transonic evaporation.

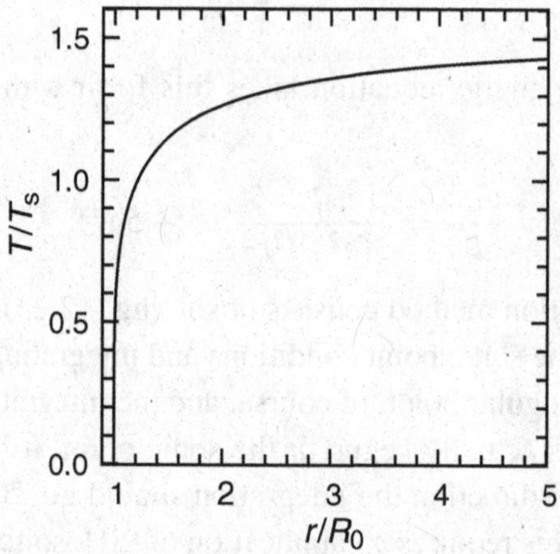

Fig. 12.5 Scaled temperature near the sonic point for transonic evaporation.

Now we simply do the numerical integrations and tabulate u^2 and a^2 as functions of radius. We find that u^2 and a^2 vanish at $r = 0.823$ in sonic radius units, which means that the true location of the sonic point is

$$\frac{r_s}{R_0} = \frac{1}{0.823} = 1.215. \tag{12.30}$$

The distributions of velocity and temperature are shown in Figures 12.4 and 12.5.

The squared sound speed has the asymptotic value at large r of 2.380, which means that the sonic point temperature is

$$T_s = 0.420 T_\infty. \tag{12.31}$$

Restoring A to its natural units gives

$$A = \frac{1}{3} a_s^5 r_s, \tag{12.32}$$

which means that the evaporation rate is

$$\dot{\mathcal{M}} = \frac{8\pi K_S}{\mathcal{R}^{7/2}} A = \frac{8\pi (a_s/a_\infty)^5 (r_s/R_\odot) R_\odot K(T_\infty)}{3\mathcal{R}}$$
$$= 0.0927 \frac{4\pi R_\odot K(T_\infty)}{\mathcal{R}}. \tag{12.33}$$

Just for fun, we put in the radius of the sun and a rough estimate, 1.5×10^6 K, for the coronal temperature, and derive an evaporation rate of $2.9 \times 10^{-14} \, \mathcal{M}_\odot \, \mathrm{y}^{-1}$, which is in the right ballpark compared with the sun's mass-loss rate!

We also find that in the subsonic part of the flow, where $u \ll a$, the pressure tends to a constant value. This pressure is the reaction to the outflowing material, i.e., it is due to the "rocket effect". This pressure turns out to be

$$p_{\mathrm{sub}} = 1.0085 \frac{\dot{\mathcal{M}} (\mathcal{R} T_\infty)^{1/2}}{4\pi R_0^2} = 0.0935 \frac{K(T_\infty) T_\infty^{1/2}}{R_\odot \mathcal{R}^{1/2}}. \tag{12.34}$$

This has the order of magnitude of the conduction flux for temperature T_∞, if the temperature gradient is $T_\infty / R_\odot$, divided by the sound speed at T_∞. For the solar example the pressure turns out to be 4×10^{-4} dyne cm^{-2} corresponding to a particle density of 2×10^6 cm^{-3}, which is on the low side for the base of the corona, but perhaps not too bad for coronal holes. The flow speed continues to rise at large r as expected from Bernoulli's law, since in isothermal conditions the work function is the Helmholtz free energy, which goes logarithmically to $-\infty$ as $\rho \to 0$. At $r = 100 R_\odot$ the velocity is around $6a_s$, which for the corona example is 820 km s^{-1}. This is large, but not by more than a factor 2.

The fact that these estimates for the mass loss from the corona are not completely outrageous, even though gravity has been neglected in this model, demonstrates that thermal conduction is a major ingredient in the solar wind, although only one part of the total picture. The conduction-dominated model of the solar wind is the one of Chamberlain (1960, 1961), although transonic solutions were considered later by Noble and Scarf (1962). These models stand in contrast to Parker's original wind model, for which conduction is ignored (Parker, 1960). A good review of the subject is found in Brandt (1970).

12.2 Ionization fronts

A structure that occurs in stars and nebulae that is like a shock wave in several respects, but also has characteristics in common with the Marshak wave, is the ionization front. (See Kahn (1954), Axford (1961), and Mihalas and Mihalas (1984).) We are considering radiation hydrodynamic behavior of a system made up of the typical cosmic mixture, which is mostly hydrogen. We suppose that there is a quite sizable radiative flux, **F**, flowing through the system. The facts that: (1) hydrogen usually ionizes around $T = 10^4$ K in LTE; (2) the hydrogen opacity is sharply increasing with T when hydrogen is neutral, but decreases with T when hydrogen is ionized; and (3) thermal relaxation causes the radiation flux to tend to a spatially constant value combine to create a spatial profile of the temperature that has a sharp step around 10^4 K. This is a consequence of the diffusion formula, according to which $\nabla T \propto \kappa_R(T)\rho F/T^3$, but is observed even when the diffusion approximation is not applicable. The temperature has a sharp step even when the flux is not actually constant, provided F varies between two fixed positive limits. Not infrequently the region on one side that is mostly neutral will relax to one constant value of the flux while the mostly ionized region on the other side will have relaxed to a different value. The two values of F remain different since the high opacity and high specific heat (due to ionization) of the $T \approx 10^4$ K material makes this an insulating barrier.

The concept of an "ionization front" emerges if we imagine that the opacity in a certain temperature range, say 7000–15 000 K, is some enormous value, and otherwise is what it is supposed to be. This idealization makes the temperature jump discontinuously from 7000 K to 15 000 K across some surface in space, the ionization front. In reality the opacity is large but not infinite in this range, and the thickness of the front is not zero, but something that depends on the magnitude of $\kappa_R(T)\rho F/T^4$. The idealization may be useful in cases for which the true thickness is quite small compared with other length scales.

This discussion has been based on the assumption of LTE, and in particular on the validity of the Saha equation which determines the degree of ionization in terms of the temperature. This is the appropriate regime for stars, but not for nebulae. Out of LTE we have to regard the ionization fraction of hydrogen, x, as an additional degree of freedom that is determined by the equation for ionization kinetics. In this case, too, it may happen that the scale length for x to jump from a value $\ll 1$ to nearly 1 may be short compared with other length scales. We can use the ionization front picture here as well, but for this case the "neutral" and "ionized" states are not strictly tied to particular values of the temperature, although the values 7000 K and 15 000 K are reasonable.

In neither the LTE nor the non-LTE case is the ionization transition actually a phase transition in the thermodynamic sense. In the latter picture the two phases can coexist only on a certain curve in (p, T) space, whereas the neutral and ionized species are present to a greater or lesser degree at all p and T. But in a broad-brush way, when we ignore the small neutral fraction in a mostly ionized plasma or the small ionized fraction in a mostly neutral gas, there is some similarity, and the quantity analogous to the latent heat is the ionization potential of hydrogen expressed per unit mass of material. This has the large value 1.302×10^{13} erg g^{-1} for the cosmic mixture with a hydrogen mass fraction $X = 0.7$. This is equivalent to the specific kinetic energy for a flow velocity of 51 km s^{-1}. For flows with $u \ll$ 50 km s^{-1} the ionization energy and the radiative flux are the largest terms in the energy budget.

We continue the discussion of ionization fronts by considering the jump conditions that express the conservation laws for mass, momentum, and energy when there is a locally-steady flow through the front. What "locally-steady" means is that changes are small in the time required for a parcel of mass to pass through the front. We assign specific temperatures to the neutral (#1) side and the ionized (#2) side of the front, 7000 K and 15 000 K, respectively, so the isothermal sound speeds are $a_1 \approx 7$ km s^{-1} and $a_2 \approx 13$ km s^{-1}. The ideal gas law for a $\gamma = 5/3$ gas is assumed, except that the ionized gas has an internal energy that is larger than that of the neutral gas by an increment I_H, the ionization energy per unit mass $\approx 1.302 \times 10^{13}$ erg g^{-1}. The jump conditions are then

$$\rho_1 v_1 = \rho_2 v_2 = C_1, \tag{12.35}$$

$$p_1 + \rho_1 v_1^2 = p_2 + \rho_2 v_2^2, \tag{12.36}$$

$$\left(\frac{5}{2}\frac{p_2}{\rho_2} + I_H + \frac{1}{2}v_2^2\right) - \left(\frac{5}{2}\frac{p_1}{\rho_1} + \frac{1}{2}v_1^2\right) = \frac{\Delta F}{C_1}. \tag{12.37}$$

Our discussion of the magnitude of I_H suggests that for relatively low-velocity flows the kinetic energy and enthalpy terms may be neglected in the energy jump condition, which then becomes a relation that fixes the mass flux through the front in terms of the jump in the radiative flux.

The possible solutions of the mass and momentum jump conditions are represented as the relations between the compression ratio, $\eta \equiv \rho_2/\rho_1 = v_1/v_2$, and the (isothermal) Mach numbers on the neutral side $M_1 \equiv v_1/a_1$ and the ionized side $M_2 \equiv v_2/a_2$,

$$M_1^2 = \frac{\eta(\eta(a_2/a_1)^2 - 1)}{\eta - 1}, \tag{12.38}$$

$$M_2^2 = \frac{\eta - (a_1/a_2)^2}{\eta(\eta - 1)}. \tag{12.39}$$

These relations are shown in Figure 12.6.

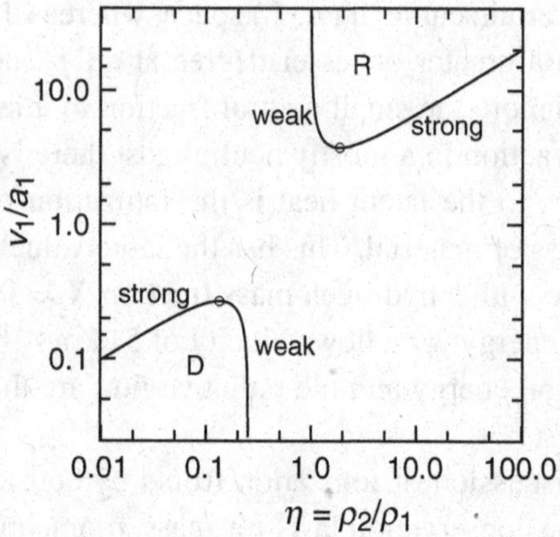

Fig. 12.6 Relation between neutral Mach number v_1/a_1 (ordinate) and compression $\eta = \rho_2/\rho_1$ (abscissa) for ionization fronts with $a_2/a_1 = 13/7 = 1.87$. The labels "R" and "D" distinguish the two branches, and each branch has a "weak" and a "strong" side, as shown. Circles mark the apex of each branch, where $M_2 = 1$.

The relation in Figure 12.6 indicates how ionization fronts are classified into four types. The upper branch corresponds to those fronts that are supersonic on the neutral side and the lower branch consists of the fronts that are subsonic on the neutral side. The "weak" fronts consist of those for which the compression is closer to unity – and the "strong" ones for which the compression is further from unity – at a given neutral Mach number. The weak fronts have the same character, supersonic or subsonic, on both sides, while for the strong fronts the supersonic/subsonic character is reversed for the two sides.

If a value for the flux jump ΔF is specified, and if ρ_1 is given, then there is an implied value of C_1 and therefore of v_1. We speak of *R-type conditions* when the implied v_1 falls in the range of R-type fronts, of *D-type conditions* when the implied v_1 falls in the range of D-type fronts, and of *M-type conditions* when the implied velocity falls in the gap between the two branches. This gap is the Mach number range 0.29–3.42 for $a_2/a_1 = 1.87$. M-type conditions cannot simply produce an ionization front with a suitable speed, but something else must happen. This is generally the creation of a shock that alters the conditions on the neutral side so that D-type or R-type conditions are met. If the shock is as weak as possible, so the D-type or R-type conditions are *barely* met, then these will be D-critical or R-critical, i.e., at one of the apex points where the ionized Mach number is unity.

There is a close parallel between ionization fronts and combustion fronts. The latter occur when the material contains chemically reacting species, and when the constituents are mostly unreacted on one side of the front while the reaction is nearly complete on the other side. If label 1 is associated with unreacted material and label 2 is associated with reacted material, then the jump conditions are the same as equations (12.35)–(12.37), except that the combination $\Delta F/C_1 - I_H$ is replaced by the heat of reaction ΔH. The solutions are qualitatively like those for the ionization front: there are supersonic combustion fronts, which are called *detonations*, and subsonic combustion fronts, which are called *deflagrations*, and each type can be weak or strong, based on the Mach number on the burnt side. Combustion processes that include a critical front, for which the flow speed is just sonic on the burnt side, are called Chapman–Jouget processes.

Because the radiation field is (presumably) negligible for combustion fronts, the second law of thermodynamics and chemical kinetics considerations lead to constraints on combustion fronts that do not apply to ionization fronts. These include the following: the flow must be in the #1–#2 direction, so the reaction goes in the exothermic direction when mass passes through the front; strong deflagrations (D-fronts) and weak detonations (R-fronts) are impossible. A weak detonation case always leads to a precompressing shock and a critical detonation. This is the Jouget hypothesis, and it makes the properties of the Chapman–Jouget detonation of prime importance in modeling explosives. This question is addressed in Landau and Lifshitz (1959), Sections 121–122, and also Zel'dovich and Raizer (1967). It is also treated from a slightly different viewpoint in Courant and Friedrichs (1948). The essential difference between ionization fronts and combustion fronts is that the latter are everywhere in thermal equilibrium, while ionization fronts are modified by the nonequilibrium radiation flux.

In order to understand better how an ionization front affects the surrounding flow, we can count the number of C_+ or C_- characteristics that can reach a point on the neutral side and on the ionized side of the front. We recall that for a shock wave, two characteristics reach the preshock side of the shock front, and just one reaches the shock from the postshock side. A C_0 characteristic also supplies the preshock entropy. There are five mechanical variables needed to describe the state of the fluid on each side of the shock and the shock's motion; these are u_1, ρ_1, u_2, ρ_2, and u_S. The data provided by the three acoustic characteristics, plus the two mechanical jump conditions, just determine these five variables. The two additional thermodynamic variables, the pre- and post-shock entropies, match the data from the C_0 characteristic and the energy jump condition. So shocks are fully determined. In particular, the details of the internal structure of the shock cannot influence the surrounding flow.

The count of just the mechanical (C_+ and C_-) characteristics reaching each side of an ionization front is given in Table 12.1, where we distinguish an advancing front, for which the flow is from the neutral side to the ionized side, from a receding front. We have the same five mechanical variables, and we have two

Table 12.1. *Count of characteristics at ionization fronts*

type		advancing	receding
R	weak	2 neutral, 0 ionized	0 neutral, 2 ionized
	strong	2 neutral, 1 ionized	0 neutral, 1 ionized
D	weak	1 neutral, 1 ionized	1 neutral, 1 ionized
	strong	1 neutral, 0 ionized	1 neutral, 2 ionized

mechanical jump conditions; the energy jump condition gives an additional constraint if the value of ΔF is imposed. This means that if there are two C_+ and C_- characteristics the flow including the ionization front with a specified ΔF is just determined; if there are less than two characteristics it is underdetermined, and if there are more than two it is overdetermined. What we see from an examination of Table 12.1 is that the weak fronts of either type, whether advancing or receding, are just determined. The advancing strong R and receding strong D fronts are overdetermined, and such flows may resolve into critical fronts followed by a rarefaction (advancing strong R) or preceded by a shock (receding strong D). The underdetermined cases, advancing strong D and receding strong R, may be truly dependent for their behavior on the details of the internal structure of the front. The advancing weak-R and both advancing and receding weak-D type ionization fronts are the ones often encountered in simulations of H II regions and stellar atmospheres, respectively.

12.3 Comptonization

In this section we consider those corrections to the statements made earlier that the frequency change in Compton (or Thomson) scattering is negligible, and that such scattering has no effect on the energy exchange between radiation and matter. These may be fair approximations, but they cannot be true in general. If this were so, we could put 10^{12} photons sampled from a blackbody distribution at 10^5 K in a $1\,\text{cm}^3$ box together with 10^{12} electrons with a kinetic temperature of 10^4 K and they would remain at their respective temperatures forever. We do not think this happens with billiard balls, why should it happen with electrons and photons? The collisions between billiard balls, when the balls bounce off each other at some angle of scattering, result in the exchange of energy and momentum between the colliders, just in order to satisfy the conservation laws. This is all it takes to result in equilibration of the two species.

It is just so with scattering of photons by electrons. We will look more closely at the relations for scattering of an arbitrary intensity field by a thermal-equilibrium electron gas. We will introduce a new function $R(\nu, \nu')$, the redistribution function for Compton scattering. Actually, we saw such a function in the discussion of partial redistribution in line scattering, and many of the properties of R are similar for the two kinds of scattering. The definition of R is this: a radiation field of low intensity (which means: ignore stimulated emission) that has an angle-averaged intensity $J_{\nu'}$, and therefore for which the energy density in the bandwidth $d\nu'$ is $4\pi J_{\nu'} d\nu'/c$, produces an emissivity due to this band in a band $d\nu$ at a distinct frequency ν equal to

$$N_e \sigma_T R(\nu, \nu') J_{\nu'} d\nu d\nu'. \tag{12.40}$$

The choice of σ_T here as the scaling cross section is just a convenience. The frequency dependence of the cross section is contained in R. The definition of R involves the *energy* of the source photons and the *energy* of the emitted photons rather than the photon numbers. Therefore R is related to the differential cross section with respect to final photon frequency as follows:

$$\sigma_T R(\nu, \nu') = \frac{\nu}{\nu'} \frac{d\sigma(\nu', \nu)}{d\nu}. \tag{12.41}$$

Notice that in $R(\nu, \nu')$ the first argument is the frequency of the photon after scattering, and the second argument is the frequency before scattering.

Recall the definition of the modal photon density n_ν: $n_\nu = (2h\nu^3/c^2)^{-1} J_\nu$, for the mean number per mode after integrating over solid angle. The equation for the rate of change of the modal density due just to scattering, when account is taken of the in-scatterings as well as the out-scatterings and the stimulated scattering factors are now included as in the discussion in Section 8.3, is this (cf. Section 4.2, (4.24)):

$$\frac{1}{N_e \sigma_T c} \frac{\partial n_\nu}{\partial t} = \frac{c^2}{2h\nu^3} \int_0^\infty d\nu' \left[- R(\nu', \nu) \frac{\nu}{\nu'} J_\nu \left(1 + \frac{c^2 J_{\nu'}}{2h\nu'^3}\right) + R(\nu, \nu') J_{\nu'} \left(1 + \frac{c^2 J_\nu}{2h\nu^3}\right)\right]. \tag{12.42}$$

Before proceeding we have to consider what kind of symmetry is possessed by the function R. Earlier it was stated that R is symmetric under exchange of ν and ν'. We will see that this is *almost* true, by the following argument. Consider a thought experiment in which some radiation is introduced into a sealed box, with perfectly reflecting walls, containing thermal electrons at a temperature T, and the radiation is allowed to come to equilibrium with them; however, no sources or sinks of radiation exist within the box – the same billiard-ball picture mentioned above. When the radiation is at equilibrium, it must be described by a Bose–Einstein distribution at the temperature T of the scatterers, with a fugacity z

that depends on the fixed number of photons in the box:

$$J_\nu = \frac{2h\nu^3/c^2}{z\exp(h\nu/kT) - 1}. \tag{12.43}$$

For this value of the intensity, the stimulation factor $1 + n$ becomes $z/[z - \exp(-h\nu/kT)]$. Therefore, if the intensity has the form given by (12.43), the photon gains and losses must exactly balance. The equation that results is

$$\int d\nu' R(\nu, \nu') \frac{\nu'^3}{[z\exp(h\nu'/kT) - 1][z - \exp(-h\nu/kT)]}$$
$$= \int d\nu' R(\nu', \nu) \frac{\nu^4}{\nu'} \frac{1}{[z\exp(h\nu/kT) - 1][z - \exp(-h\nu'/kT)]}. \tag{12.44}$$

A rearrangement yields

$$\int d\nu' \left[R(\nu, \nu') - \left(\frac{\nu}{\nu'}\right)^4 \exp\left(\frac{h(\nu' - \nu)}{kT}\right) R(\nu', \nu) \right]$$
$$\times \frac{\nu'^3}{[z\exp(h\nu'/kT) - 1][z - \exp(-h\nu/kT)]} = 0, \tag{12.45}$$

which must be true for all z. Therefore the redistribution function has to have this symmetry:

$$R(\nu, \nu') = \left(\frac{\nu}{\nu'}\right)^4 \exp\left[\frac{h(\nu' - \nu)}{kT}\right] R(\nu', \nu). \tag{12.46}$$

In other words, R is symmetric apart from a small bias that is related to the difference between the two frequencies. At high temperature the redistribution favors upshifting the photon frequencies because of the ν^4 factors, but if the temperature is low in comparison with $h\nu/k$ then downshifting is favored. In essence, there will be a small bias tending to bring the photons toward the middle of the Planck distribution at $h\nu = 4kT$.

Now we can put the two R functions in (12.42) in terms of a single one and express the result in terms of modal densities in this form

$$\frac{1}{N_e \sigma_T c} \frac{\partial n_\nu}{\partial t} = \int_0^\infty d\nu' \frac{\nu}{\nu'} R(\nu', \nu)$$
$$\times \left\{ -n_\nu(1 + n_{\nu'}) + \exp\left[\frac{h(\nu' - \nu)}{kT}\right] n_{\nu'}(1 + n_\nu) \right\}. \tag{12.47}$$

From our earlier discussion we expect $R(\nu', \nu)$ to be sharply peaked around $\nu' = \nu$, and therefore provided that the radiation field varies smoothly with ν it should be possible to introduce a Taylor series in $\nu' - \nu$ for the function in braces, do the integrals over ν', which will be moments of $(\nu/\nu')R(\nu, \nu')$, and thus express the right-hand side of (12.47) in terms of the radiation field and its derivatives. This is the Fokker–Planck method. It may be compared with Harrington's method for PRD, described in Section 9.2.1. When carried to the second order in $\nu' - \nu$ the equation that results is the Kompaneets equation (Kompaneets, 1957). The following discussion most nearly follows Pomraning (1973), with parts adapted from Rybicki and Lightman (1979).

We let $F(\nu, \nu')$ be the function in braces in (12.47),

$$F(\nu, \nu') = -n_\nu(1 + n_{\nu'}) + \exp\left[\frac{h(\nu' - \nu)}{kT}\right] n_{\nu'}(1 + n_\nu). \tag{12.48}$$

We form the Taylor expansion of F in the variable ν' around the point ν,

$$F(\nu, \nu') = F_{\nu'}(\nu, \nu)(\nu' - \nu) + \frac{1}{2}F_{\nu'\nu'}(\nu, \nu)(\nu' - \nu)^2 + \cdots, \tag{12.49}$$

where the constant term is seen to vanish. After doing the integrations we find

$$\frac{1}{N_e \sigma_T c}\frac{\partial n_\nu}{\partial t} = A(\nu)F_{\nu'}(\nu, \nu) + \frac{1}{2}B(\nu)F_{\nu'\nu'}(\nu, \nu) + \cdots, \tag{12.50}$$

where the derivatives are

$$F_{\nu'}(\nu, \nu) = \frac{\partial n_\nu}{\partial \nu} + \frac{h}{kT}n_\nu(1 + n_\nu), \tag{12.51}$$

$$F_{\nu'\nu'}(\nu, \nu) = \frac{\partial^2 n_\nu}{\partial \nu^2} + \frac{2h}{kT}(1 + n_\nu)\frac{\partial n_\nu}{\partial \nu} + \left(\frac{h}{kT}\right)^2 n_\nu(1 + n_\nu). \tag{12.52}$$

The quantities $A(\nu)$ and $B(\nu)$ are the moments of the differential cross section,

$$A(\nu) = \frac{1}{\sigma_T}\int_0^\infty d\nu'(\nu' - \nu)\frac{d\sigma(\nu, \nu')}{d\nu'}, \tag{12.53}$$

$$B(\nu) = \frac{1}{\sigma_T}\int_0^\infty d\nu'\,(\nu' - \nu)^2\frac{d\sigma(\nu, \nu')}{d\nu'}. \tag{12.54}$$

The exact, relativistic differential cross section is rather ugly. It begins with the Klein–Nishina differential cross section versus initial photon frequency and scattering angle, which apply in the rest frame of the initial electron, which then must be averaged over the relativistic Maxwellian distribution. In order to obtain the differential cross section for specific initial and final frequencies as a function of scattering angle, all measured in the fixed frame, the Doppler shift and aberration constraints for initial and final photons must be applied, and the electron-frame photon coordinates may be integrated out. This becomes a nine-dimensional integral, although it includes seven delta functions. The final two integrations are over an azimuthal angle for the electron velocity, which is easy, and the speed distribution, which can only be done numerically. Efficient routines for doing this calculation have been developed by Kershaw, Prasad, and Beason (1986). Power-series expansions in the two small quantities $h\nu/mc^2$ and kT/mc^2 have been developed for the differential cross section from which it has been shown that to first order in the two small parameters the moments are given by

$$A(\nu) = \frac{kT}{mc^2}\left(4 - \frac{h\nu}{kT}\right)\nu \tag{12.55}$$

and

$$B(\nu) = \frac{2kT}{mc^2}\nu^2. \tag{12.56}$$

These expressions could be improved using the results for the relativistic regime obtained by Cooper (1971).

The value for B, the mean square frequency shift, is not hard to derive from the nonrelativistic Doppler effect formula without worrying about the Klein–Nishina corrections. As it happens, A can then be calculated from it using the requirement of thermodynamic consistency. When these values for A and B are used with the derivatives from (12.51) and (12.53) in the Fokker–Planck equation (12.50) it becomes

$$\frac{1}{N_e\sigma_T c}\frac{\partial n_\nu}{\partial t} = \frac{kT}{mc^2}\left\{\nu^2\frac{\partial^2 n_\nu}{\partial \nu^2} + \left[4 + \frac{h\nu}{kT}(2n_\nu + 1)\right]\nu\frac{\partial n_\nu}{\partial \nu} + \frac{4h\nu}{kT}n_\nu(n_\nu + 1)\right\}, \tag{12.57}$$

which can be rearranged into the standard Kompaneets form (Kompaneets, 1957)

$$\frac{1}{N_e\sigma_T c}\frac{\partial n_x}{\partial t} = \frac{kT}{mc^2}\frac{1}{x^2}\frac{\partial}{\partial x}\left\{x^4\left[\frac{\partial n_x}{\partial x} + n_x(n_x + 1)\right]\right\} \tag{12.58}$$

in which the frequency variable has been scaled by the temperature,

$$x = \frac{h\nu}{kT}. \tag{12.59}$$

Kompaneets' equation has the expected desirable properties: it is written to precisely conserve photons, since the integral $\int n_x x^2\, dx$ is the photon number density. It must be remembered that this equation is only for the local scattering term; it must be supplemented by the transport term, the $\mathbf{n} \cdot \nabla$ part, in the complete transport equation. Besides conserving photons the Kompaneets equation will give back an exact Bose–Einstein distribution in equilibrium. In that case the frequency-flux vanishes because the equation

$$\frac{\partial n_x}{\partial x} + n_x(n_x + 1) = 0 \tag{12.60}$$

is obeyed exactly by functions

$$n_x = \frac{1}{z\exp(x) - 1}, \tag{12.61}$$

i.e., any Bose–Einstein distribution with the local electron temperature.

If we multiply the Kompaneets equation by x^3 and integrate we get an equation that describes how the mean frequency of the photons evolves with time. It is

$$\frac{d\langle x\rangle}{dt} = N_e \sigma_T c \frac{kT}{mc^2}\left(4\langle x\rangle - \langle x^2\rangle\right). \tag{12.62}$$

So a distribution concentrated initially at low frequency will be boosted exponentially with time according to

$$\langle x\rangle \propto e^y, \tag{12.63}$$

where the parameter y is

$$y = \frac{4N_e\sigma_T ctkT}{mc^2} = \frac{4kTN}{mc^2}, \tag{12.64}$$

where N is the mean number of scatterings that have occurred.

An illustration of the use of the Kompaneets equation is to find the spectral distribution of the photons that begin with a certain frequency in an infinite volume of electrons at a fixed temperature T after a large number N of scatterings. This is also a chance to illustrate some ideas about probability distributions of the number of scatterings and generating functions.

The modal photon density will obey (12.58). The left-hand side is the change of n_x in the mean time between scatterings, so we can regard the right-hand side as the change in n_x per scattering. We write the right-hand side as

$$-n_x + (n_x + K[n_x]), \tag{12.65}$$

where the first term represents the disappearance of the radiation as it was before the event, and the second term is the reappearance of radiation as modified by the Compton scattering process. That is, the first term is the rate of absorption and the second is the rate of reemission. The operator K is the Kompaneets operator, and we will drop the stimulated term, thus making it a linear operator,

$$K[n] \equiv \frac{kT}{mc^2}\frac{1}{x^2}\frac{\partial}{\partial x}\left[x^4\left(\frac{\partial n}{\partial x} + n\right)\right]. \tag{12.66}$$

Now, let us keep track of the photons according to how many times they have scattered. Let $n_0(x,t)$ be what remains of the initial distribution of photons that have survived unscattered until time t. Let $n_1(x,t)$ be those photons that have had one and only one scattering up to time t, and so on. We can write evolution equations for the groups of photons that look like this:

$$\frac{1}{N_e \sigma_T c} \frac{\partial n_0}{\partial t} = -n_0, \tag{12.67}$$

$$\frac{1}{N_e \sigma_T c} \frac{\partial n_r}{\partial t} = -n_r + n_{r-1} + K[n_{r-1}] \qquad \text{for } r \geq 1. \tag{12.68}$$

The photons in group 0 can only be destroyed, but the photons in groups $1, 2, \ldots$ can be destroyed or created by the action of scattering on the photons in group $r - 1$. Now the spectral distribution of photons is determined by how many times they have scattered, but does not change after the last scattering that qualifies them to join that group. Thus the modal density function for photons that have scattered a definite number of times can be expressed as the product of one factor that depends on time and varies as the population of this group grows and decays, and a second factor that depends on frequency, and is the spectrum for all the photons in that group:

$$n_r(x, t) = a_r(t) u_r(x). \tag{12.69}$$

We have noted that the scattering operation does not change the total number of photons, only their frequency distribution, so every function $u_r(x)$ is normalized the same as the initial distribution,

$$\int_0^\infty x^2\, dx\, u_r(x) = \int_0^\infty x^2\, dx\, u_0(x). \tag{12.70}$$

Multiplying (12.68) by x^2 and integrating over x leads to this set of equations for the populations of the groups:

$$\frac{1}{N_e \sigma_T c} \frac{da_0}{dt} = -a_0, \tag{12.71}$$

$$\frac{1}{N_e \sigma_T c} \frac{da_r}{dt} = -a_r + a_{r-1}. \tag{12.72}$$

The K term goes away in the integration since it conserves photons. This set of differential equations is easily solved by recursion, and we find

$$a_r(t) = \exp(-N_e \sigma_T c t) \frac{(N_e \sigma_T c t)^r}{r!}, \tag{12.73}$$

so that at any time t the relative numbers of photons with different numbers of scatterings follow Poisson statistics. The mean number of scatterings in time t is just $N_e \sigma_T c t$.

We are really more interested in the spectral distribution, so we integrate over time in (12.68). We verify that the integral of every $a_r(t)$ is the same as the mean time between scatterings, $1/N_e\sigma_T c$. The integral of the equation for $r = 0$ vanishes identically, and the other equations give

$$u_r = u_{r-1} + K[u_{r-1}]. \tag{12.74}$$

We want to know about large numbers of scatterings, so this equation would be painful to use for one r after another directly. This is where the generating function concept comes in. We define the generating function, or ϖ transform if you prefer, by

$$\tilde{u} = \sum_{r=0}^{\infty} \varpi^r u_r. \tag{12.75}$$

This would be the spectral distribution of all the photons together if the scattering albedo were ϖ instead of unity. But we do not have to think of ϖ as a real albedo, but can instead handle it as a free parameter. When the typical number of scatterings is so large that we can treat r like a continuous variable, this sum can be approximated by an integral, which is then the Laplace transform of u with respect to r, and the transform variable p is $-\log\varpi$:

$$\tilde{u} = \sum_{r=0}^{\infty} \varpi^r u_r \approx \mathcal{L}_{-\log\varpi}[u]. \tag{12.76}$$

When it is desired to invert the transform to obtain u_r from $\tilde{u}$ this relation can be used:

$$u_r = \frac{1}{2\pi i}\oint \frac{\tilde{u}\, d\varpi}{\varpi^{r+1}}. \tag{12.77}$$

(This is the Cauchy integral formula for the rth derivative of $\tilde{u}$, at $\varpi = 0$, divided by $r!$.) The contour should enclose the origin and not enclose singularities of $\tilde{u}$, such as the point $\varpi = 1$.

Applying the sum operation indicated in (12.75) to (12.74) and noting that the $r = 0$ term is missing leads to

$$\tilde{u} - u_0 = \varpi(\tilde{u} + K[\tilde{u}]). \tag{12.78}$$

We do a little rearrangement of this equation and it takes this form,

$$\tilde{u} = \frac{u_0}{1-\varpi} + \frac{\varpi}{1-\varpi}K[\tilde{u}], \tag{12.79}$$

so the generating function satisfies an inhomogeneous equation with the Kompaneets operator, and the initial distribution is the source.

Working out the derivatives puts the equation in this form:

$$x^2\frac{d^2\tilde{u}}{dx^2}+x(x+4)\frac{d\tilde{u}}{dx}+\left[4x-\frac{mc^2(1-\varpi)}{kT\varpi}\right]\tilde{u}=-\frac{mc^2}{kT}\frac{u_0}{\varpi}. \quad (12.80)$$

With one or two transformations this equation becomes a confluent hypergeometric equation, and we will just quote the answer for the response in the case that the initial distribution is a delta function at $x = x_0$:

$$\tilde{u}=\frac{\Gamma(s)x_0^{s+2}x^s e^{-x}}{\varpi\Gamma(2s+4)}M(s,2s+4,x_<)U(s,2s+4,x_>), \quad (12.81)$$

where $x_< = \min(x, x_0)$ and $x_> = \max(x, x_0)$, M and U are the regular and irregular confluent hypergeometric functions, and s, the root of the indicial equation, is

$$s=-\frac{3}{2}+\sqrt{\frac{9}{4}+\frac{mc^2(1-\varpi)}{kT\varpi}}. \quad (12.82)$$

The value of s depends on the Compton y parameter defined in (12.64) in terms of the number of scatterings N. Here we associate N with $\varpi/(1-\varpi)$, so s is related to y by

$$s=-\frac{3}{2}+\sqrt{\frac{9}{4}+\frac{4}{y}}. \quad (12.83)$$

When y is very small, as when the number of scatterings is small, then s is large, and in that case $\tilde{u}$ is very sharply peaked at $x = x_0$. When the y parameter is moderate, then s is moderate also, and an interesting result in that case is if we consider $h\nu_0 \ll kT$. This is the inverse Compton scattering case, because instead of hot photons being degraded by scattering on cold electrons, here cold photons are boosted by scattering on hot electrons. The limiting forms for the M and U functions produce the result that $\tilde{u} \propto x_0^{s+2}x^{-s-3}$ when $x > x_0$. So we get a declining power-law distribution for the scattered spectrum that is flatter and flatter as y becomes larger and s becomes smaller. When $y \gg 1$ then s is small, and the M and U functions tend to unity. What happens in that case is that, whatever the original distribution was, those photons are distributed over a Wien distribution $u \propto e^{-x}$ in the final spectrum. The progression of the spectra of $\tilde{u}$ with y is illustrated in Figure 12.7.

The evolution of $\tilde{u}$ with y is suggestive of how the inverse Compton spectrum changes with time or the number of scatterings, but for a complete picture the inverse ϖ-transform has to be taken. Figure 12.8 shows the time-dependent spectrum for this same case, obtained from a numerical integration of the Kompaneets equation. The suggestion of a power-law spectral distribution in the range

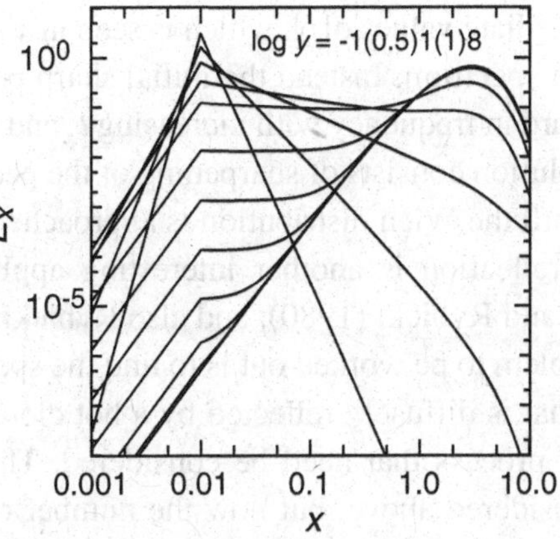

Fig. 12.7 Spectral distributions of $x^3(1 - \varpi)$ times the ϖ-transform of $u(x)$ for several values of $y = (4kT/mc^2)\varpi/(1 - \varpi)$.

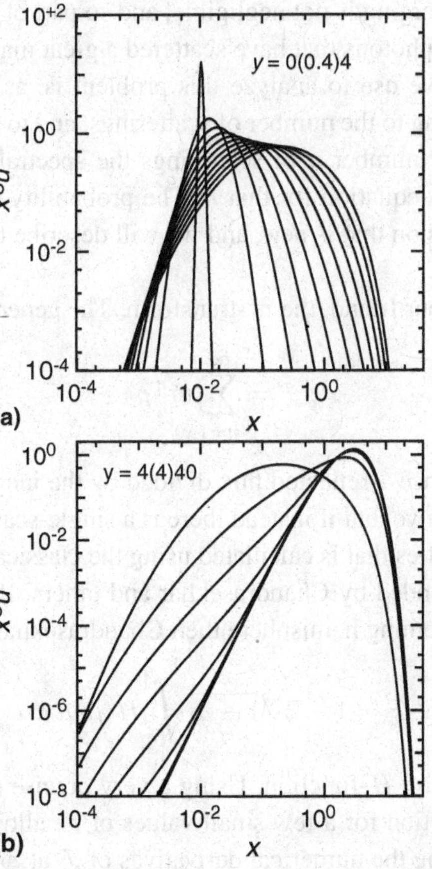

Fig. 12.8 Time-dependent spectra $x^3 n(x, t)$ vs x for Comptonization of radiation with $x_0 = 10^{-2}$. Time is expressed as values of $y = (4kT/mc^2)N_e\sigma_T ct$ ranging: (a) from 0 to 4 and (b) from 4 to 40.

$x_0 \ll x \ll 4$ at intermediate values of y, which is seen in $x^3\tilde{u}$, is not evident in the actual time-dependent spectrum. Instead the initial sharp peak at x_0 is just broadened and shifted upward in frequency with increasing t, and when the peak reaches $x \approx 1$, the further evolution consists of sharpening of the peak and depletion of the low-frequency tail, until the Wien distribution is approached.

Inverse Compton reflection is another interesting application of these same ideas. (See Lightman and Rybicki (1980), and also Rybicki and Lightman (1979), Section 7.5.) The problem to be worked out is to find the spectrum of initially low-frequency radiation that is diffusely reflected by a hot cloud, for which Compton scattering is the only process that need be considered. This is the same inverse Compton process considered above, but now the number of scatterings is not an independent variable, but has a probability distribution determined by the diffuse reflection process. Although the photons that are reflected after just a single scattering are the most numerous, the number that scatter several times, or even hundreds of times, before reemerging is not negligible, and so part of the reflected spectrum does consist of those photons that have scattered a great many times.

The method that we use to analyze this problem is, as above, to itemize the reflected flux according to the number of scatterings, and to apply to those photons that have a particular number n of scatterings the spectral distribution expected from the Kompaneets equation for that n. The probability distribution p_n of n is the part of the discussion that is new, and we will describe that in somewhat more detail.

We come back to our friend, the ϖ-transform. The generating function for p_n

$$\mathcal{F} = \sum_{n=1}^{\infty} \varpi^n p_n \tag{12.84}$$

is the diffuse reflectance – reflected flux divided by the incident flux – if the scattering is not conservative, but if instead there is a single-scattering albedo ϖ. This is one of those quantities that is calculated using the classical methods of radiative transfer theory expounded by Chandrasekhar and others. If the incident intensity is isotropic in the incoming hemisphere then Chandrasekhar's result is

$$\mathcal{F} = 1 - 2\sqrt{1-\varpi} \int_0^1 H(\mu)\mu \, d\mu \tag{12.85}$$

in terms of the standard H-function. Using a very precise subroutine to generate a table of the H-function for a few small values of ϖ allows the values of p_n to be found by estimating the numerical derivatives of $\mathcal{F}$ at $\varpi = 0$; in fact p_n is the nth derivative at $\varpi = 0$, divided by $n!$. Owing to cancellation, this method is good only for $n < 5$ or so. The values for larger n can be estimated by replacing $H(\mu)$

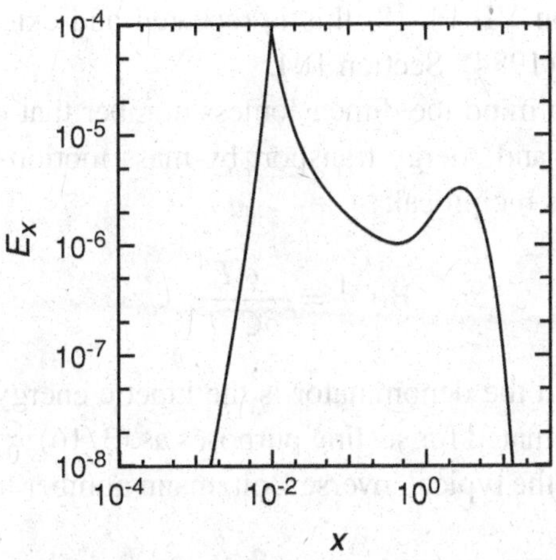

Fig. 12.9 Spectral distribution for inverse Compton reflection. Ordinate: spectral energy density $x^3 n_x$. Abscissa: $h\nu/kT$. For this case $kT = 5.11$ keV.

in (12.85) by the function for $\varpi = 1$, for which the integral has the exact value $2/\sqrt{3}$, then using the binomial theorem to expand $\sqrt{1-\varpi}$ in powers of ϖ. This gives this result for $n \gg 1$:

$$p_n \sim \frac{2}{\sqrt{3\pi}\, n^{3/2}}. \tag{12.86}$$

This has the interesting implication that something like 1% of the photons will have been scattered 10^4 or more times! The details of the numerical differentiation and of the asymptotic law can be checked using a very simple Monte Carlo code. An example of the final result is the spectrum in Figure 12.9, which is, as before, for an initial frequency $x_0 = 10^{-2}$, and here kT has the value $10^{-2}mc^2 = 5.11$ keV. In this spectrum there is a hint of a power law around $x = 10^{-1}$, and a pronounced bump at $x = 3$, the Wien peak. In this problem $4kT/mc^2$ is 0.04, so for photons to be boosted from $x_0 = 10^{-2}$ to $x = 4$ takes $\ln(400)/0.04 \approx 150$ scatterings, and 10.6% of the photons scatter at least that many times, thus populating the Wien peak.

12.4 Radiating shock waves

One of the most important situations in which the radiative energy flux works with the gas dynamics equation to produce a complex structure is in radiative shock waves. These are shocks in which the material is heated sufficiently for the radiative flux to be comparable to the flux of kinetic energy on the upwind side of the front. The classic discussion of radiating shock waves is in Zel'dovich and

Raizer (1967), Section VII.14–18; this is repeated and extended very clearly in Mihalas and Mihalas (1984), Section 104.

We have to keep in mind the dimensionless number that gives the relative importance of radiation and energy transport by mass motion. It is the Boltzmann number, or actually its reciprocal:

$$\mathrm{Bo}^{-1} = \frac{\sigma T^4}{\rho C_v T V}. \tag{12.87}$$

For the shock problem the denominator is the kinetic energy flux $(1/2)\rho_0 v_0^3$ and the temperature is estimated for scaling purposes as $(3/16)v_0^2/\mathcal{R}$, the strong shock limit for $\gamma = 5/3$. So the typical inverse Boltzmann number for a shock is $\mathrm{Bo}^{-1} = 0.0025\sigma v_0^5/\rho\mathcal{R}^4$.

Just to illustrate the numbers we quote the inverse Boltzmann number for a $100\,\mathrm{km\,s^{-1}}$ shock in gas with a preshock density of $10^{-9}\,\mathrm{g\,cm^{-3}}$; it is about 6000. Under these circumstances the influence of radiation on the shock structure is profound.

The diagnostic diagram that aids understanding the influence of radiation on the shock is the mechanical flux vs temperature diagram. This is constructed from the steady conservation laws (2.91), (2.92), and (2.96) in which viscosity is discarded but heat conduction is replaced by a more general radiation flux:

$$\rho v = C_1, \tag{12.88}$$

$$\rho v^2 + p = C_2, \tag{12.89}$$

$$\rho v e + \frac{1}{2}\rho v^3 + p v + F = C_5. \tag{12.90}$$

We divide the second equation by the first to get

$$\frac{p}{\rho} = v\left(\frac{C_2}{C_1} - v\right) = \mathcal{R}T. \tag{12.91}$$

In the third equation we assume the ideal gamma-law relation $e = (3/2)p/\rho$, so pressure and density can be eliminated to express the flux in terms of v:

$$v\left(\frac{5}{2}\frac{C_2}{C_1} - 2v\right) = \frac{C_5 - F}{C_1}. \tag{12.92}$$

Far upstream from the shock F may have an asymptotic value F_∞. The value of C_2/C_1 is v_*, the stagnation velocity. The value of C_5 is F_∞ plus an amount that is close to $v_0^2/2 \approx C_2^2/2C_1^2$ for a large Mach number. The diagnostic diagram is the plot of the left-hand side of (12.92) vs the left-hand side of (12.91).

Figure 12.10(a) illustrates the diagnostic diagram in the absence of radiation. The initial upstream state is point A at zero temperature – negligible compared

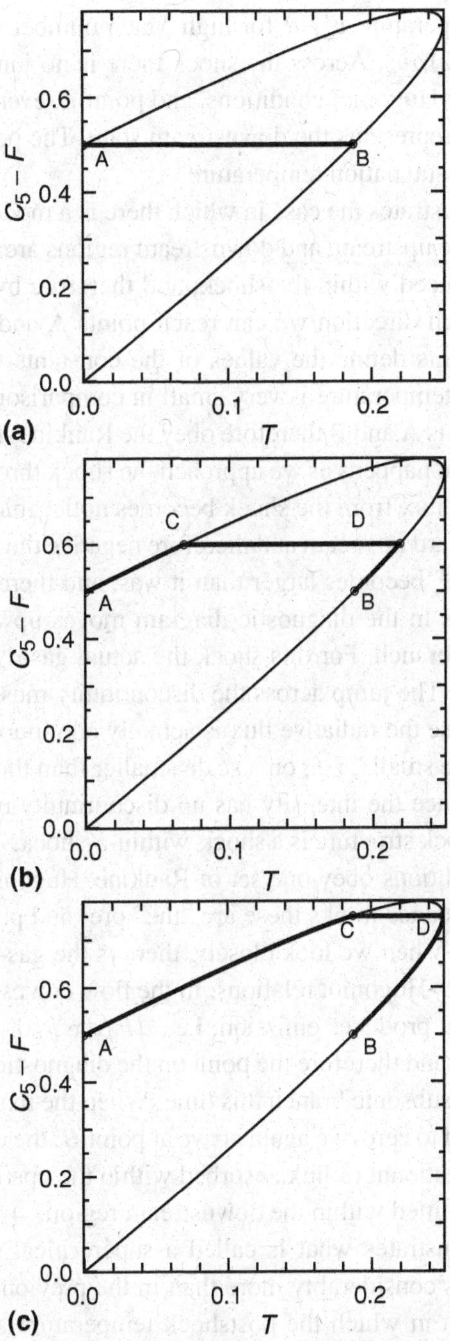

Fig. 12.10 (a) Flux vs T within a nonradiating shock. Bold lines are traversed; (b) similar for a subcritical radiating shock; (c) similar for a supercritical shock.

to the stagnation temperature $v_*^2/\mathcal{R}$ for high Mach number – and at a flux value corresponding to $(1/2)\rho v_*^3$. Across the shock there is no jump in flux, since that is one of the Rankine–Hugoniot conditions, and point B, even with point A but on the subsonic branch, represents the downstream state. The postshock temperature is 3/16 in units of the stagnation temperature.

Figure 12.10(b) illustrates the case in which there is a moderate amount of radiation present. Both the upstream and downstream regions are optically thick to the radiation that is produced within the shock, and therefore by moving far enough from the shock in each direction we can reach points A and B where the flux is negligible. These points define the values of the constants C_1 and C_2, and it is at point A where the temperature is very small in comparison with the stagnation temperature. The points A and B therefore obey the Rankine–Hugoniot conditions. But consider now what happens as we approach the shock through the preshock region. As the radiative flux from the shock becomes noticeable, representing a flux from downstream toward upstream and therefore negative the way we have defined it, the value of $C_5 - F$ becomes larger than it was, and therefore the point representing the conditions in the diagnostic diagram moves upward and to the right along the supersonic branch. For this shock the actual gas-dynamic discontinuity occurs at the point C. The jump across the discontinuity must again be horizontal in the diagram, because the radiative flux is actually continuous. This assumes we look at the flux "in the small", i.e., on a scale smaller than the radiation mean free path, and therefore since the intensity has no discontinuity neither does the flux. Thus the radiating shock structure is a shock-within-a-shock. The remote upstream and downstream conditions obey one set of Rankine–Hugoniot relations, and the observer at a great distance thinks these are "the" pre- and post-shock conditions. But within the shock, when we look closely, there is the gas-dynamic shock with another set of Rankine–Hugoniot relations. In the flow downstream from the inner shock the hot material produces emission, i.e., $dF/dx > 0$, which means that F goes up (toward zero) and therefore the point on the diagnostic diagram goes down and to the left, on the subsonic branch this time. When the material stops emitting, and the flux is restored to zero, we again arrive at point B, the eventual downstream state. We see that the amount of flux absorbed within the upstream part of the flow just equals the flux emitted within the downstream region.

Figure 12.10(c) illustrates what is called a supercritical shock, in which the amount of radiation is considerably more than in the previous case. This actually produces the situation in which the postshock temperature at the inner shock is driven up to almost exactly the maximum possible value, 1/4 of the stagnation temperature. Point C cannot rise any higher than that on the supersonic branch, since there would be nowhere for point D to go. What this does to the shock structure we will see shortly. In the postshock cooling region the temperature recovers to point B as before. Now you will notice that point C is at exactly the same temperature as point B. This is remarkable. The precursor heating due to the radiation produced within the cooling zone raises the preshock temperature up to *just what the downstream temperature would have been without radiation*. The actual postshock temperature is made hotter than this, and only after the postshock cooling has occurred does the downstream material return to this temperature. If any attempt is made to get more radiation than this out of the shock by turning up various parameters or whatever, there is no further change in these temperatures. All that happens is that the shock adjusts the structure of the precursor region and the cooling region so that the right amounts of energy are produced and absorbed to give these temperatures.

This radiating shock structure can be analyzed in a rough way that necessarily ignores the important nonlinearities but catches the main qualitative features. The idea is to observe that the supersonic and subsonic branches of the diagnostic curve are both vaguely linear with a positive slope. In fact the slopes, $-dF/d(p/\rho\mathcal{R})$, are equal to C_1C_v and C_1C_p in the supersonic and subsonic limits, respectively, where C_v is the specific heat at constant volume and C_p is the specific heat at constant pressure. So let us forget about the nonlinearity and use a linear formula connecting F and T, but perhaps with different slopes if we like. We do also need to put in the offset: the (radiative) flux goes to zero at $T = 0$ in the supersonic branch, but it goes to zero at point B in the subsonic branch, where $T = T_0$ is large. We will use the value $aT_0^4 = E_0$ below. What we actually want is a linear relation between the flux and the equilibrium energy density, aT^4. We express that slope as

$$\frac{daT^4}{dF} = 4aT^3\frac{dT}{dF} = -\frac{4aT^3}{C_1C} = -\frac{16\sigma T^4}{cC_1CT}. \tag{12.93}$$

We adopt the Eddington approximation (*not* thermal diffusion) for which

$$F = -\frac{c}{3\kappa_R\rho}\frac{dE}{dx}, \tag{12.94}$$

so the coefficient of the relation between aT^4 and $(1/\kappa_R\rho)dE/dx$ is $16\sigma T^4/3C_1CT$. Apart from the factor 16/3 this is the inverse of the Boltzmann number. As mentioned above, the specific heat C is smaller in the supersonic region than it is in the subsonic region by a factor $\gamma = 5/3$; we will not worry about this minor inaccuracy since the nonlinearity of the relation between aT^4 and T is much more serious. Finally we adopt these linear relations between aT^4 and F:

$$aT^4 = \begin{cases} \dfrac{\mathrm{Bo}^{-1}}{\kappa_R \rho} \dfrac{dE}{dx} & \text{upstream} \\ E_0 + \dfrac{\mathrm{Bo}^{-1}}{\kappa_R \rho} \dfrac{dE}{dx} & \text{downstream} \end{cases}. \tag{12.95}$$

The next step is to solve the Eddington approximation equation of transfer given that aT^4 is obtained from these linear relations. Keeping to the spirit of this crude model we will not be concerned with the variation of the absorptivity with location within the shock – a large effect in reality – and thus make all the coefficients spatially constant. The Eddington equation for E is then

$$\frac{d^2 E}{dx^2} = \begin{cases} 3(\kappa_R \rho)^2 \left(E - \dfrac{\mathrm{Bo}^{-1}}{\kappa_R \rho} \dfrac{dE}{dx} \right) & \text{upstream} \\ 3(\kappa_R \rho)^2 \left(E - E_0 - \dfrac{\mathrm{Bo}^{-1}}{\kappa_R \rho} \dfrac{dE}{dx} \right) & \text{downstream} \end{cases}. \tag{12.96}$$

The homogeneous equation has exponential solutions $E \sim \exp(px)$, and the roots for p are found to be

$$\frac{p}{\kappa_R \rho} = -\frac{3}{2\mathrm{Bo}} \pm \left[3 + \left(\frac{3}{2\mathrm{Bo}} \right)^2 \right]^{1/2}. \tag{12.97}$$

These roots tell an interesting story. The positive one has to be used upstream and the negative one downstream. When the Boltzmann number is large – weak radiation – the roots are the usual $\pm 1/\sqrt{3}$ and the radiation extends symmetrically one mean free path or so in each direction. But in the strong-radiation case the roots are very unequal. The positive root goes to $\mathrm{Bo}\kappa_R \rho$ while the negative one goes to $-3\kappa_R \rho/\mathrm{Bo}$. So on the upstream side the extent of the radiation is very large, of order Bo^{-1} mean free paths, but on the downstream side the radiation drops very abruptly, in Bo mean free paths. The very sharp cooling zone behind the shock means that the thermal diffusion approximation is hopeless there; it just gives the wrong answer.

We can find the correct solutions now. We put in adjustable coefficients for the exponentials, and try these solutions:

$$E = \begin{cases} E_a \exp(p_a x) & \text{upstream} \\ E_0 + E_b \exp(p_b x) & \text{downstream} \end{cases}, \tag{12.98}$$

where p_a is the positive root and p_b is the negative root. We have to find the constants E_a and E_b by matching the values of E and F at the shock. (The second

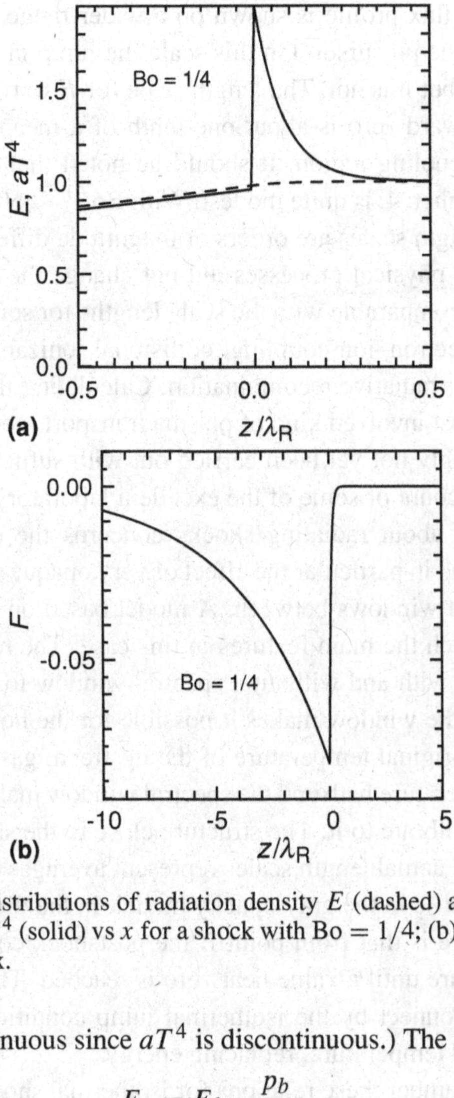

Fig. 12.11 (a) Distributions of radiation density E (dashed) and equilibrium radiation density aT^4 (solid) vs x for a shock with Bo $= 1/4$; (b) the corresponding distribution of flux.

derivative is discontinuous since aT^4 is discontinuous.) The results are

$$E_a = E_0 \frac{p_b}{p_b - p_a}, \tag{12.99}$$

$$E_b = E_0 \frac{p_a}{p_b - p_a}. \tag{12.100}$$

Since p_a is positive and p_b is negative, E_a will be positive and E_b will be negative. This means that E increases monotonically with x through the shock, so the flux is in the $-x$ direction everywhere. The flux follows from (12.94) and the distribution of aT^4 (i.e, of temperature) follows from (12.95).

A sample of what the solutions look like is shown in Figure 12.11. The runs of E and aT^4 track fairly closely with the notable exception of the temperature spike behind the shock, which, because it is quite optically thin, has almost no influence on E. The flux profile is shown on a wider range of x to illustrate the very great extent of the precursor. On this scale the jump in flux at the shock appears discontinuous, but it is not. The length scale for F to rise from its minimum value at the front toward zero is about one tenth of a mean free path, the same as the width of the cooling region. It should be noted that this value of the inverse Boltzmann number, 4, is quite modest. With $\text{Bo}^{-1} \approx 6000$ the precursor and temperature spike length scales are orders of magnitude different. Or, they would be if some neglected physical processes did not change the scale. The predicted width of the spike is comparable with the scale lengths for some of the very fastest plasma processes: electron–ion coupling, collisional ionization, and shorter than some, such as that for radiative recombination. Calculating the shock structure in detail requires the most involved kind of plasma transport modeling, and the modeling work has probably not yet been carried out with sufficient thoroughness to give a satisfactory account of some of the excellent laboratory experiments.

The final remarks about radiating shocks concerns the effect of frequency-dependent opacity, and in particular the effect of very opaque parts of the spectrum with quite transparent windows between. A model based on the Rosseland mean is not adequate to catch the main features in this case. The reason there can be a qualitative difference with and without a spectral window to the outside world is that the existence of the window makes it possible for the hot postshock material to cool down to the original temperature of the upstream gas. That is, there is an "outside world" temperature bath and the spectral window makes it possible for the postshock gas to equilibrate to it. The structure close to the shock is not modified as much, although the actual length scales represent averages over different opacities and differ from those in the gray opacity model. In the diagnostic diagram the new qualitative feature is that from point B the postshock cooling flow proceeds on down in temperature until a value near zero is reached. The final state and the starting state A are connect by the isothermal jump conditions, which conserve mass, momentum and temperature, replacing energy.

It is helpful to remember these relations for isothermal shocks:

$$\rho_0 v_0 = \rho_1 v_1, \tag{12.101}$$

$$p_0 + \rho_0 v_0^2 = p_1 + \rho_1 v_1^2, \tag{12.102}$$

$$\frac{p_0}{\rho_0} = \frac{p_1}{\rho_1} = a^2, \tag{12.103}$$

from which we see that

$$\frac{v_0}{a} = \frac{a}{v_1} = M, \tag{12.104}$$

the Mach number (relative to the isothermal sound speed), so the density ratio is

$$\frac{\rho_1}{\rho_0} = M^2. \tag{12.105}$$

The formation of isothermal shocks in the relatively dense atmospheres of pulsating stars and the difference in the way they develop with height compared with adiabatic shocks is a significant topic for pulsating star physics, which we cannot discuss here. It should be noted that not all shocks in material that has spectral windows will be isothermal. This requires that the emissivity of the hot shocked matter be sufficiently large that the length scale for the cooling is short compared with the larger problem dimensions. This is a statement about the density of the shocked material. As a shock propagates upward through a density gradient spanning orders of magnitude it may first be adiabatic because even the spectral windows are opaque enough to block the escape of radiation; then there may be an isothermal propagation phase, because the radiation escapes but the cooling length scale is still short compared with other dimensions; finally the shock is adiabatic again because the cooling processes are inefficient. These transitions in shock behavior form one of the most interesting aspects of studying pulsating stars.

12.5 Radiatively driven stellar winds

The final example to be discussed is the theory of radiatively-driven stellar winds. The astrophysical motivation is the existence in some stars, and perhaps in active galactic nuclei, molecular clouds and some circumstellar outflows, of supersonic motions that are combined with spectral line transport. The momentum deposited by the line radiation is certainly important in some of these cases. We have already studied how to approximately solve the transport equation in the Sobolev, large velocity gradient, approximation, and we found an expression for the body force on the material due to radiation pressure in spectral lines treated in this approximation.

The theory for radiatively-driven winds presented here is the CAK theory, from Castor *et al.* (1975). There has been considerable elaboration of the theory since 1975, and the current state can be found in the articles presented in Howarth (1998). A particularly rich vein has been the study of the instability of radiative driving. Some of the current work was reviewed at the conference in Isle-aux-Coudres, Quebec (Moffat, Owocki, Fullerton, and St. Louis, 1995) The hydrodynamics of the instabilities has been studied with increasing precision, and now the association between radiation-driven instabilities and the x-ray emission from the winds is secure (Feldmeier *et al.*, 1997).

The application of the Sobolev line force result (6.117) to stellar winds is made much easier than it would otherwise be by a remarkable empirical fact about large databases of line opacity data: *line opacities are distributed approximately as a power law.* The values of k_L that determine the Sobolev optical depth have been found to follow this law for the number of lines stronger than a given value of k_L:

$$N(k_L) = N_0 k_L^{\alpha-1}. \tag{12.106}$$

The constant α in the exponent ranges between 0.5 and 0.7 depending on the database. When the lines are all added up it is found that the total body force, or radiative acceleration, varies in this way in the radial beaming approximation:

$$g_R = \frac{\kappa_T F}{c} k \left(\frac{1}{v_{\mathrm{th}} \kappa_T \rho} \frac{du}{dr} \right)^{\alpha}. \tag{12.107}$$

The proportionality factor k is related to the normalizing constant N_0 in the line distribution. The other factors is this formula have been introduced to make certain of the variables dimensionless. Among these, κ_T is the Thomson opacity $N_e \sigma_T / \rho$ and v_{th} is a representative value of the rms velocity of the absorbing ions. F is the total flux. This α-power law is a result of blending optically thin lines for which the force is proportional to F with thick lines for which it is proportional to $(F/\rho) du/dr$.

The stellar wind equations including the radiative driving force are variations of (12.16) and (12.17) in which the temperature gradient term in the latter is omitted (the thermal pressure turns out to be a relatively small effect in setting the structure of the wind), and the radiative acceleration is included instead. When the density is eliminated this equation is found for the flow speed:

$$\left(u - \frac{a^2}{u} \right) \frac{du}{dr} - \frac{2a^2}{r} + \frac{G\mathcal{M}}{r^2} - \frac{C}{r^2} \left(r^2 u \frac{du}{dr} \right)^{\alpha} = 0. \tag{12.108}$$

The factor C is a collection of constants,

$$C = \frac{\kappa_T L k}{4\pi c} \left(\frac{4\pi}{\kappa_T v_{\mathrm{th}} \dot{\mathcal{M}}} \right)^{\alpha}; \tag{12.109}$$

it is the eigenvalue that is related to the mass-loss rate. The isothermal sound speed is treated as a constant since the flow is expected to be nearly isothermal for the same reasons discussed in the previous section. The boundary conditions are that the velocity should be sonic – small in comparison with the final velocity – somewhere near the stellar photosphere, and the flow should keep going beyond the Parker radius $r = G\mathcal{M}/2a^2$. The analysis is somewhat tricky, and has become generally accepted only after being corroborated by much more elaborate calculations.

The analysis in question examines the locus of points in the r–u plane where this nonlinear differential equation is singular. If the equation is written in this form, in which du/dr is denoted by u',

$$F(r, u, u') = 0, \tag{12.110}$$

then a singular point is a point (r, u), where this algebraic equation is not solvable for u'. The condition for solvability is a nonvanishing partial derivative with respect to the variable to be solved for, so singular points are those points that obey

$$\frac{\partial F}{\partial u'} = 0. \tag{12.111}$$

When u' is eliminated between (12.110) and (12.111) the result is a single equation that connects u and r. This defines the *singular locus* for this differential equation. This is more general than the sonic point that occurs in Parker's wind theory (Parker, 1960), for example, or in the transonic evaporation model discussed earlier. In fact, the sonic point is *not* one of the singular points since (12.110) is solvable there. The rest of the mathematical argument is based on the assertion that an acceptable solution can only begin with a small u near the photosphere and reach large radius with a large velocity if it grazes the singular locus at one point. The solutions that miss the singular locus entirely either do not exist with $u < a$ or cease to exist at $r > G\mathcal{M}/2a^2$. "Grazing" the singular locus means touching it at a point of tangency. Solution curves that meet it at a nonzero angle form a cusp at that point and stop. The condition for tangency is found by differentiating (12.110) with respect to radius and then substituting (12.111):

$$\frac{dF}{dr} = \frac{\partial F}{\partial r} + u'\frac{\partial F}{\partial u} + u''\frac{\partial F}{\partial u'} = \frac{\partial F}{\partial r} + u'\frac{\partial F}{\partial u} = 0. \tag{12.112}$$

We call this the regularity condition.

We can now pick any value for the radius of this regular critical point, *not* the sonic point, and solve the three equations (12.110), (12.111), and (12.112) for the three unknowns u, u', and C. Then numerical integrations beginning from this point give the complete solution, including the run of density since C fixes the mass-loss rate. If the photospheric radius computed using this flow model is not the desired one, the hypothesized critical point radius can be adjusted until it is.

We will give the simplest version of these steps, as in Castor *et al.* (1975). The equation for $F = 0$ is

$$\left(u - \frac{a^2}{u}\right) r^2 \frac{du}{dr} - 2a^2 r + G\mathcal{M} - C\left(r^2 u \frac{du}{dr}\right)^{\alpha} = 0. \tag{12.113}$$

The derivative with respect to du/dr is

$$\left(u - \frac{a^2}{u}\right) r^2 - \frac{\alpha}{du/dr} C\left(r^2 u \frac{du}{dr}\right)^{\alpha} = 0, \tag{12.114}$$

and the regularity equation is

$$\left(u - \frac{a^2}{u}\right) 2r\frac{du}{dr} - 2a^2 - \frac{2\alpha}{r}C\left(r^2 u\frac{du}{dr}\right)^\alpha + \left(1 + \frac{a^2}{u^2}\right) r^2 \left(\frac{du}{dr}\right)^2 - \frac{\alpha(du/dr)}{u} C\left(r^2 u\frac{du}{dr}\right)^\alpha = 0. \quad (12.115)$$

We consider these equations first in the hypersonic $a = 0$ limit. The first two equations become

$$r^2 u\frac{du}{dr} + G\mathcal{M} - C\left(r^2 u\frac{du}{dr}\right)^\alpha = 0, \quad (12.116)$$

$$r^2 u\frac{du}{dr} - \alpha C\left(r^2 u\frac{du}{dr}\right)^\alpha = 0, \quad (12.117)$$

which have the solution

$$r^2 u\frac{du}{dr} = \frac{\alpha}{1-\alpha} G\mathcal{M}, \quad (12.118)$$

$$C\left(r^2 u\frac{du}{dr}\right)^\alpha = \frac{1}{1-\alpha} G\mathcal{M}. \quad (12.119)$$

To locate the actual singular point we need to consider the small quantities proportional to a^2. We first express the second of the three original equations as

$$r^2 u\frac{du}{dr} - \alpha C\left(r^2 u\frac{du}{dr}\right)^\alpha = a^2 r\frac{r}{u}\frac{du}{dr}. \quad (12.120)$$

The regularity equation is written as

$$\left[r^2 u\frac{du}{dr} - \alpha C\left(r^2 u\frac{du}{dr}\right)^\alpha\right]\left[\frac{2}{r} + \frac{1}{u}\frac{du}{dr}\right] = 2a^2\left(\frac{r}{u}\frac{du}{dr} + 1\right) - a^2\left(\frac{r}{u}\frac{du}{dr}\right)^2. \quad (12.121)$$

Combining the last two equations leads to

$$r\frac{r}{u}\frac{du}{dr}\left[\frac{2}{r} + \frac{1}{u}\frac{du}{dr}\right] = 2\left(\frac{r}{u}\frac{du}{dr} + 1\right) - \left(\frac{r}{u}\frac{du}{dr}\right)^2, \quad (12.122)$$

or

$$\left(\frac{r}{u}\frac{du}{dr}\right)^2 = 1. \quad (12.123)$$

We take the positive root,

$$\frac{r}{u}\frac{du}{dr} = 1. \tag{12.124}$$

All the critical point data are now determined, given a choice of the critical point radius. It remains to relate this to the photospheric radius. We consider again the $a = 0$ limit. The differential equation (12.116) in that limit *has only a single root for $r^2 u du/dr$* that is independent of what r and u may be. In other words, the entire solution for $u \gg a$ is described by

$$u = \left[\frac{2\alpha G\mathcal{M}}{(1-\alpha)R_*}\left(1 - \frac{R_*}{r}\right)\right]^{1/2}. \tag{12.125}$$

With this solution the critical point, which is where the slope obeys (12.123), is at

$$\frac{r_C}{R_*} = \frac{3}{2}. \tag{12.126}$$

It has been discovered that mild changes in the way the temperature is distributed with height have hardly any effect on the velocity law (12.125) but change the location of the critical point.

Equation (12.125) provides the scaling law for the terminal velocity of the wind: it is proportional to the star's escape velocity. The formula for C (derived from (12.118) and (12.119))

$$C = \alpha^{-\alpha}(1-\alpha)^{-(1-\alpha)}(G\mathcal{M})^{1-\alpha} \tag{12.127}$$

gives the mass-loss rate

$$\dot{\mathcal{M}} = \left(\frac{4\pi G\mathcal{M}}{\kappa_T v_{\text{th}}}\right)\alpha(1-\alpha)^{(1-\alpha)/\alpha}\left(\frac{k\kappa_T L}{4\pi G\mathcal{M}c}\right)^{1/\alpha}, \tag{12.128}$$

in which L is the stellar luminosity.

The predictions of this theory are found to be in fair quantitative agreement with the winds of hot stars. Two notable features are the scaling of u_∞ with escape velocity and the dependence of the mass-loss rate on a power somewhat larger than unity of the stellar luminosity. The agreement is improved if the radial-beaming approximation for g_R is replaced with a proper integration over the cone of photospheric radiation. (See Friend and Castor (1983), Pauldrach, Puls, and Kudritzki (1986), Friend and Abbott (1986) and Kudritzki, Pauldrach, Puls, and Abbott (1989).) This increases the terminal velocity and softens the steepness of the velocity law at $r \gtrsim R_*$, both of which improve the agreement with observations. The more recent detailed calculations of excitation and ionization balance in the stellar wind models have made the calculations of g_R more reliable without changing the basic results much. The most recent work in stellar wind theory concerns the instability of radiative driving and the large-scale high velocity shocks that are produced as a result. Unfortunately space does not allow a discussion of that work here, but see the IAU Colloquium volume *Variable and Non-spherical Stellar Winds in Luminous Hot Stars* (Wolf, Stahl, and Fullerton, 1998) for reports of the status in 1998.

References

Abramowitz, M. and Stegun, I. A. (1964) *Handbook of Mathematical Functions* (Washington, DC: National Bureau of Standards)

Adams, M. L. (1997) *Trans. Theory and Stat. Phys.*, **26**, 385–431

Adams, M. L. and Larsen, E. W. (2002) *Prog. Nucl. Energy*, **40**, 3–159

Adams, T. F. (1972) *Astrophys. J.*, **174**, 439–448

Adams, T. F., Hummer, D. G., and Rybicki, G. B. (1971) *J. Quant. Spectr. Rad. Transfer*, **11**, 1365–1376

Ahrens, C. and Larsen, E. W. (2000) *Trans. Amer. Nucl. Soc.*, **83**, 340–341

Alcouffe, R. E. (1976) *Trans. Amer. Nucl. Soc.*, **23**, 203

Alcouffe, R. E. (1977) *Nucl. Sci. Engr.*, **64**, 344

Allen, C. W. (1973) *Astrophysical Quantities*, 3rd edn (London: Athlone)

Allis, W. P., Buchsbaum, S. J., and Bers, A. (1963) *Waves in Anisotropic Plasmas* (Cambridge, MA.: MIT Press)

Athay, R. G. (1972a) *Astrophys. J.*, **176**, 659–669

Athay, R. G. (1972b) *Radiation Transport in Spectral Lines* (Dordrecht: Reidel)

Auer, L. H. (1976) *Astrophys. J.*, **16**, 931–937

Auer, L. H. (1984) in *Methods in Radiative Transfer*, ed. W. Kalkofen (Cambridge, UK: Cambridge University Press), pp. 237–279

Auer, L. H. (1987) in *Numerical Radiative Transfer*, ed. W. Kalkofen (Cambridge, UK: Cambridge University Press), pp. 101–109

Auer, L. H. (1991) in *Stellar Atmospheres: Beyond Classical Models*, eds. L. Crivellari, I. Hubeny, and D. G. Hummer (Dordrecht: Kluwer), pp. 9–17

Auer, L. H. and Mihalas, D. (1969) *Astrophys. J.*, **158**, 641–655

Avrett, E. H. (1965) *Astrophys. J.*, **144**, 59–65

Avrett, E. H. and Hummer, D. G. (1965) *Mon. Not. Roy. Astr. Soc.*, **130**, 295–331

Avrett, E. H. and Kalkofen, W. (1968) *J. Quant. Spectr. Rad. Transfer*, **8**, 219–250

Avrett, E. H. and Loeser, R. (1987) in *Numerical Radiative Transfer*, ed. W. Kalkofen (Cambridge, UK: Cambridge University Press) pp. 135–161

Axford, W. I. (1961) *Phil. Trans. Roy. Soc. (London)*, **A253**, 301

Baldwin, C., Brown, P. N., Falgout, R., Graziani, F., and Jones, J. (1999) *J. Comput. Phys.*, **154**, 1–40

Balsara, D. and Shu, C. W. (2000) *J. Comput. Phys.*, **160**, 405–452

Baltrusaitis, R. M., Gittings, M. L., Weaver, R. P., Benjamin, R. F., and Budzinski, J. M. (1996) *Phys. Fluids*, **8**, 2471–2483

Barbier, D. (1943) *Ann. d'Astrophys.*, **6**, 113
Baron, E. and Hauschildt, P. H. (1998) *Astrophys. J.*, **495**, 370–376
Bauche, J. and Bauche-Arnoult, C. (1990) *Comput. Phys. Rep.*, **12**, 1–28
Baym, G. (1997) in *XXXVII Cracow School of Theoretical Physics*, Zakopane, Poland, May 30–June 10 1997
Bazan, G. (1998) in *Proceeding from the 2nd International Workshop on Laboratory Astrophysics with Intense Lasers*, UCRL-ID-131978, ed. B. A. Remington (Livermore, CA: Lawrence Livermore National Laboratory) pp. 42–63
Beckers, J. M. (1969) *Solar Phys.*, **9**, 372–386
Bell, J., Berger, M., Saltzman, J., and Welcome, M. (1994) *SIAM J. Sci. Comput.*, **15**, 127–138
Bell, J. B., Colella, P., and Glaz, H. M. (1989) *J. Comput. Phys.*, **85**, 257–283
Berger, M. J. and Colella, P. (1989) *J. Comput. Phys.*, **82**, 64–84
Berrington, K. A. (1995) in *Astrophysical Applications of Powerful New Databases*, eds. S. J. Adelman and W. L. Wiese (San Francisco, CA: Astronomical Society of the Pacific), pp. 19–30
Biberman, L. M. (1947) *Zh. Eksperim. i Teor. Fiz.*, **17**, 416; also (1949) *Sov. Phys. JETP*, **19**, 584
Blinnikov, S. I. (1996) *Astron. Lett.*, **22**, 79–84, translation of *Pis'ma v. Astr. Zh.*, **22**, 92–98
Born, M. and Wolf, E. (1989) *Principles of Optics: Electromagnetic Theory of Propagation, Interference, and Diffraction of Light* (Oxford: Pergamon)
Bowers, R. L. and Wilson, J. R. (1991) *Numerical Modeling in Applied Physics and Astrophysics* (Boston, MA: Jones and Bartlett Publishers)
Boyd, J. P. (2001) *Chebyshev and Fourier Spectral Methods*, 2nd edn (New York, NY: Dover Publications)
Brandt, J. C. (1970) *Introduction to the Solar Wind* (San Francisco, CA: W. H. Freeman and Co.)
Brown, P. N. and Saad, Y. (1990) *SIAM J. Sci. Stat. Comput.*, **11**, 450–481
Broyden, C. G. (1965) *Math. Comp.*, **19**, 577
Buchler, J.-R. (1979) *J. Quant. Spectr. Rad. Transfer*, **22**, 293
Buchler, J.-R. (1983) *J. Quant. Spectr. Rad. Transfer*, **30**, 395
Calvetti, D., Golub, G. H. and Reichel, L. (1994) *Num. Math.*, **67**, 21–40
Canfield, R. C. and Puetter, R. C. (1981) *Astrophys. J.*, **243**, 381–389
Canfield, R. C., McClymont, A. N., and Puetter, R. C. (1984) in *Methods in Radiative Transfer*, ed. W. Kalkofen (Cambridge, UK: Cambridge University Press) pp. 101–129
Canfield, R. C., Puetter, R. C., and Ricchiazzi, P. J. (1981) *Astrophys. J.*, **248**, 82–86
Cannon, C. J. (1973) *Astrophys. J.*, **185**, 621–630
Canuto C., Hussaini, M. Y., Quarteroni, A., and Zang, T. A. (1988) *Spectral Methods in Fluid Dynamics* (Heidelberg: Springer-Verlag)
Caramana, E. J. and Shashkov, M. J. (1998) *J. Comput. Phys.*, **142**, 521
Caramana, E. J. and Whalen, P. P. (1998) *J. Comput. Phys.*, **141**, 174
Caramana, E. J., Burton, D. E., Shashkov, M. J., and Whalen, P. P. (1998), *J. Comp. Phys.*, **146**, 227
Caramana, E. J., Shashkov, M. J., and Whalen, P. P. (1998), *J. Comput. Phys.*, **144**, 70
Carlson, B. G. (1963) in *Methods in Computational Physics*, Vol. I, eds. B. Alder and S. Fernbach (New York, NY: Academic Press), pp. 1–42
Carlson, B. G. (1970) *Transport Theory: Discrete Ordinates Quadrature over the Unit Sphere*, Report LA-4554, (Los Alamos, NM: Los Alamos National Laboratory)

Carlsson, M. (1986) *A Computer Program for Solving Multi-Level Non-LTE Radiative Transfer Problems in Moving or Statis Atmospheres*, Report No. 33 (Uppsala: Uppsala Astronomical Observatory); see also URL `http://www.astro.uio.no/~matsc/mul22/mul22.html`
Caroff, L. J., Noerdlinger, P. D., and Scargle, J. D. (1972) *Astrophys. J.*, **176**, 439–461
Castor, J. I. (1970) *Mon. Not. Roy. Astr. Soc.*, **149**, 111–127
Castor, J. I. (1972) *Astrophys. J.*, **178**, 779
Castor, J. I. (1974a) *Astrophys. J.*, **189**, 273–283
Castor, J. I. (1974b) *Mon. Not. Roy. Astr. Soc.*, **169**, 279–306
Castor, J. I., Abbott, D. C., and Klein, R. I. (1975) *Astrophys. J.*, **195**, 157–174
Castor, J. I., Dykema, P. G., and Klein, R. I. (1991) in *Stellar Atmospheres: Beyond Classical Models*, eds. L. Crivellari, I. Hubeny, and D. G. Hummer (Dordrecht: Kluwer), pp. 49–59
Castor, J. I., Dykema, P. G., and Klein, R. I. (1992) *Astrophys. J.*, **387**, 561–571
Castrianni, C. L. and Adams, M. L. (1998) *Nucl. Sci. Eng.*, **128**, 278–296
Chamberlain, J. W. (1960) *Astrophys. J.*, **131**, 47
Chamberlain, J. W. (1961) *Astrophys. J.*, **133**, 675
Chandrasekhar, S. (1960) *Radiative Transfer* (New York: Dover)
Christy, R. F. (1966) *Astrophys. J.*, **144**, 108
Colella, P. (1985) *SIAM J. Sci. Comput.*, **6**, 104–117
Colella, P. and Glaz, H. M. (1985) *J. Comput. Phys.*, **59**, 264–289
Colella, P. and Woodward, P. R. (1984) *J. Comput. Phys.*, **54**, 174–201
Colgate, S. A. and White, R. H. (1966) *Astrophys. J.*, **143**, 626
Condon, E. U. and Shortley, G. H. (1951) *The Theory of Atomic Spectra*, reprinted with corrections (Cambridge, UK: Cambridge University Press)
Cooper, G. (1971) *Phys. Rev.*, **D3**, 2312–2316
Cooper, J., Ballagh, R. J., Burnett, K., and Hummer, D. G. (1982) *Astrophys. J.*, **260**, 299–316
Courant, R. and Friedrichs, K. O. (1948) *Supersonic Flow and Shock Waves* (New York, NY: Wiley-Interscience)
Courant, R., Friedrichs, K. O. and Lewy, H. (1928) *Math. Ann.*, **100**, 32
Cowie, L. L. and McKee, C. F. (1977) *Astrophys. J.*, **211**, 135–146
Cox, A. N. and Stewart, J. N. (1965) *Astrophys. J. Suppl.*, **11**, 22–46; **19**, 243–259, 261–279
Cox, A. N., Stewart, J. N., and Eilers, D. D. (1965) *Astrophys. J. Suppl.*, **11**, 1–21
Cox, J. P., Cox, A. N., Olsen, K. H., King, D. S., and Eilers, D. D. (1966) *Astrophys. J.*, **144**, 1038
Cox, J. P. and Giuli, R. T. (1968) *Principles of Stellar Structure*, 2 vols. (New York, NY: Gordon and Breach)
Crivellari, L. Hubeny, I. and Hummer, D. G., eds. (1991) *Stellar Atmospheres: Beyond Classical Models* (Dordrecht: Kluwer)
Dawson, J. and Oberman, C. (1962) *Phys. Fluids*, **5**, 517
Dreizler, S. and Werner, K. (1991) in *Stellar Atmospheres: Beyond Classical Models*, eds. L. Crivellari, I. Hubeny, and D. G. Hummer (Dordrecht: Kluwer), pp. 155–164
Dykema, P. G., Klein, R. I., and Castor, J. I. (1996) *Astrophys. J.*, **457**, 892–921
Eastman, R. G. and Pinto, P. A. (1993) *Astrophys. J.*, **412**, 731–751
Feautrier, P. (1964) *Compt. Rend. Acad. Sci. Paris*, **258**, 3189–3191
Feldmeier, A., *et al.* (1997) *Astron. Astrophys.*, **320**, 899–912
Fleck, J. A., Jr., and Canfield, E. H. (1984) *J. Comput. Phys.*, **54**, 508–523
Fleck, J. A., Jr., and Cummings, J. D. (1971) *J. Quant. Spectr. Rad. Transfer*, **8**, 313–342

Fraser, A. R. (1966) *Atomic Weapons Research Establishment Report No. O-82/65* (Aldermaston: UK Atomic Energy Authority)
Freeman, B. E., Hauser, L. E., Palmer, J. T., Pickard, S. O., Simmons, G. M., Williston, D. G., and Zerkle, J. E. (1968) *The VERA Code. Defense Atomic Support Agency Report No. DASA 2135*, Vol. 1 (La Jolla: Systems, Science and Software, Inc.)
Friend, D. B. and Abbott, D. C. (1986) *Astrophys. J.*, **311**, 701–707
Friend, D. B. and Castor, J. I. (1983) *Astrophys. J.*, **272**, 259–272
Frisch, H. (1980) *Astron. Astrophys.*, **83**, 166–183
Frisch, U. and Frisch, H. (1975) *Mon. Not. Roy. Astron. Doc.*, **173**, 167–182
Fryxell, B., Olson, K., Ricker, P., Timmes, F. X., Zingale, M., Lamb, D. Q., MacNeice, P., Rosner, R., and Tufo, H. (2000) *Astrophys. J.*, **131**, 273
Ganapol, B. D. and Pomraning, G. C. (1983) *J. Quant. Spectr. Rad. Transfer*, **29**, 311–320
Gingold, R. A. and Monaghan, J. J. (1977) *Mon. Not. Roy. Astr. Soc.*, **181**, 375–389
Gittings, M. L. (1992) *Defense Nuclear Agency Numerical Methods Symposium, 28–30 April 1992*
Godunov, S. K. (1959) *Mat. Sbornik*, **47**, 271–306
Godunov, S. K., Zabrodyn, A. W., and Prokopov, G. P. (1962) *J. Comp. Math. Math. Phys. USSR*, **1**,1187
Gol'din, V. Ya. (1964) *Zh. Vych. Mat. i Mat. Fiz.*, **4**, 1078; trans. (1967) *USSR Comm. Math. Math. Phys.*, **4**, 136
Grauer, R. and Germaschewski, K. (2001) in *Space Plasma Simulation*, Proceedings of the Sixth International School/Symposium, Garching, Germany, eds. J. Büchner, C. T. Dums and M. Scholer (Katlenburg-Lindau, Germany: Copernicus GmbH), pp. 1–4
Grove, J. W., Holmes, R. L., Sharp, D. H., Yang, Y., and Zhang, Q. (1993) *Phys. Rev. Lett.*, **71**, 3473–3476
Hallquist, J.O. (1982) "Theoretical Manual for DYNA3D", UCID-19401, (Livermore, CA: University of California, Lawrence Livermore National Laboratory)
Hamann, W.-R. (1985) *Astron. Astrophys.* **148**, 364–368
Hamann, W.-R. (1986) *Astron. Astrophys.*, **160**, 347–351
Hammer, J. H. and Rosen, M. D. (2003) *Phys. Plasmas*, **10**, 1829–1845
Hanbury Brown, R. and Twiss, R. Q. (1954) *Phil. Mag. Ser.* **45**, 663
Hanbury Brown, R. and Twiss, R. Q. (1956a) *Nature*, **177**, 27
Hanbury Brown, R. and Twiss, R. Q. (1956b) *Nature*, **178**, 1046
Harrington, J. P. (1973) *Mon. Not. Roy. Astron. Soc.*, **162**, 43–52
Hauschildt, P. H., Baron, E., and Allard, F. (1997) *Astrophys. J.*, **483**, 390–398
Hauschildt, P. H., Storzer, H., and Baron, E. (1994) *J. Quant. Spectr. Rad. Transfer*, **51**, 875–891
Hearn, A. G. (1972) *Astron. Astrophys.*, **19**, 417–426
Hearn, A. G. (1973) *Astron. Astrophys.*, **23**, 97–103
Heitler, W. (1954) *Quantum Theory of Radiation*, 3rd edn (Oxford: Clarendon Press), p. 198
Henson, V. E. (1987) "A Multigrid Tutorial," *Lawrence Livermore National Laboratory Report UCRL-VG-136819*
Henson, V. E. (1999) An Algebraic Multigrid Tutorial, *Lawrence Livermore National Laboratory Report UCRL-VG-133749*; see also URL `http://www.llnl.gov/CASC/people/henson/presentations.html`
Hernquist, L., and Katz, N. (1989) *Astrophys. J. Suppl.*, **70**, 419–446
Hestenes, M. R. and Stiefel, E. L. (1952) *J. Res. Nat. Bur. Stand.*, **B49**, 409–436

Holmes, R. L., Dimonte, G., Fryxell, B., Gittings, M. L., Grove, J. W., Schneider, M. L., Sharp, D. H., Velikovich, A. L., Weaver, R. P., and Zhaing, Q. (1999) *J. Fluid Mech.*, **389**, 55

Holstein, T. (1947a) *Phys. Rev.*, **72**, 1212–1233

Holstein, T. (1947b) *Phys. Rev.*, **83**, 1159–1168

Howarth, I. D., ed. (1998) *Boulder-Munich II: Properties of Hot, Luminous Stars, Windsor, UK, 21–24 July, 1997*, Astronomical Society of the Pacific Conference Series, vol. 131 (San Francisco, CA: Astronomical Society of the Pacific)

Huard, S. (1997) *Polarization of Light*, trans. Gianni Vacca (New York, NY: Wiley)

Hubeny, I. and Lanz, T. (1995) *Astrophys. J.*, **439**, 875–904

Hubeny, I., Mihalas, D., and Werner, K., eds. (2003) *Stellar Atmosphere Modeling*, ASP Conference Series, Vol. CS-288, (San Francisco, CA: Astronomical Society of the Pacific)

Huebner, W. F., Merts, A. L., Magee, N. H., and Argo, M. F. (1977) *Report LA-6760-M*, (Los Alamos, NM: Los Alamos Scientific Laboratory)

Hummer, D. G. (1962) *Mon. Not. Roy. Astron. Soc.*, **125**, 21

Hummer, D. G. (1964) *Astrophys. J.*, **140**, 276–281

Hummer, D. G. and Hubeny, I. (1991) in *Stellar Atmospheres: Beyond Classical Models*, eds. L. Crivellari, I. Hubeny, and D. G. Hummer (Dordrecht: Kluwer), pp. 119–124

Hummer, D. G. and Rybicki, G. B. (1982) *Astrophys. J.*, **254**, 767–779

Hummer, D. G. and Rybicki, G. B. (1985) *Astrophys. J.*, **293**, 258–267

Hyman, J. M. and Manteuffel, T. A. (1984) in *Elliptic Problem Solvers 2*, Proceedings of Elliptic Problem Solvers Conference, Monterey, CA, 1983, eds. G. Birkhoff and A. Schoenstadt (New York, NY: Academic Press), pp. 301–313

Iglesias, C. A. and Rogers, F. J. (1996) *Astrophys. J.*, **464**, 943–953

Ivanov, V. V. (1973) *Transfer of Radiation in Spectral Lines*, trans. D. G. Hummer from *Radiative Transfer and the Spectra of Celestial Bodies*, published in Moscow 1969 (Washington, DC: National Bureau of Standards)

Jefferies, J., Lites, B. W., and Skumanich, A. (1989) *Astrophys. J.*, **343**, 920–935

Jiang, G. S. and Shu, C. W. (1996) *J. Comput. Phys.*, **126**, 202–228

Jones, J. E. and Woodward, C. S. (2001) *Adv. Water Res.*, **24**, 763–774

Kahn, F. D. (1954) *Bull. Astron. Inst. Netherlands*, **12**, 187

Kalkofen, W., ed. (1984) *Methods in Radiative Transfer* (Cambridge, UK: Cambridge University Press)

Kalkofen, W., ed. (1987) *Numerical Radiative Transfer* (Cambridge, UK: Cambridge University Press)

Karp A. H., Lasher, G., Chan, K. L., and Salpeter, E. E. (1977) *Astrophys. J.*, **214**, 161–78

Keller, G. and Meyerott, R. E. (1955) *Astrophys. J.*, **122**, 32–42

Kershaw, D. S. (1978) *J. Comput. Phys.*, **26**, 43–65

Kershaw, D. S., Prasad, K. K., and Beason, J. D. (1986) *J. Quant. Spectr. Rad. Transfer*, **36**, 273–282

Klein, R. I. (1999) *J. Comput. Appl. Math.*, **109**, 123–152

Klein, R. I., Castor, J. I., Greenbaum, A., Taylor, D., and Dykema, P. G. (1989) *J. Quant. Spectr. Rad. Transfer*, **41**, 199–219

Klein, R. I., Fisher, R. T., Krumholz, M. R., and McKee, C. F. (2003) *Rev. Mex. Astron. Astrophys.*, Conf. Ser. **15**, 92–96

Klein, R. I., Fisher, R. T., McKee, C. F., and Krumholz, M. R. (2004) *Chicago Workshop on Adaptive Mesh Refinement Methods*, Proceeding of a Conference, Chicago, Illinois, September 3–5, 2003 (Berlin: Springer-Verlag), in press

Kompaneets, A. S. (1957) *Sov. Phys. JETP*, **4**, 730

Kopp, H. J. (1963) *Nucl. Sci. Engr.*, **17**, 65
Kourganoff, V. (1963) *Basic Methods in Transfer Problems* (New York, NY: Dover)
Kudritzki, R.-P., Pauldrach, A., Puls, J., and Abbott, D. (1989) *Astron Astrophys.*, **219**, 205–218
Kunasz, P. B. and Auer, L. H. (1988) *J. Quant. Spectr. Rad. Transfer*, **39**, 67–79
Kurucz, R. L. and Bell, B. (1995) *Atomic Line Data*, Kurucz CD-ROM No. 23 (Cambridge, MA: Smithsonian Astrophysical Observatory)
Lamb, H. (1945) *Hydrodynamics* (New York, NY: Dover Publications)
Lanczos, C. (1952) *J. Res. Nat. Bur. Stand.*, **B49**, 33–53
Landau, L. D. and Lifshitz, E. M. (1959) *Fluid Mechanics*, Vol. 6 of *Course of Theoretical Physics* (London: Pergamon); 2nd edn (Oxford: Reed Educational and Professional Publishing Ltd)
Landau, L. D. and Lifshitz, E. M. (1960) *Electrodynamics of Continuous Media*, Vol. 8 of *Course of Theoretical Physics* (London: Pergamon)
Landi Degl'Innocenti, E. (1983) *Solar Phys.*, **85**, 3–31
Landi Degl'Innocenti, E. (1987) in *Numerical Radiative Transfer*, ed. W. Kalkofen (Cambridge, UK: Cambridge University Press)
Landi Degl'Innocenti, E. and Landi Degl'Innocenti, M. (1972) *Solar Phys.*, **27**, 319
Larsen, E. W. and Mercier, B. (1987) *J. Comput. Phys.*, **71**, 50–64
Larsen, E. W. and Pomraning, G. C. (1980) *SIAM J. Appl. Math.*, **39**, 201–212
Lathrop, K. D. and Carlson, B. G. (1965) *Discrete Ordinates Angular Quadrature of the Neutron Transport Equation*, Report LA-3186, (Los Alamos, NM: Los Alamos National Laboratory)
Lax, P. D. (1954) *Comm. Pure Appl. Math.*, **7**, 159–193
Lax, P. D. and Wendroff, B. (1960) *Comm. Pure Appl. Math.*, **13**, 217
Lenoir, W. B. (1967) *J. Appl. Phys.*, **38**, 5283–5290
Lenoir, W. B. (1968) *J. Geophys. Res.*, **73**, 361–376
Levermore, C. D. (1979) *A Chapman-Enskog Approach to Flux-Limited Diffusion Theory, Report No. UCID-18229* (Livermore, CA: Lawrence Livermore National Laboratory)
Levermore, C. D. and Pomraning, G. C. (1981) *Astrophys. J.*, **248**, 321–334
Lewis, E. E. and Miller, W. F., Jr. (1984) *Computational Methods of Neutron Transport* (New York: John Wiley & Sons); also (1993) (La Grange Park, IL: American Nuclear Society)
Liepmann, H. W. and Roshko, A. (1957) *Elements of Gas Dynamics* (New York: Wiley)
Lightman, A. P. and Rybicki, G. B. (1980) *Astrophys. J.*, **236**, 928–944
Lindquist, R. W. (1966) *Ann. Phys.*, **37**, 487
Liska, R. and Wendroff, B. (2003), submitted to *SIAM J. Sci. Comput.*; also report LA-UR-01-6225 (Los Alamos, NM: Los Alamos National Laboratory)
Lucy, L. B. (1977) *Astron. J.*, **82**, 1013–1024
MacFarlane, J. J. (1993) *Collisional-Radiative Equilibrium (CRE) Model for the CONRAD Radiation-Hydrodynamics Code, Fusion Technology Institute Report UWFDM-937* (Madison, WI: University of Wisconsin Fusion Technology Institute)
Manteuffel, T. A. (1977) *Num. Math.*, **28**, 307–327
Manteuffel, T. A. (1978) *Num. Math.*, **31**, 183–208
Marcus, P. S. (1993) *Ann. Rev. Astr. Astrophys.*, **31**, 523–573
Marshak, R. E. (1958) *Phys. Fluids*, **1**, 24
Martin, W. R. and Brown, F. B. (2001) *Trans. Amer. Nucl. Soc.*, **85**, 329–332
McKee, C. F. and Cowie, L. L. (1977) *Astrophys. J.*, **215**, 213–225
Meijerink, J. A. and van der Vorst, H. A. (1977) *Math. Comp.*, **31**, 148–162

Mercier, R. P. (1964) *Proc. Phys. Soc.*, **83**, 819
Messiah, A. (1961) *Quantum Mechanics*, volume 1 (Amsterdam: North-Holland)
Messiah, A. (1962) *Quantum Mechanics*, volume 2 (Amsterdam: North-Holland)
Mészáros, P. and Nagel, W. (1985) *Astrophys. J.*, **298**, 147–160
Mihalas, D. (1978) *Stellar Atmospheres*, 2nd edition (San Francisco, CA: Freeman)
Mihalas, D. (2003) *A Guide to the Literature on Quantitative Spectroscopy in Astrophysics, Report No. LA-14062-MS*, (Los Alamos, NM: Los Alamos National Laboratory)
Mihalas, D., Auer, L. H., and Mihalas, B. W. (1978) *Astrophys. J.*, **220**, 1001–1023
Mihalas, D. and Mihalas, B. W. (1984) *Foundations of Radiation Hydrodynamics* (New York: Oxford University Press); reprinted (New York: Dover Publications)
Mirin, A. A., Cohen, R. H., Curtis, B. C., Dannevik, W. P., Dimits, A. M., Duchaineau, M. A., Eliason, D. E., Schikore, D. R., Anderson, S. E., Porter, D. H., Woodward, P. R., Shieh, L. J., and White, S. W. (1999) Very High Resolution Simulation of Compressible Turbulence on the IBM-SP System, Supercomputing 99 Conference, Portland, OR, November 1999. Also available as Lawrence Livermore National Laboratory technical report UCRL-JC-134237
Moffat, A., Owocki, S., Fullerton, A., and St. Louis, N., eds., (1995) *Instability and Variability of Hot Star Winds – Proceedings of an International Workshop held at Isle-aux-Coudres, Quebec, 23–27 August, 1993*, *Astrophys. Sp. Sci.*, **221**, 11ff
Monaghan, J. J. (1992) *Ann. Rev. Astron. Astrophys.*, **30**, 543–574
Morel, J. E. and Montry, G. R. (1984) *Trans. Theory and Stat. Phys.*, **13**, 615–633
Ng, K. C. (1974) *J. Chem. Phys.*, **61**, 2680
Noble, L. M. and Scarf, F. L. (1962) *J. Geophys. Res.*, **67**, 4577
Noh, W. F. (1987) *J. Comput. Phys.*, **72**, 78–120
Norman, M. L., ed. (1996) *Computational Astrophysics. 12th Kingston Meeting on Theoretical Astrophysics*, Astronomical Society of the Pacific Conference Series v. 123 (San Francisco, CA: Astronomical Society of the Pacific)
Olson, G. L. and Kunasz, P. B. (1987) *J. Quant. Spectr. Rad. Transfer*, **38**, 325–336
Olson, G. L., Auer, L. H., and Buchler, J.-R. (1986) *J. Quant. Spectr. Rad. Transfer*, **35**, 431–442
Omont, A., Smith, E. W., and Cooper, J. (1972) *Astrophys. J.*, **175**, 185–199
Panofsky, W. and Philips, M. (1962) *Classical Electricity and Magnetism*, 2nd edn, (Reading, MA: Addision-Wesley Publishing Co.)
Parker, E. N. (1960) *Astrophys. J.*, **132**, 821
Pauldrach, A., Puls, J., and Kudritzki, R.-P. (1986) *Astron. Astrophys.*, **164**, 86–100
Payne, M. G. and Cook, J. D. (1970) *Phys. Rev.*, **A2**, 1238–1248
Pinto, P. A. and Eastman, R. G. (2000) *Astrophys. J.*, **530**, 757–776
Pomraning, G. C. (1973) *The Equations of Radiation Hydrodynamics* (Oxford: Pergamon)
Pomraning, G. C. (1979) *J. Quant. Spectr. Rad. Transfer*, **21**, 249–261
Pomraning, G. C. (1982) *J. Quant. Spectr. Rad. Transfer*, **27**, 517–530
Pozdnyakov, L. A., Sobol', I. M., and Syunyaev, R. A. (1979) *Soviet Astronomy Letters*, **5**, 279–284
Rachkovsky, D. N. (1962) *Izv. Krymsk. Astrofiz. Obs.*, **27**, 148
Ralston, A. (1965) *First Course in Numerical Analysis* (New York, NY: McGraw-Hill), problem 17(d), p. 220
Ramoné, G. L., Adams, M. L., and Nowak, P. F. (1997) *Nucl. Sci. Engr.*, **125**, 257

Rasio, F. A. (2000) in "Proceedings of the 5th International Conference on Computational Physics (ICCP5)", eds. Y. Hiwatari *et al.*, *Prog. Theo. Phys. Suppl.*, No. 138, 609–621

Rathkopf, J. A., Miller, D. S., Owen, J. M., Stuart, L. M., Zika, M. R., Eltgroth, P. G., Madsen, N. K., McCandless, K. P., Nowak, P. F., Nemanic, M. K., Gentile, N. A., Keen, N. D., and Palmer, T. S. (2000) in "Physor 2000 American Nuclear Society Topical Meeting on Advances in Reactor Physics and Mathematics and Computation into the Next Millennium Pittsburgh, PA May 7–11, 2000" (US Department of Energy)

Reed, W. H. (1971) *Nucl. Sci. Engr.*, **45**, 245

Rees, D. E. (1987) in *Numerical Radiative Transfer*, ed. W. Kalkofen (Cambridge, UK: Cambridge University Press)

Richtmyer, R. D. and Morton, K. W. (1967) *Difference Methods for Initial-Value Problems*, 2nd edn (New York: Interscience)

Rider, W. J., Greenough, J. A., and Kamm, J. R. (2003) submitted to *Trans. Amer. Inst. Aero. Astronaut.*

Rider, W. J., Knoll, D. A., and Olson, G. L. (1999) *J. Comput. Phys.*, **152**, 164–191

Rogers, F. J. and Iglesias, C. A. (1995) in *Astrophysical Applications of Powerful New Databases*, eds. S. J. Adelman and W. L. Wiese (San Francisco, CA: Astronomical Society of the Pacific), 31–50

Rybicki, G. B. (1971) in Symposium on interdisciplinary applications of transport theory, Oxford, UK, 1–4 Sept. 1970, *J. Quant. Spectr. Rad. Transfer*, **11**, 589–595

Rybicki, G. B. (1984) in *Methods in Radiative Transfer*, ed. W. Kalkofen (Cambridge, UK: Cambridge University Press), pp. 21–64

Rybicki, G. B. and Hummer, D. G. (1978) *Astrophys. J.*, **219**, 654–675

Rybicki, G. B. and Hummer, D. G. (1983) *Astrophys. J.*, **274**, 380–398

Rybicki, G. B. and Hummer, D. G. (1991) *Astron. Astrophys.*, **245**, 171–181

Rybicki, G. B. and Hummer, D. G. (1992) *Astron. Astrophys.*, **262**, 209–215

Rybicki, G. B. and Lightman, A. P. (1979) *Radiative Processes in Astrophysics* (Chichester, UK: Wiley)

Saad, Y. (1996) *Iterative Methods for Sparse Linear Systems* (Boston, MA: PWS Publishing Co.); (2003) 2nd edn (Philadelphia, PA: SIAM)

Sampson, D. H. and Zhang, H. L. (1992) *Phys. Rev.*, **A45**, 1556–1561

Sampson, D. H. and Zhang, H. L. (1996) *J. Quant. Spectr. Rad. Transfer*, **55**, 279–284

Scharmer, G. B. (1981) *Astrophys. J.*, **249**, 720–730

Scharmer, G. B. and Carlsson, M. (1985) J. Comput. Phys., **59**, 56–80

Schwarzschild, M. (1958) *Structure and Evolution of the Stars* (Princeton, NJ: Princeton University Press); reprinted (New York, NY: Dover Publications)

Seaton, M. J. (1995) in *Astrophysical Applications of Powerful New Databases*, ed. S. J. Adelman and W. L. Wiese (San Francisco, CA: Astronomical Society of the Pacific), pp. 1–17

Seaton, M. J., Yu Yan, Mihalas, D., and Pradhan, A. K. (1994) *Mon. Not. Roy. Astr. Soc.*, **266**, 805–827

Sedov, L. I. (1959) *Similarity and Dimensional Methods in Mechanics*, trans. M. Friedman (New York, NY: Academic)

Shi, J., Zhang, Y. T., and Shu, C. W. (2003) *J. Comput. Phys.*, **186**, 690–696

Shu, C. W. and Osher, S. (1989) *J. Comput. Phys.*, **83**, 32–78.

Sobel'man, I. I. (1979) *Atomic Spectra and Radiative Transitions* (Berlin: Springer)

Sobolev, V. V. (1960) *Moving Envelopes of Stars*, trans. of *Dvishushchuesia obolochki zvezd.* published in Leningrad 1947, by S. Gaposchkin (Cambridge, MA: Harvard University Press)

Sobolev, V. V. (1963) *A Treatise on Radiative Transfer,* trans. S. I. Gaposchkin from *Perenosluchistoi energii v atmosferakh zvezd i planet.*, published in Moscow 1956 (Princeton, NJ: Van Nostrand)
Sod, G. A. (1978) *J. Comput. Phys.*, **27**, 1–31
Spitzer, L. and Härm, R. (1953) *Phys. Rev.*, **89**, 977
Stone, J. M. and Norman, M. L. (1992a) *Astrophys. J. Suppl.*, **80**, 753–790
Stone, J. M. and Norman, M. L. (1992b) *Astrophys. J. Suppl.*, **80**, 791–818
Stone, J. M., Mihalas, D., and Norman, M. L. (1992) *Astrophys. J. Suppl.*, **80**, 819–845
Strang, W. G. (1968) *SIAM J. Numer. Anal.*, **5**, 506
Su, B. and Olson, G. L. (1996) *J. Quant. Spectr. Rad. Transfer*, **56**, 337-351
Synge, J. L. (1957) *The Relativistic Gas* (Amsterdam: North-Holland)
Thomas, L. H. (1930) *Quart. J. Math. (Oxford)*, **1**, 239
Thomas, R. N. (1960) *Astrophys. J.*, **131**, 429–437
Thomas, R. N. and Athay, R. G. (1961) *Physics of the Solar Chromosphere* (New York, NY: Interscience)
Tipton, R. (1991) "CALE Users Manual, Version 920701", (Livermore, CA: Lawrence Livermore National Laboratory)
Truelove, J. K., Klein, R. I., McKee, C. F., Holliman, J. H. II, Howell, L. H., and Greenough, J. A. (1997) *Astrophys. J. Letters*, **489**, L179–183
Truelove, J. K., Klein, R. I., McKee, C. F., Holliman, J. H., II, Howell, L. H., Greenough, J. A., and Woods, D. T. (1998) *Astrophys. J.*, **495**, 821
Unno, W. (1956) *Publ. Astron. Soc. Japan*, **8**, 108–125
van Leer, B. (1977) *J. Comput. Phys.*, **23**, 276
van Leer, B. (1979) *J. Comput. Phys.*, **32**, 101–136
Van Regemorter, H. (1962) *Astrophys. J.*, **136**, 906–915
Vinsome, P. K. W. (1976) in *Proceedings of the Fourth Symposium on Reservoir Simulation*, Society of Petroleum Engineers, pp. 149–159
Wehrse, R., Baschek, B., and von Waldenfels, W. (2003) *Astron. Astrophys.*, **401**, 43–56
Weisskopf, V. F. (1933) *Observatory*, **56**, 291–308
Werner, K. (1987) in *Numerical Radiative Transfer*, ed. W. Kalkofen (Cambridge, UK: Cambridge University Press), pp. 67–99
Werner, K. and Husfeld, D. (1985) *Astron. Astrophys.*, **148**, 417–422
Wesseling, P. (1992) *An Introduction to Multigrid Methods* (Chichester, England: John Wiley and Sons); also URL `http://casper.cs.yale.edu/mgnet/www/mgnet-books-wesseling.html`
Whitham, G. B. (1974) *Linear and Nonlinear Waves* (New York, NY: Wiley)
Winkler, K.-H. A. and Norman, M. L., eds. (1982) *NATO Advanced Research Workshop on Astrophysical Radiation Hydrodynamics* (Dordrecht: Kluwer)
Wolf, B., Stahl, O., and Fullerton, A. W., eds. (1998) *Variable and Non-spherical Stellar Winds in Luminous Hot Stars*, IAU Colloquium No. 169, Heidelberg, Germany, June 15–19, 1998 (Berlin: Springer-Verlag)
Woodward, P. and Colella, P. (1984) *J. Comput. Phys.*, **54**, 115–173
Wooley, R. v. d. R. and Stibbs, D. W. N. (1953) *The Outer Layers of a Star* (Oxford: Clarendon Press)
Yokokawa, M., Itakura, K., Uno, A., Ishihara, T., and Kaneda, Y. (2002), in *Conference on High Performance Networking and Computing*, archive Proceedings of the 2002 ACM/IEEE Conference on Supercomputing (Baltimore, Maryland: Association for Computing Machinery), 1–17
Zel'dovich, Ya. B. and Raizer, Yu. P. (1967) *Physics of Shock Waves and High-Temperature Hydrodynamic Phenomena* (New York, NY: Academic); reprinted (New York, NY: Dover Publications)

Index

Printed in the United States
By Bookmasters

Printed in the United States
By Bookmasters